Yan Pei · Hao Shang Ma · Yu-Wei Chan ·
Hwa-Young Jeong
Editors

Proceedings of Innovative Computing 2024 Vol. 1

Proceedings of The 7th International Conference on Innovative Computing Vol. 1 (IC 2024)

Springer

Editors
Yan Pei
Computer Science and Engineering
University of Aizu
Aizuwakamatsu Shi, Japan

Yu-Wei Chan
Department of Computer Science
and Information Management
Providence University
Taichung, Taiwan

Hao Shang Ma
Department of Computer Science
and Information Engineering
National Taichung University of Science
North District, Taiwan

Hwa-Young Jeong
Humanitas College
Kyung Hee University
Dongdaemun-gu, Korea (Republic of)

ISSN 1876-1100 ISSN 1876-1119 (electronic)
Lecture Notes in Electrical Engineering
ISBN 978-981-97-4195-3 ISBN 978-981-97-4193-9 (eBook)
https://doi.org/10.1007/978-981-97-4193-9

This Springer imprint is published by the registered company Springer Nature Singapore Pte Ltd.
The registered company address is: 152 Beach Road, #21-01/04 Gateway East, Singapore 189721, Singapore

Lecture Notes in Electrical Engineering 1214

The book series *Lecture Notes in Electrical Engineering* (LNEE) publishes the latest developments in Electrical Engineering—quickly, informally and in high quality. While original research reported in proceedings and monographs has traditionally formed the core of LNEE, we also encourage authors to submit books devoted to supporting student education and professional training in the various fields and applications areas of electrical engineering. The series cover classical and emerging topics concerning:

- Communication Engineering, Information Theory and Networks
- Electronics Engineering and Microelectronics
- Signal, Image and Speech Processing
- Wireless and Mobile Communication
- Circuits and Systems
- Energy Systems, Power Electronics and Electrical Machines
- Electro-optical Engineering
- Instrumentation Engineering
- Avionics Engineering
- Control Systems
- Internet-of-Things and Cybersecurity
- Biomedical Devices, MEMS and NEMS

For general information about this book series, comments or suggestions, please contact leontina.dicecco@springer.com.

To submit a proposal or request further information, please contact the Publishing Editor in your country:

China

Jasmine Dou, Editor (jasmine.dou@springer.com)

India, Japan, Rest of Asia

Swati Meherishi, Editorial Director (Swati.Meherishi@springer.com)

Southeast Asia, Australia, New Zealand

Ramesh Nath Premnath, Editor (ramesh.premnath@springernature.com)

USA, Canada

Michael Luby, Senior Editor (michael.luby@springer.com)

All other Countries

Leontina Di Cecco, Senior Editor (leontina.dicecco@springer.com)

**** This series is indexed by EI Compendex and Scopus databases. ****

Contents

Layout Control Algorithm of Interior Lighting Design Based on ATMEGA16

Jinmin Cui(✉)

Department of Information and Art Design, Anhui Zhong-ao Institute of Technology, Hefei, China
cuijinmin0621@163.com

Abstract. Aiming at the shortage of traditional lighting system, an improved design scheme of indoor lighting system based on intelligent detection technology is proposed. The brightness of the room and environment is sufficient to meet people's needs. ATMEGA16 MCU is used for signal processing, analysis and decision-making. Therefore, the control of lighting equipment is more accurate, convenient and user-friendly. The test results show that the system performance is stable and reliable. This paper introduces an intelligent indoor lighting control system based on ATMEGA16 microcontroller. The system collects ambient light parameters through traditional photoresist and color sensor, and transmits the current light intensity and color to multiple parts of Zigbeed. ZigBee wireless communication technology is used to transmit data to the shared ZigBee node, and then to the processor. Then the corresponding CPU control unit intelligently sets the headlights.

Keywords: ATMEGA16 · Control Algorithm · Design · Interior Lighting · Layout

1 Introduction

The lighting system plays an important role in people's production and life, but at present, most people use ordinary switches to control the on/off of the lighting system, while the lighting and dark modulator are just ordinary buttons or modulators, which control street lighting. Most of them are double switches, and users can work independently. Hotels or pensions usually require many counters. Compared with the comfort people bring through these lights, there are many areas that need to be improved. For a long time, the combination of natural light and intelligent indoor lighting system has been ignored by designers. Most rooms still use the traditional lighting mode. In some public places, excessive lighting time will not only lead to energy loss, but also accelerate the aging of equipment.

Y. Pei et al. (Eds.): IC 2024, LNEE 1214, pp. 1–10, 2024.
https://doi.org/10.1007/978-981-97-4193-9_1

2 System Structure and Working Principle

With the development of science and technology and the acceleration of social informatization process, the relationship between people's work, life and intelligence is getting closer and closer. Informatization has changed people's way of life, but also challenges people's living conditions. People in different environments need different lighting. At present, most living spaces use non compensated light sources, which can easily lead to visual fatigue and even eye damage. The system collects optical data through multiple branch nodes, transmits it to the cpu wirelessly, and then responds to lighting settings through a group of devices to design a closed-loop feedback connection to make the system more stable and intelligent. The system completes the infrared spectrum of human, stops the activities of UAV, and saves energy [1].

2.1 System Structure

The indoor lighting control system mainly uses ATMEGA16 single chip microcomputer as an independent controller, and combines LED display technology, light detection technology, key detection processing technology, infrared detection technology, delay technology and other technologies to intelligently control the indoor lighting equipment. The system structure is shown in Fig. 1 [2, 3].

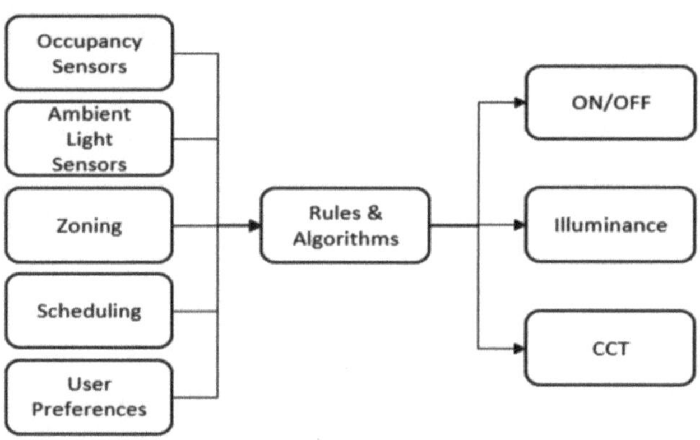

Fig. 1. Indoor Lighting Control Scheme

2.2 Light Data Acquisition

At present, the widely used optical data acquisition technologies mainly include photocell detection and electronic sensor detection. Photocell is a photocell that directly converts light into electrical energy. When the light reaches the surface of the selenium photoelectric element, the incident light reaches the interface between the selenium semiconductor layer and the metal film through the metal film. The photoelectric effect occurs at the voltage difference between the interface and the two ends. The resulting potential difference is proportional to the illumination of the receiving surface of the photoelectric element, and the current measured in the circuit reflects the illumination data. The latter mainly uses high-precision electronic sensors to change the resistance and voltage of components by changing the light; Light intensity depends on the functional relationship between voltage and illumination. The system uses optical resistance and color sensor to collect information. The sensitivity of general photoresistors (spectral characteristics) is very close to the response of human eyes to visible light (0.40–0.76 μm). As long as the human eye can perceive light, the resistance value of the photoresistor will change, so we can use the photoresistor to detect effective light intensity information; The TCS3200 color sensor is used to identify the three main colors. The circuit converts the three color illumination intensity information into electric pulse signal, and realizes color recognition effect through certain information processing [4–6].

2.3 Working Principle

(1) Single chip microcomputer controls the switch of indoor lighting equipment through relay. (2) The light detection circuit around the lighting device detects the brightness around the device. The signal reaches the microcontroller through the ambient brightness sensor module. If the brightness can meet the needs of life, the microcontroller keeps the lighting device off through a relay. If the brightness is not enough, the microcontroller relays.

Continue to detect whether the pyroelectric inductance signal of human body is collected. (3) If the passive pyroelectric infrared sensor detects the human body signal, the microcontroller will immediately control the lighting device and turn it on; If no human body signal is detected, the microcontroller will keep the lighting device off. (4) When the lighting device is turned on, if the microcontroller does not detect the human body signal at a certain time, the lighting device will be turned off after a period of time. If the human body signal is detected during the delay period, the delay will end and the lighting device will continue to turn on. (5) The PA0–PA3 pins of the single chip microcomputer can be set to different delay time values according to the different lighting application occasions and different users. (6) When the lighting circuit works normally, the lighting equipment can be forced to control as long as the forced button is pressed, and the circuit can also be switched to the automatic control state through the button [7, 8].

The machine controls the indoor lighting switch through the relay. The lighting controls around the lighting fixture determine the external brightness of the fixture. The signal reaches the microcontroller through the ambient brightness detection module. If the brightness meets the needs of life, the microcontroller supports turning off the lighting device through a relay. If the brightness is not enough, the automatic relay continues to check whether someone is sending a thermal signal. When the passive infrared heat dissipation sensor detects an artificial signal, the microcontroller immediately monitors the lighting device and turns on the lighting device; If no human signal is detected, the entire machine will remain off. If the lighting device is turned on and the monitor does not detect the human body signal for a certain time, the lighting device will turn off after a long time. If a manual signal is detected during the delay, the delay ends and the lighting continues to turn on. PA0-PA3 connector can set different delay values according to lighting fixtures and quantity. If the lighting circuit works normally, just press the button to violently control the lighting equipment, and turn the circuit into automatic control state through the button. As shown in Fig. 2, indoor lighting design effect.

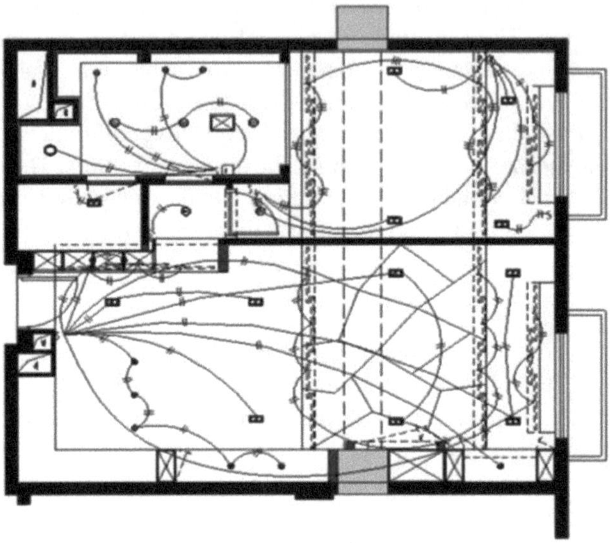

Fig. 2. Interior Lighting Design Rendering

2.4 Light Control Procedure

In the design process of the lighting control system, a single-chip microcomputer is used to control the two-way switch between transistors to adjust the brightness and darkness of the lamp. The working effect of the ON/OFF light is that when the remote controller presses the button, the SCM sends out the crystal control signal light ON/OFF, and then presses the button, the SCM sends out the crystal control signal light OFF, that is, after receiving the infrared signal, the indicator light signal turns. If the indicator light is 0, the indicator light will light up, and then press the button again, and the indicator light will go out at the same time. In the whole design, the light and dark adjustment of the lighting level, that is, the lamp with dimming function, is the key. At present, many residential and office buildings are installed with rotary dimmers. These light modulator switches are implemented by hardware. However, the infrared remote controller should use an intelligent lighting system to adjust the brightness and darkness of the lamp through the infrared signal of the remote controller. Therefore, it is necessary to find a way to achieve light modulation. In order to understand this method, starting from the principle of the bulb itself, you can adjust the light and shade of the bulb by changing the voltage on the bulb. On the hardware handle, adjust the resistance to change the voltage at both ends of the lamp. In the intelligent lighting system developed by us, we use SCM to change the function of bidirectional transistors and the line angle voltage of lamps. The principle of changing the shutter voltage by changing the shutter angle is shown in Fig. 3 [9–11].

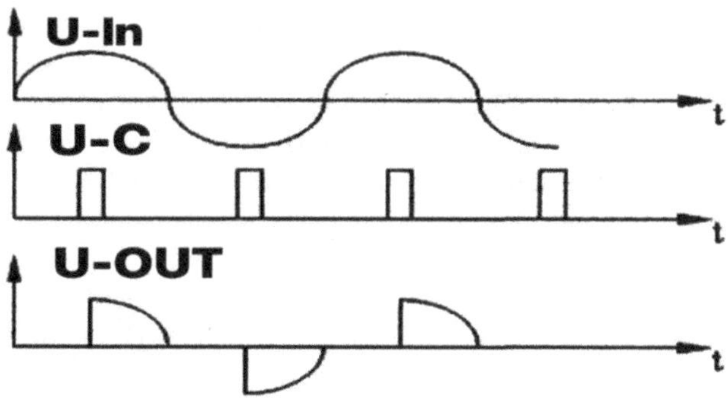

Fig. 3. Change of power supply voltage by crystal tube control

3 Sensor Design

3.1 Design of Environmental Brightness Sensor Module

The main component of the ambient light sensor module is photoresist. The peak spectral response of photoresist is close to the visual sensitive area of the human eye, and the wavelength is 555mm. On the other hand, photoresist uses the principle of photoelectric semiconductor effect, and its resistance varies with the intensity of incident light. The lower the incident light intensity is, the lower the resistance value is; The lower the incident light intensity is, the longer the reaction time is, and the resistance value increases. The detection module is encapsulated with epoxy resin, featuring small size, high sensitivity, fast response and high reliability. If the ambient brightness is detected when the detected light intensity exceeds the set value, the output level is low, otherwise the output level is high. The function of photoresist is to convert the brightness of ambient light into analog voltage, and then use a computer amplifier to transmit accurate digital signals to the microcontroller. In order to avoid the disadvantage of small light area, the detector must be reasonably placed around the building to accurately determine the intensity of natural light [12].

The design of passive thermal infrared sensor. The infrared heat dissipation sensor sends out electrical signals to detect infrared radiation from humans or animals. This.is an infrared energy sensor, which can detect the radiation of objects. The sensor of this circuit is a P2288 passive infrared heat release sensor, which is mainly composed of three parts: (1) Fresnel filter chip, as a bandpass filter, can match human radiation with a wavelength of 9 to 10 microns μ M infrared axis, which distinguishes people from animals. Fresnell lens can decompose thermal infrared signal into thermal infrared sensor, and divide alarm into several bright areas and dark areas. When the object moves within the warning range, the electrical signal of the infrared heat sink sensor can be changed by changing the temperature, thus changing the electrical signal of the infrared heat sink. The experiment shows that if there is no Fresnel lens, the infrared heat dissipation sensor can detect a distance of about two meters. With Fresnel lens, the detection distance can be increased to more than ten meters. Thermoelectric ceramic materials and pyrothermic infrared sensors convert changes in infrared radiation energy into electrical signals, that is, thermoelectric conversion is carried out through filter chips.

In the design, the human body can transmit about 10mV infrared ray by filtering and amplifying M with Fresnel filter chip. The infrared signal is detected by the Fresnel lens on the thermal infrared sensor. Then, the thermal infrared sensor converts the infrared energy into electrical signals through the thermoelectric conversion of the filter chip. Finally, the signal processor and logic circuit of the infrared detector are used for processing. If the value of output logic is 1, it indicates that someone is present; Otherwise, it means no one [13].

3.2 Control Part Design

(1) Delay time selection circuit

The time of the delay chain is determined by the timer in the microcontroller. The delay is selected to be set in the PA output of the microcontroller, mainly because the lighting must be delayed when the ambient light is dark. Set PA0–PA3 outputs with different delays. When it is detected that PA0–PA3 switches are not closed, the system monitors the lighting delay according to the set initial value. When PA0–PA3 switches are closed, the program starts to check the status of each port in the order of PA0–PA2, and the system sets different port delays according to different signal inputs [14].

(2) Single wing control unit

ATMEGA 16 uses RISC architecture as the main controller. This series of chips has the advantages of high efficiency, low power consumption, strong RISC processing capacity, large storage capacity, high cost performance, etc. The chip has 32 programmable I/O ports, 1KB SRAM and 8-way 10 bit ADC, meeting the system requirements. The photoelectric sensor converts the light intensity into electrical signal through a/d converter, and then converts it into digital signal. The system controller calculates the control variable according to the detected light intensity and the system input command to meet the light control requirements. The minimum schematic diagram is shown in Fig. 4.

(3) Display circuit

The system adopts external display configuration to display the operation status and changes of operation parameters of each device in real time. In order to save system cost, this design uses LED as external display to display the current state of the system. Based on the principle of photoelectric effect, the semiconductor photoresist was prepared. General materials include selenium, cadmium sulfide, sulfide pencil, sulfide secretion, cadmium selenide, zinc selenide, gallium arsenide, silicon, etc., among which the most important is the photosensitive resistance of photosensitive cadmium sulfide. Cured CDS photoresist, GL5516 series, 5–10 Ω photoresist and 0 Ω black resistor are used in the design [15]. The design consists of 5.1 Ω resistance, and the voltage fluctuation range is shown in Fig. 5 below:

The specific scope is shown in the following formula:

$$V1 = 5 * (5/(5 + 5.1)) = 2.47\ \text{V} \tag{1}$$

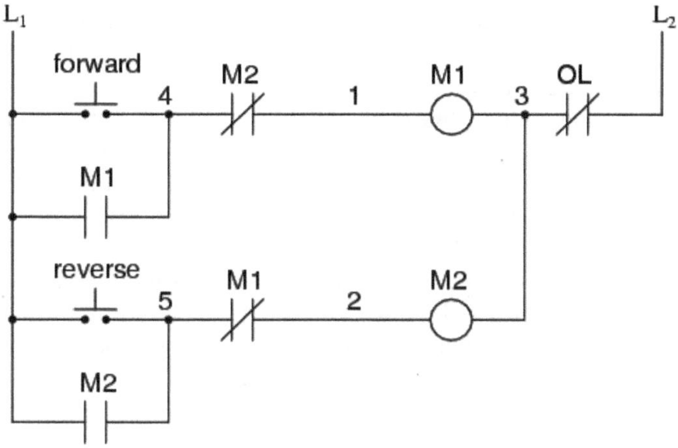

Fig. 4. Minimum circuit system

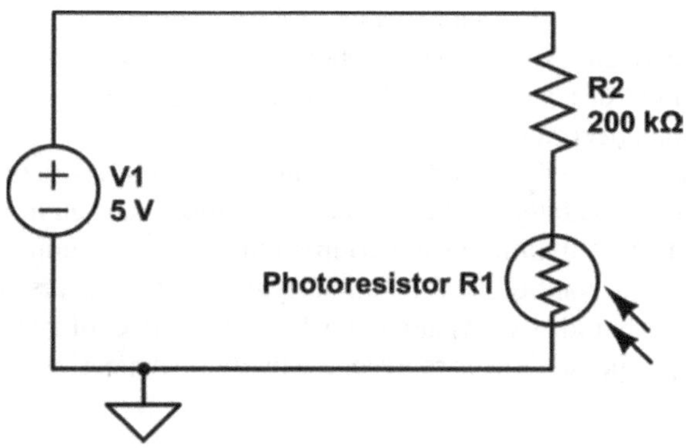

Fig. 5. Photoresistor control circuit

$$V2 = 5 * (10/(10 + 5.1)) = 3.31\,\text{V} \tag{2}$$

$$V3 = 5 * (100/(100 + 5.1)) = 4.87\,\text{V} \tag{3}$$

The working principle of the output control circuit is that a machine recognizes the output signals of the light detection circuit and the sensor processing circuit. PC0 microcontroller outputs control signal. If there is strong light or weak light in the room, and no one is in the room, the PC0 microcontroller output is high, and the relay does not work. The light is off. When the indoor light is dim and there is someone in the room, the PC0 pin output is low, the relay works and the lighting is normal.

3.3 Shape of Interior Lighting

The basic form of lighting is the design of points, lines and surfaces, and the point light source is the reference of the phase. For relatively small single tube lighting, such as coil, ceiling light, light bulb, etc. D. In the middle of lighting design. Light source is the most common, but the target is very clear. Stable and orderly, centralized lighting, clear alarm effect; Light source mainly refers to:

Appropriate combination of linear light sources, including linear and curved light sources, can produce straight lines. The feeling of seriousness and tranquility has the effect of expansion and guidance, and the curve gives a feeling of flowing beauty. It has a strong decorative effect; Surface light source refers to the light flowing through the surface of the large surface cladding.

Linear light source can make lighting more uniform, softer and transparent, creating more artistic effects for people. Intelligent use is not required for indoor lighting, depending on the shape of the room. Lamps of the same shape make the atmosphere of the room an ideal place for people to be comfortable.

3.4 Ambient Lighting, Decorative Effect of 4-Lamps

The artistic treatment of lighting has produced a decorative effect that no decorative tool can match. Suggestion. But an excellent lighting effect is not easy to achieve, but through everything related to it. Analyze the means sensibly and use them correctly. With the development of social productivity. Also the appearance changes. Reflecting its exquisite style and decoration.

The beauty and fashion of modern society. In modern society people live and work, As physical objects in the interior, the existence of lamps plays an important role. It's a decorative element. As a lighting designer, in addition to the requirements for interior lighting, Environmental conditions, decoration style and other factors affect the choice of lamps, and attention should also be paid to the decorative effect of lamps. The decorative personality of lamps and lanterns is represented by their shape, size, material, color and other components.

4 Conclusion

With the development of science and technology, people's lives have changed, and the requirements for buildings and housing are becoming higher and higher. This system is suitable for building intelligent lighting. Modular design, simple and feasible; It can be used in accommodation, jewelry, clothing stores and other places with high lighting requirements. The integrated circuit module is small in size, does not affect the appearance, and has a wide range of applications. The optical matching system developed by the system uses multi-sensor light collection, ZigBee transmission frame, external light source and equipment to supplement and slow down light in the system. Compared with the previous intelligent control system, the system uses multiple sensors to collect complete information. The system passes the physical test, and each circuit module has low power consumption, high optical accuracy, simple detection circuit and fast information acquisition speed. Aiming at the shortcomings of traditional lighting system, an improved design of indoor lighting system based on intelligent detection technology is proposed. Using infrared and pyrolytic human body temperature control systems, combined with the two conditions of human body and environmental brightness adequacy evaluation, signal processing, analysis and decision-making of a single ATMEGA16 machine can control lighting equipment more accurately. Strong humanization. The system is designed to control the lighting of large indoor spaces such as schools and shopping centers.

Acknowledgments. This work was financially supported by Key Project of Philosophy and Social Sciences in Anhui Province: Design and Implementation of Multi-view Visualization for Dynamic Monitoring of Epidemic Information (No. 2022AH053126); Key Project of Philosophy and Social Sciences in Universities of Anhui Province: Research and Implementation of Virtual Reality Display for Anhui Architectural Cultural Heritage Information (No. 2022AH053124); Key Project of Natural Science Research in Universities of Anhui Province: Research on Cortex_A9 Formaldehyde "Superposition" Pollution Detection Mechanism for Indoor Space (No. KJ2021A1507).

References

1. Prajakta, P., Priyanka, P., Tanuja, N., Mohan, T., Nayana, J.: Performance enhancement of dual axis solar tracker system for solar panels using proteus ISIS 7.6 software package. Glob. Trans. Proc. **2**(2) (2021)
2. Hudha, W.N., Kholilul, M.A., Budi, S.A., Kuat, S.: Design and development of heart rate per minutes based on atmega16 microcontroller with alarm warning. IOP Conf. Ser.: Mater. Sci. Eng. **835**(1) (2020)
3. Kumar, D.A., Mrinal, M.: Experimental realization of 2D chaos synchronization using microcontroller ATMEGA 16 and applied in a secure PIN transaction. Int. J. Simulat.: Syst. Sci. Technol. (2020)
4. Cahya, N.L.E., Bayu, F.: Pengembangan trainer pembelajaran aplikasi mikrokontroler atmega 16 menggunakan bahasa pemrograman C dengan codevision AVR DAN downloader USB ASP. Teknologi dan Kejuruan: Jurnal Teknologi, Kejuruan, dan Pengajarannya **41**(2) (2018)
5. Alkapon, I.F., Nandang, T., Decy, N.: Implementasi Sistem Perparkiran Otomatis dengan Menentukan Posisi Parkir Berbasis RFId. ELKOMIKA: Jurnal Teknik Energi Elektrik, Teknik Telekomunikasi, Teknik Elektronika, **4**(1) (2018)
6. Naga Mallik, R.S., Neeraja, S.: Bluetooth based home robotization by using atmega16 microcontroller. Int. J. Smart Home **12**(1) (2018)
7. Mufti, F., Zakki, F.A.: Pemanfaatan Teknologi Eco Push Button Sebagai Pengaman Box Brankas Berbasis ATMEGA 16. LINK **24**(1) (2018)
8. Sriwati, N.I.I., Musrawati, Baco, S., Achmad, A.S.A., Umrianah, E.: Early leakage protection system of LPG (liquefied petroleum gas) based on ATMega 16 microcontroller. IOP Conf. Ser.: Mater. Sci. Eng. **336**(1) (2018)
9. Suprianto, B., Nidzom, S.: Wireless sensor network based Atmega16 microcontroller as temperature and current monitoring system on distribution network transformator. IOP Conf. Ser.: Mater. Sci. Eng. **288**(1) (2018)
10. Kadek Agus Sara Sawita, I., Wayan Supardi, I., Gusti Agung Putra Adnyana, I.: Alat Monitoring Suhu Melalui Aplikasi Android Menggunakan Sensor LM35 dan Modul SIM800L Berbasis Mikrokontroler ATMega16. Buletin Fisika **18**(2) (2017)
11. Gede Surya Merta, I., Gusti Agung Widagda, I., Bagus Alit Paramarta, I.: Perancangan Alat Ukur Kadar Alkohol Menggunakan Sensor Mq-3 Berbasis Mikrokontroler Atmega16. Buletin Fisika **18**(2) (2017)

12. Ruzianto, R.S., Setiyono, B., Sumardi, S.: Perancangan plant pencampur air menggunakan kontrol pid untuk pengaturan suhu cairan berbasis ATMEGA16. Transmisi **19**(2) (2017)
13. Jir, A.H., Ferry, H., Syaifurrahman, S.: Designing of robot gamelan music using ATmega 16 microcontroller. IAES Int. J. Robot. Automat. **6**(2) (2017)
14. Nazilah, C.A.N., Muhammad, H.G., Maharani, R.N., Iswanto, I.: Uninterruptable power supply based on switching regulator and modified sine wave. Int. J. Electric. Comput. Eng. **7**(3) (2017)
15. Susanti, V., Martides, E., Mirdanies, M., Prawara, B., Kristi, A.A., Junianto, E.: Design system of high-velocity oxygen fuel (HVOF) thermal spray coating based on computerization. Key Eng. Mater. 4263 (2017)

Pricing Basis System Planning of New Electric Power System

Ye Ke(✉), Fangshun Xiao, Cong Zeng, Xuemei Zhu, and Ying Wang

State Grid Fujian Economic Research Institute, Fuzhou, China
keye901008@163.com

Abstract. Based on SWOT analysis of the current power engineering pricing basis system, SWOT analysis method is used to objectively analyze the current power engineering pricing basis system, and analyze the impact of new technology and new forms of development in the power field on the power engineering pricing basis and the opportunities. The construction of the pricing system framework for the electric power construction project of the new electric power system led by electric power enterprises not only takes into account the inheritance of the original system, guarantees the universality and pertinence of the pricing basis, but also combines the current dual-track system of list pricing and quota pricing in China. The valuation system framework based on the three level model of "foundation - rule - application", comprehensive quota and bill of quantities, and through the estimation of the preliminary settlement is established, and the working method is proposed to improve the system framework in the future.

Keywords: New Electric Power System · Electric Power Engineering · Pricing Basis System · Accurate Control · Project Valuation

1 Introduction

The new electric power system is based on the premise of carrying the internal requirements of achieving carbon peak carbon neutrality, implementing the new development concept, building a new development pattern and promoting high-quality development, ensuring the safety of energy and electricity as the basic premise, meeting the power demand of economic and social development as the primary goal, maximizing the consumption of new energy as the main task, and taking the strong smart grid as the hub platform.

Supported by the interaction of source and network, charge and storage and multi-energy complementarity, it has the basic characteristics of clean and low-carbon, safe and controllable, flexible and efficient, intelligent and friendly, open and interactive power system. However, due to the rapid technological development of the new electric power system, the pricing basis system of electric power engineering has some problems, such as small coverage, incomplete specialties, inconsistent preparation methods and actual conditions, disconnection between price and market, excessive dependence on industry quota, and inability to realize market pricing mechanism.

Y. Pei et al. (Eds.): IC 2024, LNEE 1214, pp. 11–19, 2024.
https://doi.org/10.1007/978-981-97-4193-9_2

In the face of the new demands of marketization, refinement, informatization and internationalization of cost management in the power industry, it is necessary to further innovate and improve the pricing basis system of power engineering, better guide the whole life cycle investment and cost control of power projects, and provide the basic basis for the optimal allocation of resources and value management of power enterprises [1–3].

2 SWOT Analysis of Current Power Engineering Pricing Basis System

2.1 Analysis on the Advantage of Current Power Engineering Pricing System

According to the content of the project pricing, the quota pricing system is relatively perfect, thermal power and power grid engineering categories are complete, and renewable energy is gradually improving; The list pricing basis system has been developed successively since 2003. Thermal power and power grid projects basically meet the demand, but there is a temporary lack of list pricing basis for wind power, photovoltaic and other types. In terms of the compilation method of engineering valuation basis and the measurement of resource consumption, a relatively mature theoretical method system has been formed at present, and there is a relatively clear price formation and price adjustment mechanism. In terms of engineering valuation basis application, the electric power industry has basically formed a relatively standardized pricing basis application behavior, so that the investment control, project cost management and engineering audit of electric power enterprises can rely on the basis of evidence.

2.2 Analysis of Disadvantages of Current Power Engineering Valuation Basis System

In terms of the content of the engineering valuation basis, the quota valuation system of the whole life cycle has not been established, and the engineering types covered by the electric power engineering valuation basis are not comprehensive. The mode of bill of quantities valuation has shortcomings in short execution time, whole life cycle, whole project type, new technology and new process in electric power industry. In terms of the compilation method of engineering valuation basis, the reform of electric power system makes the traditional resource consumption measurement method not applicable, and the current expert experience method has some problems such as quota content, consumption and actual mismatch. Factor price formation and adjustment mechanism are not in line with the market. In terms of the application of engineering valuation basis, engineering bidding and engineering settlement audit over-rely on industry quota, and cannot realize the mechanism of market pricing.

2.3 Opportunity of Current Pricing Basis System for Power Engineering

Energy transformation and innovative development will gradually realize green, low-carbon and efficient clean energy in the future. The development of energy and electricity provides soil and opportunities for the development of engineering pricing basis.

Whole process engineering consulting, whole life cycle cost management and other management technologies, big data, artificial intelligence. Blockchain and other new technologies provide methods and means for the innovation of pricing basis for power engineering. The successful experience of international cooperation projects of electric power enterprises lays a foundation for improving the wide applicability of the pricing basis of electric power engineering and helps the multi-angle development of the cost management standard of electric power engineering in the international electric power field [4].

2.4 The Challenge of Current Pricing Basis System for Power Engineering

The pricing basis of power engineering needs to be further improved in order to adapt to the control of investment cost of various types and stages of energy and power engineering. On the one hand, the pricing basis of electric power engineering should meet the requirements of the new management mode, on the other hand, the compilation method of the pricing basis of engineering is different from the advanced technical means. At present, the pricing basis of electric power engineering in China's electric power industry is not well adapted to the differences between the management of international engineering and overseas investment projects and the foreign pricing basis, and no adjustment method has been proposed to adapt to the foreign pricing basis.

3 The Influence of the Development of New Power Technology and New Business Form on the Basis of Project Valuation

The technological innovation and application in the field of electric power promote the constant adjustment of the pricing basis of electric power engineering to adapt to the development trend of new technologies in the electric power industry and timely enrich the element types of the pricing basis system of electric power engineering. The rapid development of new business forms provides an opportunity for the expansion of the application scope of the pricing basis system of electric power engineering and the optimization of the overall structure. The pricing basis of electric power engineering needs to be supplemented and perfected to effectively meet the pricing demand of new business forms of electric power.

3.1 Electric Power Technology Innovation

The power industry has made continuous breakthroughs in science and technology and promoted the improvement of the technical level of the industry. In the field of hydropower, it has mastered the world's leading manufacturing technology of giant hydro-generator sets with a capacity of one million kilowatts. In the field of thermal power, major breakthroughs have been made in the technology of secondary reheating boilers and heavy-duty gas turbines. In the field of nuclear power, third-generation nuclear power technology continues to advance; In the field of new energy, China's first large-scale offshore wind farm was built, and the demonstration and application of solar photovoltaic thermal power generation technology were carried out. Important progress was made in research on new efficient solar cell technology. In the field of power grid, the company has comprehensively conquered key core technologies such as ±1100 kV UHVDC transmission, successfully developed the world's first set of UHV GIL equipment and achieved mass production [5, 6].

3.2 Optimization of Power Structure

To optimize the power structure and promote the clean development of power is the trend of global energy transformation and development. In the power sector, efforts have been made to explore and practice diversified energy and power investment, giving birth to new forms of energy and power business aimed at improving the quality and efficiency of energy development and enhancing the capacity to use clean energy. New forms of electric power development, such as integrated energy, electric energy storage, distributed energy, micro-grid, and electric vehicle charging, are developing rapidly in terms of technology and commercial application [7].

3.3 Application of Electric Digital Information Technology

With the continuous development of Internet, Internet of Things, blockchain, big data, artificial intelligence, edge computing and other technologies, as well as breakthroughs in quantum computer and 5G/6G technologies, modern information technology is applied in the field of power engineering construction to build smart construction sites, smart grids and smart power plants, and gradually realize remote monitoring, real-time recording and data sharing for construction and operation of engineering projects. We will promote the digitalization, intellectualization and intelligentization of the power industry. The rise and comprehensive promotion of big data, cloud computing and blockchain technologies will effectively realize the sharing and exchange of industrial cost data. BIM, as a platform carrier, can realize the information sharing of the whole life cycle of buildings, connect data, processes and resources at different stages of the whole life cycle, and support the integration of the industrial chain of the construction industry. It provides the basis for the establishment of the valuation basis system of the whole life cycle of the project [8–10].

4 New Electric Power System Valuation Based on System Model Top-Level Design

Based on the development of the above new electric technology and new business forms and the improvement of the pricing basis system of electric power engineering, this paper, on the basis of the original pricing basis, takes solving the existing pricing problems as the fundamental starting point and establishing the pricing system in accordance with law and compliance as the ultimate goal, and sets up a three-level framework model of "base layer - rule layer - application layer". A new system of pricing basis for power grid construction projects conforming to the characteristics of power grid cost management and convenient for application is formed.

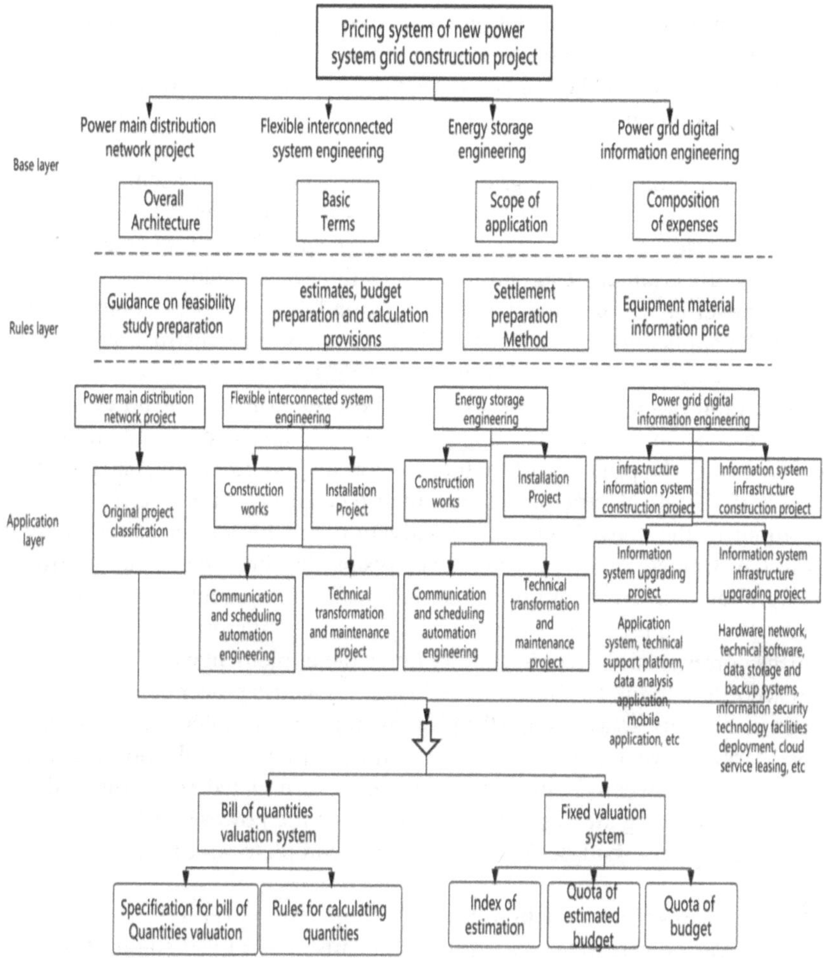

Fig. 1. Pricing system of new power system grid construction project

As show in Fig. 1, the base layer pricing basis is at the top of the pricing system framework of the new electric power system grid construction project, and it coordinates the standardization and supports the basic standardization of the pricing of the new electric power system grid construction project. It mainly includes the overall structure of the pricing system, basic terms, scope of application, cost structure and other basic instructions.

The pricing basis of the rule layer is located in the middle layer of the pricing system framework of the new electric power system grid construction project, connecting the pricing basis of the base layer and the application layer, and standardizing and supporting the estimation, budget estimate and budgeting of the new electric power system grid construction project. It mainly includes general standards such as research and development guidelines, budgeting and calculation provisions, information prices of equipment and materials, and settlement preparation methods.

The valuation basis of the application layer is at the bottom of the valuation system framework of the grid construction project of the new electric power system. It is divided into the main distribution network project, the flexible interconnection project, the energy storage project and the grid digital information project according to the engineering types. The corresponding valuation basis is set respectively according to the list and quota. Including bill of quantities valuation specification, engineering quantity calculation specification and estimation index, estimate quota and budget quota. Based on the content setting of the model and application layer, this paper puts forward suggestions on the application layer of the pricing system of the construction project of the new power system, as shown in the following Table 1.

The valuation basis at different levels of the valuation system framework of new electric power system grid construction project should be standardized and formulated in accordance with the principle of mutual support and coordination, and applied to the valuation work of new electric power system grid construction project.

According to the above content, a relatively complete and clear hierarchy of the pricing system framework of the new power system has been built, but the framework should be gradually filled according to the technical development of power engineering. It is suggested that in recent 1–2 years, based on the existing pricing basis in the power industry, the corresponding supplementary quota or supplementary cost standard can be prepared according to the characteristics of various electric power construction projects of new power systems, so as to solve the prominent pricing problems at the present stage. In the long-term 3–5 years, the valuation basis fully applicable to the power construction project of the new power system shall be gradually formulated and formed, that is, the valuation basis covering all stages of the estimate of the power construction project of the new power system shall be formed to form a complete set of independent pricing basis.

The electric power construction project of the new power system involves many industries such as power, water conservancy, chemistry, heat and so on. It is suggested that the country or related industries should take the lead, and all industries should jointly promote the establishment of the pricing system. In particular, it is necessary to establish unified project pricing rules from the national level: including improving the construction cost item composition, to meet the construction project cost control and general project contracting needs; Perfect bill of quantities format, project composition, cost composition, compilation method, unified national bill of quantities valuation method and valuation rules; To unify the compilation rules of consumption quota, the comprehensive index index of project cost and the release standards of labor and material price information, and promote the formation of a unified and open construction market.

Table 1. Application layer of pricing system of new power system grid construction project

No.	Type of project	Valuation basis name
	Main and distribution network engineering	Specification for valuation of bill of quantities for main and distribution network engineering
		Specification for calculation of quantities of main and distribution network engineering
		Estimation index of main and distribution network engineering
		Estimated quota of main and distribution network projects
		Budget quota for main and distribution network projects
		Technical reform and maintenance quota of main and distribution network engineering
	Flexible interconnection system engineering	Specification for valuation of bill of quantities for flexible interconnected systems engineering
		Specification for engineering quantity calculation of flexible interconnected system engineering
		Flexible interconnected system engineering estimation index
		The budget estimate quota of flexible interconnected system engineering
		Flexible interconnected system engineering budget quota
		The technical reform and maintenance quota of flexible interconnected system engineering
	Energy storage engineering	Specification for valuation of bill of quantities for energy storage engineering
		Specification for engineering quantity calculation of energy storage engineering
		Energy storage engineering estimation index
		The estimate quota for energy storage projects
		Budget quota for energy storage projects

(*continued*)

Table 1. (*continued*)

No.	Type of project	Valuation basis name
		Technical reform and maintenance quota of energy storage engineering
	Power grid digital information engineering	Specification for valuation of bill of quantities of power grid digital information engineering
		Specification for the quantity calculation of power grid digital information engineering
		Estimation index of power grid digital information engineering
		The estimate quota of power grid digital information engineering
		The budget quota of power grid digital information engineering
		The technical reform and maintenance quota of power grid digital information engineering

5 Conclusion

This paper uses SWOT analysis to analyze the advantages, disadvantages, opportunities and challenges of the current power engineering pricing basis system. It is urgent to establish a unified and standardized pricing basis system for the new electric power system. Combined with the current situation of electric power engineering construction, the influence of new technology and new business forms on the electric power engineering pricing basis system is analyzed, and the top-level design framework of the new electric power system pricing basis system model is studied and formulated, which provides a reference and reference for the accurate control of engineering cost in the future power field and the improvement and development of more applicable electric power engineering pricing standard by enterprises.

Acknowledgment. This work is supported by the management consulting project of Fujian Economic Research Institute of State Grid. The name of the project is the Research on the grid construction Pricing Basis system Architecture suitable for new electric power system.

References

1. Jia, H.-y.: Research on the development strategy of power generation enterprises entering the electric selling market under new electric power system reform. In: Proceedings of 2018 3rd International Conference on Education and Management Science (ICEMS 2018), pp. 182–186 (2018)

2. He, Y.-G., Li, Y.: Evaluation of power demand-side management factors in the new electric power system reform based on FAHP. In: Proceedings of the 2016 5th International Conference on Energy and Environmental Protection (ICEEP 2016) (2016)
3. Liu, B., Chen, B.-n., Li, J., Xia, R-h.: A new power system phase identification method based on difference value. In: Proceedings of 2015 3rd International Conference on Machinery, Materials and Information Technology Applications (ICMMITA 2015), pp. 1270–1275 (2015)
4. Wang, Q.: Research on development model of whole process engineering consulting under new situation. In: 2021 3rd International Conference on Economics, Marketing, and Management (2021)
5. Wang, C.-L., Ma, L., Xu, X.-F.: Decision model for "Going Global" of china's electric power technology and equipment. Procedia Comput. Sci. **162**(C) (2019)
6. Gu, Z., Wang, Y., Gao, J.: Evaluation index and concrete method of electric power technology economy. Acad. J. Eng. Technol. Sci. **3.0**(8.0) (2020)
7. Kristian, T., Michal, F.: System level simulation of micro grid power electronic system. J. Phys.: Conf. Ser. **2022**(1) (2021)
8. Shin, Y.K., et al.: Digital information technology use and transnational healthcare: a population-based study on older Russian-speaking migrants in Finland. J. Immigrant Minor. Health **24**(1) (2021)
9. Mudit, K., Rahul Dev, G.: Evaluation of optimum PV tilt angle with generated and predicted solar electric data using geospatial open source software in cloud environment. Sādhanā, **46**(2) (2021)
10. Sun, L., et al.: Implementation and verification of 5G network slicing for smart grids. In: Proceedings of 2021 5th International Conference on Electrical, Automation and Mechanical Engineering (EAME2021), pp. 145–151 (2021). https://doi.org/10.26914/c.cnkihy.2021.044372

Application of Intellectualization in Medical Beauty Space

Yu Gui[1], Shuaipeng Liu[1], Jian Tian[2], Zonghui Bo[1], Yuxin Miao[1], Yuxiao Fan[1], Shunzi Ren[1], and Lei Cui[1](✉)

[1] School of Horticulture and Landscape Architecture, Tianjin Agricultural University, Tianjin, China
36361950@qq.com
[2] Tianjin Urban Renewal Construction Development Co., Ltd., Tianjin, China

Abstract. With the advent of information technology, intelligent medical treatment will be the general trend. At present, there are still many defects in the integration of medical beauty center and intelligent technology, which is reflected in the lack of intelligent and humanized design of medical beauty space. This paper focused on the intelligent design and research of medical beauty space from the perspective of building a digital intelligent medical beauty center. Through intelligent technology, innovative and experiential design of medical beauty space is carried out, and the safety and service of the space are improved, so that the intelligent technology in the medical beauty space can be fully utilized, so as to rationally allocate medical resources and improve efficiency, reduce the medical burden and other aspects to provide a more healthy, safe, intelligent and comfortable diagnosis and treatment environment.

Keywords: Medical Beauty · Intelligence · VR Technology · Sense of Experience

1 Introduction

In recent years, with the mature development of intelligent technology, residential intelligence, medical intelligence, information technology and other new ideas have emerged, some medical devices have introduced intelligence into them, which has a certain reference value for the intelligent design of medical beauty salons [1]. Due to the constraints of the conditions, there are still some shortcomings, intelligence in the overall design of the medical space is still a very small part, only for medical equipment functions and medical technology, with the advent of information technology, the intelligent technology into the design of medical beauty interior space, medical beauty institutions and intelligent technology integration still have many defects, intelligent in the design of human nature is still inadequate performance, improve the intelligent technology. Based on intelligent technology, the design and research of medical beauty space is the focus of this paper [2].

Y. Pei et al. (Eds.): IC 2024, LNEE 1214, pp. 20–27, 2024.
https://doi.org/10.1007/978-981-97-4193-9_3

2 Project Overview

This design project is for the Lido Medical Aesthetic Centre, which is located on the first floor of Block D of Rock Times, 103 Huizhongli, Chaoyang District, Beijing, adjacent to the Bird's Nest and the Water Cube, the main venues of the Beijing Olympic Games, with easy access to the surrounding area and convenient parking. The reception area is the first area that patients come into contact with when they enter the Lido Medical Beauty Centre, and includes a reception desk, a discussion room, a VIP room, a rest area and a pharmacy, providing patients with services such as product introduction and answering questions. It includes a waiting room, consultation room, skin management room, VR technology experience area, laser treatment room, patient recovery observation room and operation room; finally, there is the medical staff office area, which is an area for discussion and research on surgical solutions and medical technology among medical staff, including offices and research rooms (as shown in Figure) This area includes offices, research rooms, etc. (as shown in Fig. 1). Indoor flow analysis: Indoor pedestrian flow lines include customer walking routes and central medical staff walking routes (as shown in Fig. 2). The space can be navigated according to the user's location by means of human and infrared sensors in the intelligent system, based on the intensity of light.

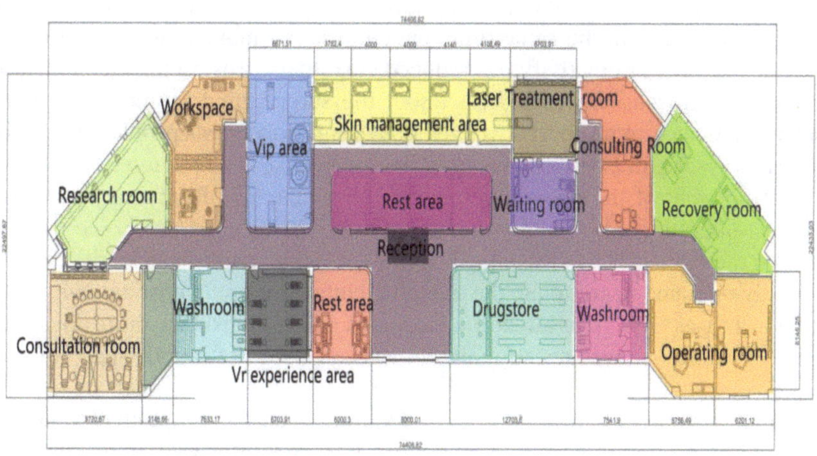

Fig. 1. Functional zoning map

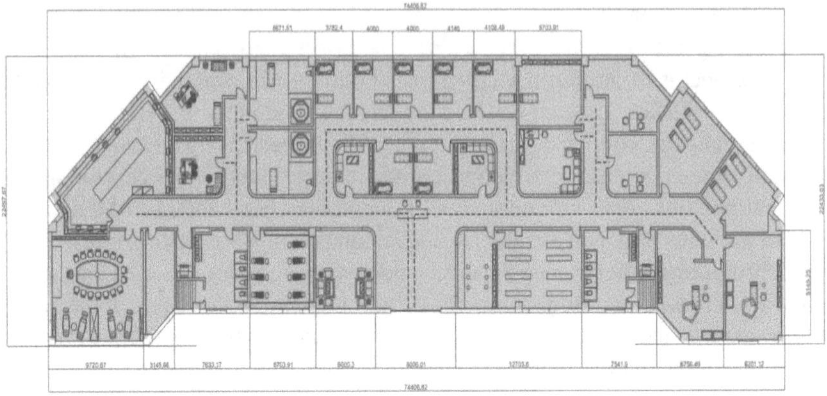

Fig. 2. Movement line analysis diagram

3 Analysis of the Current State of the Interior Space of Lido Medical and Beauty Hospital

Due to the increasing demand of patients, the lack of experience and creativity in the medical aesthetic interior space is gradually revealed. It is because of the poor experience that the psychological problems of pre-operative tension and anxiety of customers cannot be well addressed, and at the same time, patients cannot make a firm decision to have surgery [3]. The lack of creativity has reduced customers' trust in the medical aesthetic centre. In order to reassure customers about the surgery and moreover to make the medical aesthetic centre more trustworthy, this design solves this problem by designing a VR technology vegan experience area in the medical aesthetic centre.

With the advent of the post-epidemic era, hospitals are still facing problems such as imperfect operational protocols, imperfect work stations and no choice of hygienic disinfection methods [4]. At this stage, hospitals still use "hand-push sterilisers", which are manually operated to complete the overall disinfection of the hospital, while manual operation increases the risk of infection. Hospitals need to introduce more advanced and intelligent disinfection systems to maintain the safety of the indoor space of medical aesthetics.

For some customers with high privacy requirements, they hope that some of their information and needs can be kept confidential. The lack of privacy in the current Lido Medical Beauty Hospital can cause problems for some patients. In addition, there is not enough intelligence in the equipment, so that patients cannot understand their own body condition in a timely and effective manner, which makes customers worry about their own safety before surgery and reduces the demand for medical beauty [5].

4 Principles of Intelligent Design in Medical Interior Spaces

Firstly, the intelligent system design is used to create the experience and innovation of the medical beauty centre. VR virtual reality technology is used in the experiential area of the medical beauty centre and the mobile VR virtual APP program is designed to provide a convenient and intelligent network platform for the medical beauty centre. Centre intelligence to provide a convenient and intelligent network platform [6]. Secondly, the principle of safety, the use of intelligent disinfection equipment to disinfect and protect the entire medical aesthetic hospital, the technology has a certain application value. Finally, the service principle, in order to make customers more satisfied with the services of the medical beauty centre, the indoor environment of the centre is upgraded and transformed, and the intelligent detection system is implanted in the indoor environment to effectively regulate the indoor environment and meet the service experience of customers in the centre.

5 Intelligent Design Expressions in Interior Spaces

5.1 Intelligent Creative and Experiential Space Design

VR virtual reality is a computer simulation technology that enables the creation and experience of virtual worlds, which uses computers to generate an interactive three-dimensional dynamic vision with a simulation system of their physical behaviour to immerse the user in that environment [7].

The advantages of intelligent design can be felt to the greatest extent in this area. After the patient enters the experience area, virtual reality technology is used to pre-set the scene, a scanning system is used to analyze data on the patient's face and make a data report, and adjustments can be made on the computer according to the patient's own ideas. Finally, VR virtual reality technology is used to simulate the effect intuitively and give feedback to the patient, relieving the patient's post-plastic concerns, reassuring patients that the plastic surgery process is conducive to maintaining a good state of mind and more positive and trusting cooperation with the surgeon in receiving surgical treatment. As shown in Fig. 3.

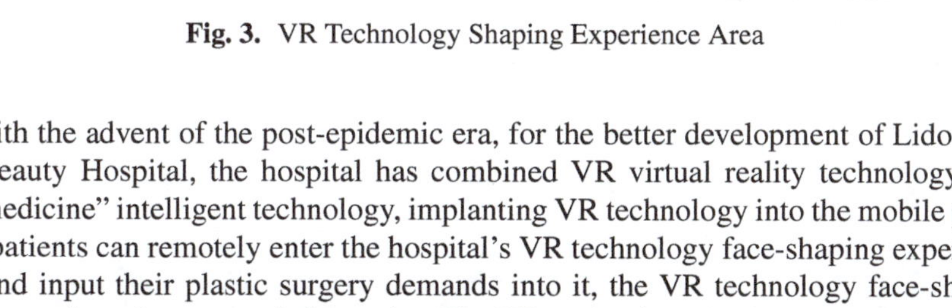

Fig. 3. VR Technology Shaping Experience Area

With the advent of the post-epidemic era, for the better development of Lido Medical Beauty Hospital, the hospital has combined VR virtual reality technology with "telemedicine" intelligent technology, implanting VR technology into the mobile phone APP, patients can remotely enter the hospital's VR technology face-shaping experience area and input their plastic surgery demands into it, the VR technology face-shaping system receives the remote information The VR technology body system receives the remote information and quickly generates simulated results for the patient's reference to better serve the patient. The adoption of this intelligent technology has expanded the business of Lido Medical Beauty Hospital to a national scale, enabling the hospital to develop better in the post-epidemic era. The so-called "telemedicine" is an "internet" service technology that connects doctors and patients, allowing patients to network with doctors at home and express their plastic surgery demands, and the Lido Aesthetic Hospital mainly uses a mobile phone app system, which patients use to Patients can use the app to consult and understand the hospital's plastic surgery program, which is easy and convenient to operate [8]. Open the mobile phone app, follow the prompts, and send their plastic surgery requirements by the system automatically to the hospital's medical aesthetician, who will give the corresponding data analysis and simulated graphics according to the patient's description through VR virtual reality technology.

5.2 Intelligent Secure Space Design

Prior to the epidemic, a hand-push type hand disinfector was used in the Lido Medical and Aesthetic Hospital, where doctors had to press its button to disinfect their hands, which increased the chances of cross infection while pressing the button. This design changes the hospital hand disinfection equipment from the original push-button type to an induction type, which only requires the hand to be placed in the inductor and the equipment will disinfect the hands. The disinfector uses a combination of disinfectant spray and drying technology and does not use UV disinfection measures, which is less harmful to human skin. And the principle of this equipment, the construction of large disinfection equipment, used at the entrance to the operating theatre, doctors and patients will be disinfected by induction of the whole body before entering the operating theatre, for the exposed part of the skin, the sensor will automatically adjust to the disinfectant spray drying session, only the clothing covered part of the light UV disinfection to ensure

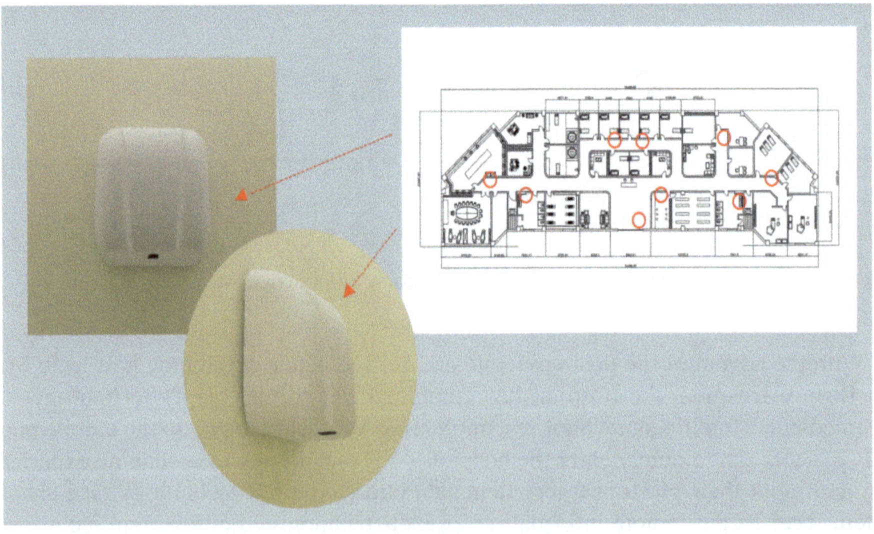

Fig. 4. Distribution of induction disinfection equipment

that the operating theatre is sterile [9]. The design and application of this equipment reduces the chances of cross-infection between doctors and patients and greatly improves the hygienic safety of the space within the medical aesthetic room (as shown in Fig. 4).

5.3 Intelligent Private Space Design

From an environmental psychology point of view, privacy is expressed as control over others, the right to choose to exchange information with others, that is, it isolates external influences when one is needed: when you want to co-exist, you can also do so with others. Therefore, the definition of "privacy" varies from discipline to discipline, but can be broadly divided into two types: the first focuses on avoiding interaction with other individuals or groups; the second definition does not emphasize withdrawal, but it implies that privacy involves the opening and closing of personal information.

The VIP space in the Medical Beauty Centre is more fully functional, with the functions of treatment, dressing, therapy, bathing and resting, and is an integrated service space design with a certain degree of privacy. The VIP room not only ensures the privacy of the patient, but also makes appropriate adjustments to the comfort of the patient, using intelligent systems for improved design, as customers inevitably experience tension and anxiety during the consultation process. The VIP room uses data information such as the location of the patient, sound perception following through intelligent sensors, and humidity control of the room space to quickly collect and adjust the environment in a timely manner. After data analysis, the system analyses the customer's mood based on their state, movements and expressions in the centre and uses the results to adjust the lighting, music and indoor air to relieve their tension, thus achieving the best treatment results and receiving good feedback from the customer, ensuring patient comfort and privacy. The VIP room is a private space in the centre where many customers do not want to be disturbed by the outside world and where they can protect their information and consumer needs [10].

5.4 Intelligent Service-Oriented Space Design

The interior space for medical aesthetics is a large service space, providing services to customers in the field of medical aesthetics. This large service space contains many small service spaces, and this design combines intelligence and service to effectively design and rationalise the space environment. First of all, when entering the Lido Medical Beauty Centre, customers can choose a robot to provide consultation according to their wishes, and then the robot will take the patient to the rest area to wait. The treatment room is a relatively private space, where customers are given a physical examination, and according to their physical condition and needs, a specific plan is reasonably formulated and an appointment is made with the customer for the operation.

An intelligent induction sterilisation area is set up at the entrance to the operating theatre space to sterilise people entering. This device is set up to greatly reduce the risk of infection for customers, and an intelligent induction system for the environment is installed in the operating theatre. During surgery, the intelligent induction system automatically detects the indoor environment and displays relevant values such as air quality on the intelligent display to regulate the environment and reduce customers' anxiety.

After the plastic surgery is completed, the medical staff will take the client into the recovery room; the recovery room is where the client will be medically observed after the surgery, and some clients who have undergone larger surgeries will need to go through the hospitalisation procedure for rehabilitation treatment. The recovery room is also equipped with an intelligent induction system, which focuses on calming down the client for a better recovery; it is intelligently regulated by light, temperature and humidity. The intelligent design of the service space not only saves manpower but also greatly enhances the comfort of the space.

5.5 The Use of Intelligent Creative Devices

Design and invention of wearable medical equipment - post-operative bracelets, post-operative bracelets are the use of a more complete information system, with the function of detecting post-operative data, patients who have done plastic surgery wear post-operative bracelets, the tiny chip in the bracelet will automatically bind to the human body, mainly to detect the human surgical site, in order to confirm the post-operative site's condition and characteristics, real-time monitoring of the post-operative site metabolism, recovery status and other data records, when the value appears to be a problem, it will automatically display a warning sign to remind the patient to go to the hospital for a review and timely treatment.

The device is particularly functional as it can not only be monitored remotely but also upgraded automatically to avoid bugs in the device and to reduce unnecessary troubles for the patient [11].

6 Conclusion

In this paper, by integrating and sorting out the development of the concept of intelligence, and by analyzing the factors related to the principle of intelligence and interior space design, we have summarized the problems that arise in the current interior space design of Lido Medical Beauty. The application of intelligent technology to medical and aesthetic institutions provides patients with high-quality, efficient and safe medical insurance services; reduces the workload of health care workers, improves work efficiency, promotes medical services and reduces hospital health care costs. Along with China's growing comprehensive national strength, the use of intelligence in the medical space enables rapid processing of information and data, making the quality and efficiency of medical care greatly improved.

References

1. Jeyanthi, S., Brian, H., Nisha, S.: Artificial intelligence improves efficiency in the one stop symptomatic breast clinic. Eur. J. Surg. Oncol. **48**(5) (2022)
2. Suman, S.: Artificial intelligence in nuclear industry: chimera or solution? J. Clean. Prod. **278** (2021)

3. Kanwar Manreet, K., Arman, K., Mehra Mandeep, R.: Machine learning, artificial intelligence and mechanical circulatory support: a primer for clinicians. J. Heart Lung Transplant. **40**(6) (2021)
4. Jeremy, S.: Expert system gradient descent style training: development of a defensible artificial intelligence technique. Knowl.-Based Syst. **228** (2021)
5. Ayres Lucas, B., Gomez Federico, J.V., Linton Jeb, R., et al.: Taking the leap between analytical chemistry and artificial intelligence: a tutorial review. Analyt. Chim. Acta **1161** (2021)
6. Biswajit, J., Sanjay, S., Nayak Gopal, K., et al.: Artificial intelligence-based hybrid deep learning models for image classification: the first narrative review. Comput. Biol. Med. **137** (2021)
7. Nishanth, G.: Artificial intelligence: potential tool to subside SARS-CoV-2 pandemic. Process Biochem. **110** (2021) (prepublish)
8. Song, B., Soria Zurita Nicolas, F., Gyory Joshua, T., et al.: Decoding the agility of artificial intelligence-assisted human design teams. Design Stud. **79** (2022)
9. Anahita, H., Andrea, S., Ashkan, E.: Gender-specific patterns in the artificial intelligence scientific ecosystem. J. Informetrics **16**(2) (2022)
10. Covadonga, D.-S., Antonio, S.-C., Amado, A.B.: The impact of artificial intelligence and big data on end-stage kidney disease treatments. Exp. Syst. Appl. **180** (2021)
11. Tomasz, C.: An estimation of motor yacht light displacement based on design parameters using computational intelligence techniques. Ocean Eng. **231** (2021)

Dynamic Value Control Index of Power Grid Project Investment Under New Policy Environment

Shiyan Mei[1], Gang Sun[1], Kai Hou[1], and Jia Li[2](✉)

[1] Grid Planning Research Center of Guangdong Power Grid Co., Ltd., Guangzhou, China
[2] China Electricity Council Electric Power Construction Technical and Economic Consulting Center, Beijing, China
lijia@cec.org.cn

Abstract. The exponential development of power technology and the policy requirements of "carbon peak" and "carbon neutrality" have got higher requirements for investment management methods aboutpower grid project. This paper analyzes the precision investment decision-making mode ofpower grid project under the new policy environment, sorts out and studies the factors have the main impact which on the cost of project investment, studies and designs the multi-level dynamic value control index, and lays the application and theoretical foundation for the development of Power Grid Project investment control technology under the new policy environment.

Keywords: Power Investment Decision · Power Grid Project · Dynamic Value Control of Project Investment

1 Introduction

The economy developed model of China has take turns from "high-speed growth" to "high-quality". As a supporting industry of national economy, the electric power industry actively responds to the strategy of high-quality development, responds to the Electricity price reform, and aims at lean engineering management. Under background of "carbon peak" and "carbon neutrality", a series of national policies such as power system reform and investment supervision of central enterprises continue to advance, and the technology development of power grid project, the expansion of investment scale, internal management requirements and external construction environment continue to be updated. Power grid enterprises urgently need to improve their power grid investment decision-making level to ensure the effectiveness and compliance of their power grid investment control. Avoid inefficient investments. Therefore, in order to further improve the quality of power grid project investment control work and promote the more standardized and orderly development of the whole process of investment control business, it is need to make a in-depth analysis and research witch about precise investment decision-making mode of power grid project under this new policy environment, so as to scientifically control cost, select best construction plan and improve quality. Promote company's power grid high-quality construction to provide strong support [1, 2].

Y. Pei et al. (Eds.): IC 2024, LNEE 1214, pp. 28–35, 2024.
https://doi.org/10.1007/978-981-97-4193-9_4

2 Multi-level Dynamic Investment Decision-Making Management pattern of Power Grid Project

2.1 Traditional Investment Control Management Concept

The traditional Power Grid Project investment control mode mainly adopts the phased management mode, and different management departments and personnel are responsible for different Phases, leading to the lack of coordination between the management units before and after. The basis used by different units and departments in the preparation of investment estimation or budget estimate is different, and there are differences in the basis of industry, enterprise and local valuation, and the design depth of the investment planning Phase does not reach the depth required by the construction, and the engineering quantity and project valuation cannot be fully refined accurately. Industry, enterprise, local quota and expense standard revision work often lasts several years, the cost of the project is subject to the influence of various policies, market, technology, management and other factors, can not achieve dynamic control, precision is insufficient [3].

2.2 Multi-level Dynamic Investment Decision Management Method

Multi-level dynamic investment decision-making can manage the project investment decision-making behavior dynamically by considering the influence factor from different angles. The investment decision management carries out reverse calculation through the historical settlement data of the project to ensure that the indexes are consistent in the caliber of planning feasibility study, construction and completion handover. The level of investment decision system structure mainly considers the influence of project type and typical project cost, partial project cost, dynamic factor adjustment and other factors on project investment; At the same time, it can be dynamically applied and adjusted year by year by setting a reasonable and universal calculation method.

2.3 Research Implications

The power grid project unit's investment should run through the whole process of construction project, including the feasibility study, engineering design phase, construction drawing design phase, construction phase, completion and production phase. Horizontal comparison of the current power grid enterprise investment control index, power grid enterprise investment in infrastructure projects generally refers to the total cost of a certain infrastructure construction, investment control index is mainly for the cost index, the whole process of investment control has a strict limit control, in the initial design phase to consider the repetition, after the implementation and completion Phase have strict supervision and control function. For the setting of investment control index system of a power grid company, to build an index system with richness, feasibility and applicability, we should take overall consideration and establish it by modules, divide the index into blocks, and implement the index evaluation to the main body [4, 5].

It is very important to electric power enterprises on development planning to make a reasonable and scientific investment decision on power grid infrastructure, which can not only meet the needs of safe operation but also get reasonable return on investment. The multi-angle and multi-level decision-making method of dynamic power grid infrastructure investment make significance to the safe operation and the improvement on economic benefits of electric power enterprises.

3 Influence Factor of Multi-level Dynamic Value Control Index of Power Grid Projects

3.1 Design Idea of Dynamic Value Control Index

Taking power transmission and transformation project investment as an organic system, this paper studies the dynamic system of investment control, and grasps the interaction and relation among the cost elements of power transmission and transformation project investment control in each phase. Among them, the dynamic factor adjustment auxiliary decision index is selected to study and develop the comprehensive adjustment index from five perspectives: the price of equipment and materials, the price level of manpower and machinery, the construction cost, the management and construction conditions, and the basic reserve cost [6, 7].

3.2 Analysis of influence factor on Power Grid Project Investment Cost

The power grid project has the characteristics of long construction time, large investment amount and strong integrity. The influence factor restrict and influence each other, and the relationship is complex. From a macro point of view, government behavior, overall construction level, economic development level, market supply and demand will have an impact on the cost of the power grid project; from a microscopic perspective, natural environment, design level, equipment selection standard, mechanized equipment level and other major factors. Therefore, in order to more accurately control the investment of Power Grid Projects and assist decision-making, it is necessary to systematically identify and carding the factors affecting of power grid projects cost, and establish an auxiliary decision-making system that can dynamically describe the fluctuation of the project investment index under the influence of various factors. After sorting and screening, the influence factor of Power Grid Project cost is shown by Table 1 below [8, 9].

3.3 Selection of influence factor of Dynamic Value Control Indicators

Based on the principle of scientific, comprehensive, systematic, independent and operable index design, this paper further analyzes the influential factors of investment cost that can be quantified and controllable in Power Grid Projects, so as to build the dynamic value control index of Power Grid Projects investment. The dynamic value control indexes of power transformation engineering and line engineering were finally confirmed by the method of sub-analysis and expert investigation, as shown in Table 2 and Table 3 below.

Table 1. Influence factor of power grid project cost

The serial number	Category of factors	factors	sub-factor
1	Scale influencing factor	Factors influencing of power grid project scale	/
2	Market influence factors	Main equipment price factors	/
3		Main material price factors	/
4		influence factor of labor cost	/
5		Factors influencing the price of auxiliary materials	/
6		Factors affecting the price of machinery	/
7		influence factor of earth material price	/
8		Other market influences	The existing quota system and the influence factor of the actual construction market level
9	Project external environmental factors	influence factor of site selection/line selection conditions	Topographic factors
10			Factors affecting altitude
11			influence factor of seismic intensity
12			influence factor of earth resistance mode/ratio of corner tower
13		Factors influencing the time of construction and production	/
14		influence factor of construction site requisition and cleanup fee	Factors affecting requisition and compensation costs
15			influence factor of construction fee balance
16		influence factor of different regional development levels	Factors influencing the level of development between cities
17			influence factor of urban inner region construction scheme
18		Social and environmental factors	/

(*continued*)

Table 1. (*continued*)

The serial number	Category of factors	factors	sub-factor
19		Government planning and local planning influence factor	/
20		Policy influence factor	Environmental Protection Policy
21			Water conservation policy
22			Land acquisition policy
23			Environmental assessment policy
24			Forest survey requirements
25			Other Assessments
26	Design and construction standard factors	influence factor of cost	Factors influencing the change of construction cost of transformer project
27			influence factor of cost change of line engineering ontology
28		Factors influencing accuracy of feasibility study report	/
29		Design level influence factor	/
30		influence factor of survey and design fees	/
31		influence factor of cost management work	influence factor of construction management cost
32			influence factor of construction management cost
33			The cost audit is based on uniform influence factor
34			The influence factors of standard difference such as quota pregauge
35			Factors affecting the difference between standard cost and contract price
36			Contract management factors

Table 2. Dynamic value control index of power transformation project investment

The serial number	First order index	Secondary index	Three level index
1	Dynamic value management index of power transformation project investment	Main equipment price factors	Main transformer/main circuit breaker
2		Main material price factors	The control cable
3			cable
4		influence factor of labor cost	/
5		Installation auxiliary price changes	Installation auxiliary price changes
6		Changes in the price of building materials	sand
7			stone
8			cement
9			steel
10		Factors affecting the price of machinery	/
11		Factors influencing substation site selection conditions	Treatment of ground
12			Resistance to earth
13		influence factor of construction site requisition and cleanup fee	Factors affecting requisition and compensation costs

Table 3. Dynamic value control index of power transformation project investment

The serial number	First order index	Secondary index	Three level index
1	Overhead project investment dynamic value management index	Main material price influence factor	conductor
2			Tower material
3		influence factor of labor cost	/
4		Factors influencing the price change of auxiliary materials	Auxiliary material price changes

(continued)

Table 3. (*continued*)

The serial number	First order index	Secondary index	Three level index
5		influence factor of earth material price change	sand
6			stone
7			cement
8			steel
9		Factors affecting the price of machinery	/
10		influence factor of line selection conditions	influence factor of terrain change
11			Factors influencing the proportion of corner tower
12		influence factor of construction site requisition and cleanup fee	Factors affecting requisition and compensation costs

4 Application Value of Multi-level Dynamic Value Control Index for Power Grid Projects

4.1 Break the Shackles of the Application of Investment Control Index System to Achieve Accurate Investment Control of Power Grid Project

Form a dynamic and adjustable cost control method for Power Grid Project investment, put forward a more scientific and reasonable, perfect and more in line with the actual situation of Power Grid Project cost management dynamic adjustment index, so as to better guide the project investment, project cost, cost control in the whole life cycle of the project, and solve the problem of traditional project investment control. For enterprises to further precise investment control to lay the application foundation.

4.2 Provide the Basis for Resource Optimization Allocation and Value Management of Electric Power Enterprises

The multi-level dynamic value control index of Power Grid Project can accurately grasp the key points of risk control of project cost, identify the main management factors, provide the basis for the optimal allocation of resources and value management of electric power enterprises, and provide data support for cost risk control and material management planning.

4.3 Lead Reform of Power Grid Investment Management and Promote Dynamic Management and Sustainable Development

The concept of multi-level dynamic control brings vitality to the investment control of electric power enterprises, expands the vision of investment management concept, and further combines the coordinated development of management technology in the fields of project management such as risk control, schedule control and quality control, and could promote the dynamic management of Power Grid Projects [10].

5 Conclusion

This paper systematically identifies and sorts out the influence factor of power grid project cost, and designs a multi-level dynamic value control index that can dynamically describe the fluctuation of project investment index under the influence of various factors. This research findings of this paper could help on control Power Grid Project investment more accurately and assist enterprises, it is useful to improve investment decision-making management means, the application of this index can improve the contradictions and problems of traditional investment cost management means, assist power enterprises to optimize the allocation of resources.

Acknowledgment. This work is supported by the management consulting project of Guangdong Power Grid Co., LTD. This topic is titled "Research on the investment index system and analysis evaluation of Power Grid Projects."

References

1. The Benefits and Drawbacks of BIM-Based Engineering Cost Management. Indust. Eng. Manag. **10**(11) (2021)
2. Wang, L., Cao, M.: Application of intelligent measuring robot in engineering cost management. Int. J. Syst. Assur. Eng. Manag. (2021)
3. Wei, G.: Research on construction engineering cost management. Acad. J. Architect. Geotech. Eng. **2.0**(1.0) (2020)
4. Duan, Y., Sun, Y., Zhang, Y., Fan, X., Dong, Q., Guo, S.: Risk evaluation of electric power grid investment in china employing a hybrid novel MCDM method. Mathematics **9**(5) (2021)
5. Wang, Y., et al.: Chinese grid investment based on transmission and distribution tariff policy: an optimal coordination between capacity and demand. Energy **219** (2021)
6. Construction and Empirical Analysis of Intelligent Prediction Model for Electric Grid Project Cost. **446**(4) (2020)
7. Peng, Y., Ding, Y., Yu, Y.: Grid project cost control research based on data mining. In: Proceedings of 2016 International Conference on Advanced Materials Science and Environmental Engineering (AMSEE2016), pp. 296–299 (2016)
8. Do, P., Chow, C.W.K., Rameezdeen, R., Gorjian, N.: Understanding the impact of spot market electricity price on wastewater asset management strategy. Water Conserv. Sci. Eng. (2022)
9. Liu, D., Wang, M., Li, X., Wu, W., Zhang, L., Li, H.: Hydropower market electricity distribution model based on large system decomposition theory. IOP Conf. Ser.: Earth Environ. Sci. **189**(5) (2018)
10. Liang, Z., Du, Y.: Analysis of human resources optimization of the electric power project management. In: Proceedings of 2016 3rd International Conference on Education, Management and Computing Technology (ICEMCT 2016), pp. 593–595 (2016)

Application of Vulnerability Detection Technology Based on Network Space Recognition

Hongxi Chen[1,2,3] and Fanfeng Lin[2,4](✉)

[1] Macao Polytechnic University, Macao, China
cc06200602@163.com
[2] Ulaanbaatar Erdem University, Ulaanbaatar, Mongolia
triones@163.com
[3] Nan'an Branch, Quanzhou Yixing Electrical Engineering Construction Co., Ltd., Quanzhou, China
[4] Xiamen Taqu Network Technology Co., Ltd., Xiamen, China

Abstract. The rapid development of the Internet has brought great convenience to people, and the impact of network attacks caused by vulnerabilities has increased the impact on enterprises and people. However, the vulnerability scanner on the market has defects with insufficient performance and effect. In order to discover and repair loopholes, this article analyzes the principle of the scanner and proposes some improvement and optimization discussions, this application of vulnerability detection technology based on network space recognition, which can discover and repair vulnerabilities early to prevent safety risks. This method effectively improves the identification efficiency and accuracy of vulnerabilities, and can effectively improve the level of network security.

Keywords: Network Security · Network Space Recognition · Vulnerability Scanning · Distributed Scan · Fingerprint Detection

1 Introduction

With the rapid development of the Internet, people enjoy the convenience and efficiency of life brought by social informatization. But followed, it also brought security risks. On the one hand, for the system exposed to public networks, the world can access the system at any corner, and the cost of attack has decreased sharply. On the other hand, the system is becoming more and more complicated, and the safety awareness of developers is uneven, causing large and small loopholes. The above two factors lead to a significant increase in the probability of system damage due to vulnerability factors, which causes economic losses to affected the business of the enterprise. At the same time, it may also cause legal risk of compliance. It may be threatened by telecommunications fraud or other aspects [1–3]. However, due to different applications, versions, and ports, it is difficult for technical personnel to position assets using the program when a certain program finds security vulnerabilities. The scanner on the current market, due to the insufficient

Y. Pei et al. (Eds.): IC 2024, LNEE 1214, pp. 36–46, 2024.
https://doi.org/10.1007/978-981-97-4193-9_5

efficiency of the scanning, the fingerprint update is not timely, and the detection script update is not timely. As a result, the company with a lack of security measures is always lagging behind. Hurry up to save the fire. The main purpose of studying space recognition and vulnerability detection technology is to investigate the scanning principle of the current open source scanner on the market and understand the technology used by the current mainstream scanner. On the other hand, for the shortcomings of the current scanner, optimization and improvement are proposed to improve the performance of the scanner.

The browser/server mode is used here. Based on the Python language, the entire system is built based on the Flask Web development framework. The main server is the core of the overall, management and execution of the vast majority of tasks. In addition, there are multiple NODE nodes as a scan node agent to share some tasks of the main server to achieve higher scanning performance. In addition, the main server provides Web services at port 80 and 443. Users log in to the main server management page through the browser to manage the task. The entire system implementation method is divided into two steps, as follows: The first step is to implement the realization of network space recognition. Use Masscan, NMAP, Python scripts to scan the Internet regularly through distributed scanning technology, and collect the online status of the target network, open port open status, and service fingerprint information, and use the local fingerprint warehouse rules to identify the asset information of the target network. In addition, for the new Internet products, we can identify assets by updating the corresponding fingerprint rules and label each asset. The second step is to perform vulnerability detection. After label the information collected by the first step above, after labeling the corresponding services of each port, they can quickly detect whether the target assets are distributed by scanning and scanning and scanning the scanning engine. In accordance with the application name, we collect the authentication script (POC) of the historical vulnerabilities of the corresponding product, and check whether each application has corresponding security vulnerabilities one by one [4, 5].

The principle of network space recognition and vulnerability detection technology described in this article. The vulnerability scanner on the market usually exists in three forms: (1) a single scanning software, mainly installed on the user's computer, scanning engine, rules, and network scanning. For example, a new generation of network security test tool Goby. (2) Based on the client (management end)/server (scan engine) mode or browser/server mode, it is usually software, deployed on different servers, and there are also ways to make the scan engine into hardware and sell them through hardware.

The above is the principle of network space recognition and vulnerability detection technology. In terms of implementation details, this article adds multi-threaded technology, distributed design, and multi-condition regular matching fingerprint library. In terms of efficiency, the scanner on the market has a better improvement.

2 Network Space Recognition

2.1 The Principle of Network Space Recognition

Use Masscan to quickly identify the port switching status of the target network IP. Because some IPs have certain protection strategies, such as adding a safe firewall, it may lead to a certain deviation in the scanning results. The problems that have been discovered are as follows:

(1) Limit IP access specified port;
(2) All ports return Open.

Due to the general host, the number of port openness is rarely more than 100. In order to ensure that the scanning results are as accurate as possible, the number of ports in the scanning result is more than 100 in the program.

Masscan scanning results export the IP address + port list, provided to NMAP for further service recognition. NMAP is also a powerful port scanning tool. It has strong service recognition capabilities and combines NMAP to further service recognition to the found ports [6–8].

Finally, we use CURL to send HTTP or HTTPS requests to the port. If there is a return HTTP status code, it means that the port is a web service. If there is no expected return information, it means that it is a non-web service. For web services, we can determine the application name used in the web service by detecting the web service icon, title, HTML, and Headers information. For non-web services, we obtain the banner information of the port through Telnet or NC tools, and judge the service used in the port based on the characteristics of banner information. The specific implementation process is shown in Fig. 1:

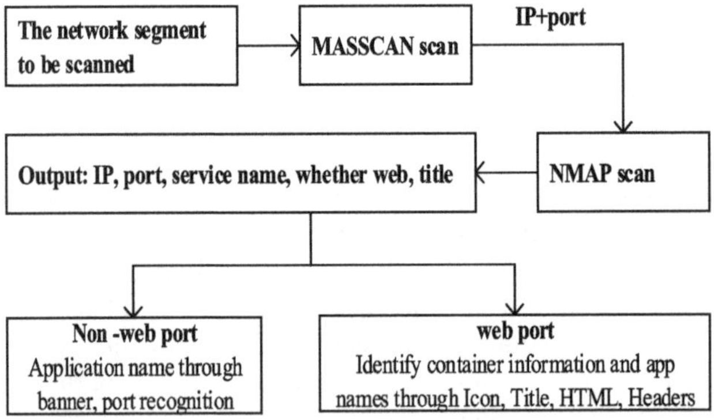

Fig. 1. The principle diagram of network space recognition

2.2 Implementation of Network Space Recognition

1) Survival host detection

When scanning the specified network segment, first of all, we need to survive the host in the network segment. If you do not do this step and directly scan the port, it will consume a lot of useless time, resulting in extremely low efficiency. There are currently two main methods for surviving host detection:

(1) Ping detection. We send ICMP (Internet Control Message Protocol ICMP (Internet Control Message Protocol) to a specific target host, testing the target host whether the target host can reach and understand the relevant status through the content of the return packets. If you receive the answering data packet with a target host type of 0, it means that the target host is survived, and then you can scan the next step for the surviving target host. Bags indicate that the target host cannot reach.
(2) Open port detection. Due to safety reasons, many hosts or firewalls are disabled ping, which has caused even the surviving host to receive the ICMP package. Therefore, the ping method cannot be all judged to judge the survival status of the host. The host is survived. We use common ports as port detection targets. For the Window system, the common ports include port 3389 (remote desktop port), port 135 (RPC port), port 139 (Netbios port), port 445 (shared printer port). System, common ports include port 22 (SSH port). In addition, there are default ports of common application software, such as port 3306 (MYSQL database default port), port 21 (FTP service default port), port 53 (DNS service port), port 80 (web service port, and 80-80, 8080, etc.), port 110 (POP3 mail port), port 389 (LDAP port), etc.

Through the above two methods, we remove the scanning task that is not online in the network segment, and only retains the actual IP integration of the actual surviving IP list to be scanned to the next port scan, which can greatly improve the scanning efficiency. Masscan scanner is an efficient port scanning tool that integrates the Ping scan detection. At the same time, it uses the most common port to assist the detection to efficiently achieve surviving host detection. We can submit the task of surviving host detection with MASSCAN.

2) Port Scanning Technology

Because each host has a maximum of 65535 ports, the port scanning steps of the host are the longest part of this program, which takes more than 60%of the total scanning task. Therefore, how to optimize port scanning and improve port scanning speed is one of the very important links of this program optimization.

This program uses Masscan to complete the implementation of this function. Masscan's advantage is that multi-threaded technology can make the task high concurrently performing scan tasks. At the same time, Masscan also uses a state-free scanning technology. No state scanning does not need to establish a complete TCP three-time handshake, thereby reducing the waiting time of the program and improving the detection efficiency of the port. Through the optimization of the above two technologies, Masscan scanning speed has improved by dozens of times the scanning efficiency compared with NMAP scanning speed. Masscan jumps into one of the indispensable software for port scanning and detection [9].

Masscan's configuration is also a key link that affects the speed of scanning. The scanning speed is determined by the throughput of the data packet according to the network and network card. Exception of the throughput performance will cause data loss and cause inaccurate scanning results. Masscan controls the package rate of the network card through–rate parameters. After testing, controlling the package rate within 2000 is to increase the scanning speed as much as possible, and to reduce the relative safety value of the packet loss as much as possible. Secondly, for the scanning port range, we provide two scanning strategies according to the actual scanning needs:

(1) The full port scan. As the name suggests, it is the full scan of more than 60,000 ports of the port of the host 0-65535. In some scenarios that need to be strictly obtained from full assets, we use this configuration to scan, and we can detect most of the asset information as much as possible. The scanning speed is relatively long, and a single IP needs to take timely time. A section C-section network requires about one hour according to the number of survival IP.
(2) Quick scan. Quick scanning is the balance of integrity and speed. Masscan can customize the port list, or you can use the built-in common port lists to scan. After testing the use of "-Top-Ports 2000" to take 2000 ports used, most of the known ports can be detected, but the number of scan ports has been reduced by 97%, which is greatly accelerated. So it is a very preferred scanning strategy. According to the actual needs, we can select the appropriate configuration option when configuration creation.

3) Distributed Scan

Quick scanning the huge Internet network is often a time-consuming task. At the same time, in order to ensure the real time of assets, assets often need to be updated in a timely manner. This makes how to shorten the network scanning as much as possible a very necessary and very challenging task. In order to further accelerate the scanning speed, we use the Docker container technology to encapsulate the scan module into agent one by one and send it to multiple servers. Each scan Agent obtains the scanning task from the master node. At the same time, the scanning results are reported to the master node to summarize the scanning results [10].

We deploy the Harbor application (an open source container mirror management tool) at the Master node, and encapsulate the scanning service into docker mirrors and upload it to Harbor. Each service node can pull the mirror in Harbor through the Docker Pull, and start the mirror with Docker Run to quickly generate the Node. This method can quickly deploy multiple NODE nodes in batches. The more nodes, the faster the scanning speed. How to distinguish between each node which tasks are scanned and which tasks are not scanned, we need to set a state for each task, we define 5 scanning status according to the entire life cycle of the scanning, as shown in Table 1:

Each Node only pulls the state of "Not scanned" and updates the task status to "start scanning", so that it will not be repeatedly scanned by other NODE nodes, and a distributed scan is achieved.

4) Fingerprint Detection Technology

Fingerprint detection includes: fingerprint detection of non-web ports, fingerprint detection of web ports. We first use NMAP to perform preliminary fingerprint detection

Table 1. Scanning task 5 scanning status of the entire life cycle

ID	State	Description
0	Not scanned	The task of scanning
1	in the line	The task waiting for scanning
2	Start scanning	The task in scanning
3	Scanning	Scanning task
4	abnormal	Abnormal interruption, error reporting, etc. leading to the task of scanning abnormalities

on the port. NMAP has a relatively powerful fingerprint library, and we can quickly identify some common port application types.

Next, use the CURL request port. If the HTTP status code is returned, it means that the port is the web port. If the status code is not returned, it means that the port is a non-web port. In non-web ports and web ports, we use different identification methods to identify the application type of ports.

For non-web ports, we usually send requests with Telnet or NC, and judge the service type of the port through the banner information returned by the port. For the web port, due to the huge third-party CMS, there are many versions, and the recognition will be relatively complicated. First, there are two parts of HTTP requests: header and body. Both Header and Body include a lot of useful information to help us identify the programs used in the web port.

Through the above characteristics, we can find that the same type of website system can be classified by the same type of system. When the application of this type of application has vulnerabilities, we can extract the detection of vulnerabilities in these types of websites as soon as possible. At the same time, the website administrator can be notified in time to repair the loopholes in time [11].

So how do we remember these characteristics? Through the joint maintenance of many technicians, the common CMS features are extracted, and the community has maintained a set of CMS fingerprint databases. With this fingerprint database, we can quickly identify the target application. At the same time, we can also update the newly discovered fingerprint to the feature library.

However, practice shows that a single characteristic is often not accurate enough, such as the fingerprint matching of "security entry verification failure" is relatively rude. When there are other applications, there are just "security entry verification failures". I hit, which leads to misunderstanding. This program introduces the function of multi-condition query. For the rules that may lead to misunderstandings, we add multiple detection conditions to enhance the accuracy of the rules recognition and reduce the error alarm.

In addition, there is also a test method that can help us identify the website. We can help judge the application used by the website through the Favicon icon of the website. The vast majority of Web sites will have a relatively special static resource file, usually the name is Favicon.ico. The Favicon.ico icon is a tap logo of the website, which can be displayed on the left side of the browser label, the left side of the address bar, and the collection clip [12]. Beautiful and more personalized, Favicon.ico is essential. Most applications will have a unique Favicon.ico (not ruled out that a few applications use Favicon.ico of others, but most of the cases are different). Favicon.ico can recognize which application is in most cases. We calculate the common application Favicon.ico through MD5 to make a fingerprint library, and to identify which application belongs to by matching the MD5 value.

Through the above technical methods, we can identify most of the common ports on the network accurately and in detail.

3 Vulnerability Detection Technology

3.1 Principles of Vulnerability Testing Technology

The vulnerability detection technology is to use vulnerabilities to test the vulnerabilities of known assets in batch of the script, so as to find safety issues. Because the current vulnerability scanning tool is not good enough for fingerprint recognition, all the vulnerabilities need to be tested again, resulting in a long time consumption. Based on network space recognition, we can identify various applications more accurately [13]. We associate the fingerprint tags each POC applicable object, and each POC only detects the detection objects suitable for them, so as to avoid useless scanning, which can achieve vulnerability scan more quickly. The specific detection process is shown below (Fig. 2):

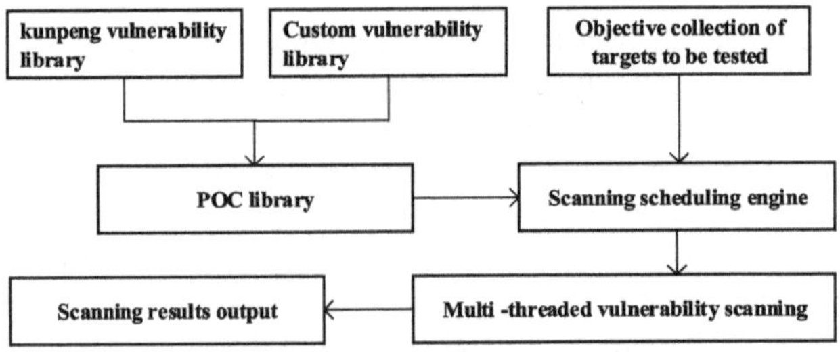

Fig. 2. Virgin detection technology schematic diagrams

3.2 Implementation of Vulnerability Detection Technology

1) Create a Scanning Task

In order to improve the scanning efficiency, we need to classify the scanning object according to the label. The assets that need to be scanned from the asset list, and classify the labels. For example, if we need to withdraw mysql assets, we can screen according to the MySQL label.

2) POC Vulnerability Library Integration

The vulnerability library is the soul of the vulnerability scanner. We must collect the loopholes as much as possible, and we need to eliminate the vulnerability detection rules that are invalid or higher. We first introduced the Kunpeng vulnerability library that the community maintained. It has contained 86 verification scripts, which enriches the vulnerability library. Contains the detection of vulnerabilities of common applications, such as common application services SSH, SMB, FTP, Redis, PostgreSQL, Mongodb, Memcache, etc., as well as common web service Discuz!, Grafana, Zabbix, WordPress, Struts2, ThinkPhp, etc.

With the development of the Internet, various new applications will continue to appear and widely apply, and its related loopholes will continue to appear. We need to continuously maintain the update vulnerability library to support the detection of new security issues in time. Therefore, it is necessary to customize vulnerability detection scripts. There are 68 known custom detecting scripts, which will gradually increase in the future. In order to enable the script to be managed in an orderly manner, the structure of the script is required reasonably. In order to make the scanning program know the type, use, and applicable application of the scanning script, in Table 2 the following fields are designed:

Table 2. Custom vulnerability detection script structure

Field Name	Meaning	Example
Name	Vulnerability name	Ftp weak password
Info	description information of Vulnerability	As a result of sensitive information leakage, serious situations can cause the server to be invaded and controlled
level	Level of Vulnerability	High risk
Type	Type of Vulnerability	Weak password
author	Author of Script	wolf@YSRC
url	Link of Vulnerability	
keyword	Keyword information, used to match fingerprint	ftp
source	source	xunfeng

Each vulnerability detection module is packaged into a Python script, which can be easily loaded into the scan engine. The script can be debugged separately, and the scanning engine can be called, which can debug the detection effect of the script. There are two special detection rules in POC. We collect applications that are often easily explosive on the market, which is shown as Table 3:

Based on these applications, try to log in with common passwords or weak passwords. If the login is successful, it indicates that there is a weak password. You need to modify the password to avoid information leakage or damage.

The other is to detect the detection rules for access restrictions. There are some applications. If there is no permissions restrictions, and open Internet access will cause

Table 3. Common applications that are easy to blast

Application name	Default port	Application
mysql	3306	Database application
ftp	21	File transmission service
smb	139 or 445	File sharing service
rdp	3389	Windows remote desktop service
imap	143	Mail port
telnet	23	Remote login protocol
pptp	1723	VPN port
Http basic	80 or 443	web simple certification protocol

information leakage or remote command to perform risks, causing information leakage or business to damage. The commonly unauthorized access application on the market is as Table 4.

Table 4. Detecting service whether to do the detection rules for access restrictions

Application name	Application	Risk
MongoDB	database application	MongoDB has unauthorized access vulnerabilities in the default configuration. Users can access the database anonymous, leading to information leakage
Redis	database application	Redis defaults to install unauthorized access vulnerabilities, and users can use empty password link databases. Using Redis file writing function can cause the server remote command to execute vulnerabilities. For Redis applications, you need to do access restrictions or open password restrictions
Memcached	High-speed cache application	There is also an unauthorized access vulnerability at MEMCACHED by default, resulting in information leakage. You need to do a good job of IP
Elasticsearch	database application	There is an unauthorized visit, causing the database to leak. Need to do IP restrictions
docker	Application container engine	The Docker Remote API executes the docker command. Unauthorized access can lead to the vulnerability of the remote command

3) Scanning Scheduling Engine

The scheduling engine loads all POCs and uses multi-threaded technology to verify whether the assets are vulnerable. This module is the core function of the vulnerability detection module.

We set the scanning rules of Kunpeng and custom POC integration, classify the KEYWORD fields, and then remove the assets of the corresponding label in the assets according to the keyword. After allocating the scanning target according to the label, use the multi-threaded class Thread in Python to create the scanning task, so that many scanning tasks can perform scanning parallel to increase the scanning time of the scanning speed.

4) Scanning Result Generation

In order to enable the results of the scanning to be displayed friendly, push it to the relevant personnel to repair the problem, you can make related scan report templates, export it into a PDF scan report, and automatically send it to the relevant personnel to repair the problem.

4 Conclusion

The vigorous development of the Internet has become an indispensable part of people's lives. How to ensure the safe operation of services and how to ensure that the privacy information of citizens is not infringed, it has become increasingly concerned. This article helps enterprise technicians identify assets, discover vulnerabilities, and repair unknown security issues by helping enterprises' technical personnel. The technical advantage of this design is compared to the existing scanning tools on the market, and there are several outstanding improvement points:

(1) Distributed scanning: The scanner on the market is mostly an independent whole, which leads to the network IO performance of the current machine. This paper is divided into an independent agent by splitting the scan module and encapsulated it into a Docker image, so that it can deploy multiple servers, realize distributed scanning, and break through the problem of the scanning performance of a single server.
(2) Fingerprint detection technology: The scanner on the market, most of the fingerprint library is collected by the initial developers, solidified in the system. This also caused the new application to be updated in a timely manner to the fingerprint library in a timely manner. Some commercial scanning software, even if it is upgraded by upgrading the firmware, it takes a long time to be updated for a longer time. In addition, there is also a disadvantage that the fingerprint library is usually uniformly managed by the system and cannot be modified separately. When a fingerprint is found, there is a mistake in a fingerprint, and the technicians cannot modify the specified fingerprint separately. The fingerprint library designed in this article can customize and modify the fingerprint library to update the fingerprint library information in a timely manner. In addition, it will also count the number of hits of each fingerprint. For the large number of fingerprints, we can focus on its accuracy, and continuously optimize the fingerprint rules to better identify assets.

(3) POC vulnerability library integration: The scanner on the market, because the fingerprint recognition fails to be very accurate, the vulnerability scan will scan each asset. A large amount of useless scanning takes up the scanning time and reduces the scanning speed. The fingerprint label designed in this thesis is labeled on each POC script and assets. The scanned asset label and the POC label pairing, eliminate the useless scanning tasks, and improve the scanning efficiency.

The above is the lack of scanners on the current market, and the design is improved and optimized.

5 Summary

We can learn how the scanners on the market can find the assets of the network space and how to use the tool to detect the loopholes on the Internet. After research and exploration, the knowledge in this area was enhanced. We have also discovered the current deficiencies and improved and optimized, so that the efficiency of asset recognition and vulnerability scanning fields has been further improved.

References

1. Ping, L., Jin, S., Yang, X.: Research on software security vulnerability detection technology. In: International Conference on Computer Science Network Technology. IEEE (2012)
2. He, Y., Ma, F.: Research on computer software security vulnerability detection technology. China Comput. Commun. (2017)
3. Guan, J.: Research on the application of computer network security and vulnerability scanning technology. Comput. Knowl. Technol. (17), 25–26+32 (2022)
4. Wang, H.: Study on the automation deployment of python-based scanning tasks of vulnerabilities. In: China Mobile "5G+AICDE" Technical Seminar, pp. 194–197+253 (2021)
5. Li, J.: Research on the research of vulnerability scanning technology in computer network security. Comput. Program. Skills Mainten. **06**, 168–169 (2021)
6. Zhao, L., Mei, S., Zhong, W.: Firewall, invasion detection system and vulnerability scanning technology based on game theory. J. Southeast Univ. (English Edition) (02), 144–147 (2011)
7. Hong, H., Zhang, Y., Hu, Y., Dai, Z.: Research on network security scanning technology. Comput. Eng. **10**, 54–56 (2004)
8. Chen, Y., Sun, J., Cheng, S., Zhang, Z., Wang, Z., Wang, B.: Safety management technology for industrial control networking equipment. Sci. Technol. Eng. **33**, 14266–14272 (2021)
9. Humphries, J.W., Carver, C.A., Pooch, U.W., Member, S.: Secure mobile agents for network vulnerability scanning. In: Proceedings of the IEEE Workshop on Information Assurance Security (2000)
10. Ali, A., Shakhatreh, A., Abdullah, M.S., Alostad, J.: SQL-injection vulnerability scanning tool for automatic creation of SQL-injection attacks. Procedia Comput. Sci. **3**, 453–458 (2011)
11. Sharmila, C., Gopalakrishnan, J., Prasath, P.S., Daniel, Y.: Multipurpose Linux tool for Wi-Fi Based Attack, Information Gathering and Web Vulnerability Scanning Automations. Springer, Cham (2022)
12. Dwivedi, A.: Vulnerability Scanning Technology on Web Applications (2021)
13. Wangsiripitak, S.: Multimodal biometrics recognition using a deep convolutional neural network with transfer learning in surveillance videos. Computation **10** (2022)

Exploration of Building a Lean Energy Service System Based on Digital Economy

Jiang Yu, Shuqi Liu(✉), Xiao Tang, Jin Yu, and Senlin Lan

State Grid Shanghai ShiNan Electric Power Supply Company, Shanghai 201109, China
18901822342@163.com

Abstract. With the rapid development of the digital economy, improving energy utilization efficiency and building a sustainable lean energy service system are crucial for achieving dual economic and environmental benefits. This study explored the construction of a lean energy service system in the digital economy and verified its potential in reducing energy consumption and improving economic efficiency through experiments. The experimental results showed that during the five month implementation period of the digital economy company, power consumption decreased from 2800 kWh to 2400 kWh, water resource consumption decreased from 1200 m^3 to 1050 m^3, and energy consumption reduction rates reached 10.2% and 6.8%, respectively. This validated the important role of lean energy service systems in achieving energy efficiency in the digital economy. Through reasonable data analysis and innovative management methods, a lean energy service system could provide sustainable development support for the digital economy and reduce economic costs.

Keywords: Digital Economy · Lean Energy Utilization Service System · Data Mining Algorithms · Digital Energy Monitoring and Management System

1 Introduction

With the rapid development of the digital economy, energy consumption has become an important concern worldwide, and achieving sustainable development and Low-carbon economy has become the common goal of all countries [1, 2]. Therefore, building an effective lean energy utilization service system has become particularly important. Lean energy utilization refers to minimizing energy consumption while achieving the same or higher level of products or services. By integrating digital technology and data analysis, monitoring, analyzing, and optimizing energy use can be achieved, thereby improving energy utilization efficiency. Building a lean energy service system based on the digital economy can bring multiple benefits [3, 4]. Firstly, through the application of digital technology, real-time monitoring and management of energy can be achieved. In this way, enterprises and individuals can better understand the use of energy and take corresponding energy-saving measures. Secondly, the digital economy can provide the ability to analyze big data. Through the mining and analysis of energy data, bottlenecks and

Y. Pei et al. (Eds.): IC 2024, LNEE 1214, pp. 47–57, 2024.
https://doi.org/10.1007/978-981-97-4193-9_6

problems of energy consumption can be found and corresponding solutions can be provided. In addition, digital technology can also support intelligent energy management systems, achieving efficient energy utilization through automation and optimized control [5].

In recent years, many scholars and experts have conducted research on the construction of a lean energy service system for the digital economy. Among them, with the increasing importance of privacy protection and data security, blockchain technology is widely applied in demand response management in intelligent grid systems. Wang C explored the requirements and challenges of demand response management in blockchain based privacy protection intelligent grids, analyzed the key issues of privacy protection, demand response, and blockchain technology in intelligent grids, and proposed corresponding solutions. The experimental results showed that the blockchain based demand response management system could ensure user privacy while achieving efficient demand response, and effectively addressed data tampering and forgery issues [6]. The combination of the Internet of Things and blockchain has brought many opportunities and challenges to various industries. Chong W analyzed the key opportunities of blockchain technology in the Internet of Things, including advantages in data security, decentralization, credibility, and smart contracts. At the same time, it also explored the challenges faced by Internet of Things application blockchain, such as scalability, privacy protection, energy consumption, and standardization. Through the review of relevant research and actual cases, it showed that the combination of the Internet of Things and blockchain could improve data security and cross organizational trust. However, in practical applications, challenges at the technical and normative levels needed to be overcome [7]. Blockchain technology provides a method of security and privacy protection, which can maintain the anonymity of smart meter data in smart grids. A smart meter is a device that can measure and communicate energy usage. In traditional power grids, meter data is usually collected and managed by energy companies. However, with the emergence of smart grids, the processing of smart meter data has become more complex, involving more participants. Ye F used blockchain technology. Smart meter data could be encrypted and stored on the blockchain, ensuring data security and integrity. Each smart meter had a unique identification identifier that could be used to identify and distinguish different devices. However, personal privacy was crucial, so maintaining the anonymity of electricity meter data in the blockchain was crucial [8].

In order to explore the construction of a lean energy utilization service system based on the digital economy, it requires joint efforts and cooperation from all parties. Governments, enterprises, and social organizations can jointly promote the application and innovation of digital technology, establish relevant policies and standards, and provide support and incentive measures. At the same time, it is also necessary to increase investment in the research and development of digital technology, improve the reliability and performance of digital technology, and promote the development of a lean energy service system.

2 Exploring Methods for Constructing a Lean Energy Service System in the Digital Economy

2.1 Building a Digital Energy Monitoring and Management System

Building a digital energy monitoring and management system is a key part of achieving a lean energy service system. Through the application of digital technology, real-time monitoring, management, and optimization of energy can be achieved, thus improving energy utilization efficiency, reducing energy waste, and achieving the goal of energy conservation and emission reduction [9, 10].

The construction of a digital energy monitoring and management system mainly includes the following aspects:

1) Introduction to data collection and monitoring technology

   In the energy monitoring and management system, it is necessary to collect and monitor data on various energy sources, including electricity, gas, water, etc. Common data collection technologies include sensor technology, smart electricity meters, smart water meters, etc. These devices can monitor energy usage in real-time and transmit data to data management platforms [11, 12].

2) Construction of data management and processing platform

   Building an efficient data management and processing platform is crucial for energy monitoring and management. This platform should have functions such as data storage, data processing, data analysis, and visual display. Through this platform, energy data can be organized, stored, and managed, while also providing the ability for data analysis and report generation.

3) Real time energy analysis and visualization display

   The digital energy monitoring and management system should be able to analyze and process real-time energy data in order to understand the use of energy in real time. Through data analysis, bottlenecks and issues in energy consumption can be identified and corresponding solutions provided. At the same time, the visualization of data is also important. Through intuitive charts and reports, users can better understand the energy consumption situation and take corresponding adjustment measures.

4) Application of automation control and intelligent equipment

   The digital energy monitoring and management system can be combined with the application of automation control and intelligent devices to achieve intelligent management and optimization of energy. By connecting to intelligent devices, remote monitoring and control of energy equipment can be achieved, improving energy utilization efficiency. For example, an intelligent control system can automatically adjust the equipment operation mode based on the analysis results of real-time energy data, achieving the optimal effect of energy consumption.

5) Data security and privacy protection

   When building a digital energy monitoring and management system, data security and privacy protection are crucial. Reasonable data security measures and privacy protection mechanisms should be adopted to ensure the security and privacy of energy data. This includes encryption of data transmission, setting of permission management, and formulation of privacy protocols [13, 14].

By building a digital energy monitoring and management system, real-time monitoring, optimization, and management of energy can be achieved, improving energy utilization efficiency and reducing energy waste. At the same time, digital technology can also provide more innovation and development space for energy management, further promoting the construction of a lean energy service system. However, the construction of digital energy monitoring and management systems also faces some challenges, including the formulation of technical standards, data security and privacy protection, system integration, and other aspects. Therefore, active participation and cooperation from all parties are needed to jointly promote the development of digital energy monitoring and management systems, as shown in Fig. 1:

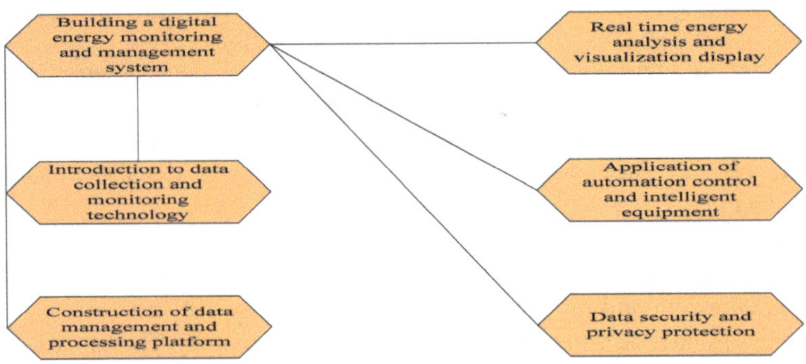

Fig. 1. Construction of a digital energy monitoring and management system

2.2 Digital Economy

The digital economy refers to an economic form based on digital technology that drives economic development and creates value through the Internet and other digital tools. It covers digital industry, digital commerce, digital finance and other aspects, and has a huge impact on the Traditional economy model and business model [15, 16]. The development of digital economy stems from the rapid development of digital technology, including cloud computing, big data, artificial intelligence, the Internet of Things and blockchain. The popularization and application of these technologies have made information transmission faster, transactions more convenient, and innovation more efficient, promoting the digitization, digitization, and intelligence of economic activities. The digital economy has many characteristics and advantages. Firstly, it can break geographical limitations and achieve cross-border communication and cooperation. Through the Internet, enterprises can conduct global business, and consumers can choose goods and services from a global perspective [11, 17]. Secondly, the digital economy focuses on data collection and analysis, mining the commercial value contained in data through big data technology to help enterprises make more accurate decisions and improve efficiency and competitiveness [18]. In addition, the digital economy has also promoted the emergence of new business models, such as the sharing economy, e-commerce and online education, to provide consumers with more choices and convenience, as shown in Fig. 2:

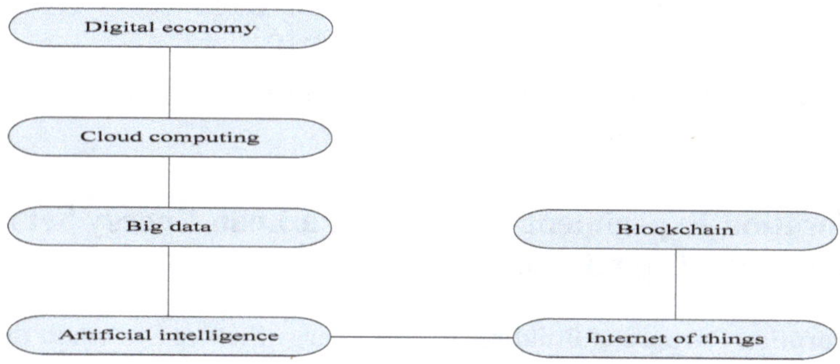

Fig. 2. Digital economy flowchart

2.3 Evaluation of Algorithms of Data Mining

For the construction of a lean energy service system in the digital economy, it is crucial to collect, organize, and preprocess data. It is particularly important in the feature processing of data mining algorithm [19] to clean the data through data mining technology, remove the Outlier and missing value in the data and normalize the data. Usually, algorithms such as association rules, clustering, and classification work together to achieve data analysis results. Taking Apriori algorithm association rules as an example, in association rule mining, the correlation between each element is usually measured through two key factors: support and confidence.

In association rules, it is assumed that there are two sets X and Y that do not intersect with each other. The probability of shared data in the sample with X and Y attributes in the overall data set Z is usually expressed as Formula (1):

$$\text{support}\ (X \rightarrow Y) = \frac{\text{count}\ (X, Y)}{\text{count}\ (Z)} \tag{1}$$

In Formula (1), count (X,Y) represents the sum of the number of samples in the sample set where X and Y attributes appear simultaneously, and count (Z) represents the total number of all samples in the sample set. Meanwhile, confidence level is usually the correlation between X and Y, representing the probability that the sample contains both attribute X and attribute Y, which can be expressed as Formula (2):

$$\text{confidence}\ (X \rightarrow Y) = \frac{\text{count}\ (X,Y)}{\text{count}\ (X)} \tag{2}$$

In Formula (2), count (X) represents the total number of samples containing attribute X. By combining the formulas, the degree of improvement can be obtained: The degree of improvement refers to how much the other attribute Y would increase when the attribute X increases. The calculation method is shown in Formula (3):

$$\text{Lift}\ (X \rightarrow Y) = \frac{\text{confidence}\ (X \rightarrow Y)}{\text{support}\ (X)} \tag{3}$$

In Formula (3), Lift (X → Y) represents the degree of improvement, and support (X) represents the support of X.

3 Exploration Experiment on Building a Lean Energy Service System in the Digital Economy

3.1 Exploration Purpose of Building a Lean Energy Service System in the Digital Economy

This experiment aims to explore the construction scheme of a lean energy service system in the digital economy, in order to improve energy utilization efficiency and promote sustainable economic development. By establishing a reasonable energy monitoring and optimization mechanism, the energy usage situation is accurately analyzed, and innovative measures to improve energy utilization efficiency are sought to provide sustainable development support for the digital economy.

3.2 Exploring and Evaluating the Construction of Lean Energy Service System in the Digital Economy

Firstly, starting with data analysis of energy usage, the existing energy efficiency status is evaluated by collecting and analyzing energy consumption data of enterprises, institutions, and individuals. At the same time, it is also necessary to consider the time, space, and industry characteristics of energy consumption in order to better develop lean energy utilization strategies. Secondly, in the process of building a lean energy service system for the digital economy, it is necessary to establish a comprehensive energy data monitoring and management system. The system should include real-time data collection, storage, processing, and analysis functions to support monitoring and prediction of energy usage. At the same time, the system should also have effective data interaction with various energy subsystems to achieve precise energy planning and optimization. On the basis of analyzing energy data, it is necessary to combine digital technology and intelligent means to explore innovative solutions for energy management. For example, artificial intelligence technology is used to develop energy consumption prediction models. Based on historical energy consumption data and external factors, future energy demand is predicted and optimization plans are formulated. In addition, by applying Internet of Things technology, remote monitoring and intelligent control of energy equipment can be achieved, saving energy while improving management efficiency. The energy consumption data of a digital economy company was collected and analyzed and optimized based on this data. The results are shown in Table 1:

In Table 1, the specific values of electricity consumption and water resource consumption, as well as the percentage reduction rate of energy consumption, can be seen.

Table 1. Energy consumption data and optimization effects

time	Electricity consumption (kWh)	Water resource consumption (m^3)	Energy consumption reduction rate (%)
Month 1	2800	1200	10.2
Month 2	2650	1150	8.7
Month 3	2700	1180	9.5
Month 4	2550	1100	7.3
Month 5	2400	1050	6.8

Taking electricity consumption as an example, the consumption for month 1 was 2800 kWh, month 2 was 2650 kWh, month 3 was 2700 kWh, month 4 was 2550 kWh, and month 5 was 2400 kWh. This meant that after building and optimizing a lean energy service system, the company achieved a stable decrease in electricity consumption within five months. Similarly, in terms of water resource consumption, the consumption for month 1 was 1200 m^3, month 2 was 1150 m^3, month 3 was 1180 m^3, month 4 was 1100 m^3, and month 5 was 1050 m^3. This indicated that through the implementation of lean energy strategy, the company successfully reduced the use of water resources within five months. The energy consumption reduction rate represented the degree of reduction in monthly consumption compared to the previous month. For example, the energy consumption reduction rate from month 1 to month 2 was 8.7%, and the reduction rate from month 2 to month 3 was 9.5%. This demonstrated the energy consumption improvement effect brought about by the continuous implementation of lean energy measures by the company over different time periods. The actual results obtained during the construction process of the lean energy service system in the company could be clearly understood. The interpretation of these data demonstrated the effectiveness of the service system and emphasizes the potential and importance of lean energy use in the digital economy.

3.3 Exploration Results of Building a Lean Energy Service System in the Digital Economy

This experiment aims to explore the construction scheme of a lean energy utilization service system in the digital economy. By analyzing energy data, establishing monitoring and management systems, and combining digital technology and intelligent means, a lean energy service system can be successfully constructed, and certain optimization effects have been achieved. This experiment provides useful practical experience for the sustainable development of the digital economy and provides a reference basis for future research and practice.

4 Exploration Results and Discussion on the Construction of Lean Energy Service System in the Digital Economy

4.1 Exploration Status of Lean Energy Service System Construction in the Digital Economy

Currently, with the rapid development of the digital economy, global energy consumption is constantly increasing, bringing enormous pressure to the balance of energy supply and demand and sustainable environmental development. There is a close connection between the rapid growth of the digital economy and energy consumption. In the digital economy, the operation of a large number of high energy consuming facilities such as data centers, communication networks, and intelligent devices has put forward higher requirements for energy consumption. Therefore, building a lean energy service system is one of the important strategies for achieving sustainable development of the digital economy. However, the energy utilization rate in the current digital economy is still relatively low, and there are several problems: firstly, many digital economy enterprises lack quantitative understanding and monitoring methods of energy consumption, making it difficult to accurately evaluate energy utilization efficiency. Secondly, the innovative nature of the digital economy leads to highly uncertain energy demand, making it difficult to establish a stable energy planning and scheduling system.

4.2 Exploration and Verification of Building a Lean Energy Service System in the Digital Economy

In order to verify the feasibility and effectiveness of the lean energy utilization service system, a digital economy enterprise was selected for validation experiments. Quantitative measurement and control measures for energy consumption would be implemented to evaluate the actual effectiveness of the system in improving energy utilization efficiency, as shown in Fig. 3:

Figure 3 shows the electricity consumption, water resource consumption, and energy costs of the digital economy enterprise over the past five years. In terms of electricity consumption (kWh), this indicator represents the total amount of electricity consumed by digital economy enterprises in each year, measured in kWh. From 2019 to 2023, electricity consumption was decreasing year by year. This decreased from 1500 kilowatt hours to 1280 kilowatt hours, which reduced by 220 kilowatt hours. Water resource consumption (cubic meter) represents the total amount of water resources consumed by digital economy enterprises in each year, measured in cubic meters. From 2019 to 2023, water resource consumption was decreasing year by year. This decreased from 1000 cubic meters to 850 cubic meters, which reduced by 150 cubic meters. Energy cost (in 10000 yuan) represents the energy procurement cost of digital economy enterprises in each year, expressed in 10000 yuan. From 2019 to 2023, energy costs was decreasing year by year. This decreased from 500000 yuan to 430000 yuan, which decreased by 70000 yuan. The changes in these values indicated that digital economy enterprises achieved certain results in building a lean energy use service system. By implementing energy management and conservation measures, companies successfully reduced electricity consumption, water resource consumption, and energy costs over the past five years, demonstrating the potential and effectiveness of lean energy use.

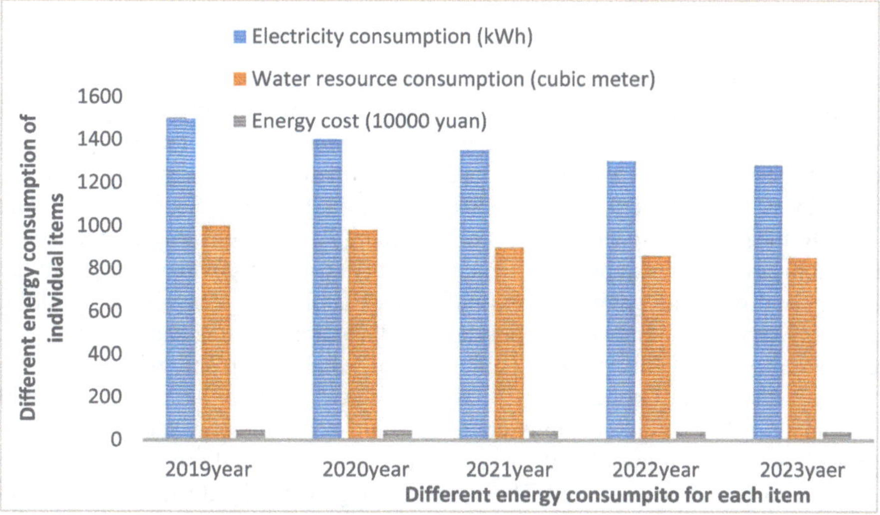

Fig. 3. Energy consumption of digital economy enterprises

4.3 Exploring Strategies for Building a Lean Energy Service System in the Digital Economy

Based on the results of the validation experiment, the following strategies are proposed for the construction of a lean energy utilization service system in the digital economy. Firstly, it is necessary to strengthen the energy management awareness of digital economy enterprises and enhance their awareness and monitoring capabilities of energy consumption. By establishing an intelligent monitoring system, energy consumption data is collected and analyzed in real-time, and energy utilization efficiency is accurately evaluated, providing decision-making support and optimization solutions for enterprises. Secondly, it is necessary to promote and apply energy conservation technologies and equipment to improve the energy efficiency level of digital economy enterprises. By introducing intelligent control systems and optimizing production processes and equipment configurations, energy waste is minimized to the maximum extent and energy utilization efficiency is improved. At the same time, it is necessary to focus on technological innovation, promote the development of new energy technologies, and reduce the carbon emissions of the digital economy. Finally, sustainable energy policies and economic incentives need to be developed to encourage digital economy enterprises to actively participate in the construction of a lean energy service system. For example, it is necessary to establish an energy efficiency ranking reward mechanism, and provide corresponding rewards and preferential policies to enterprises with low energy consumption and high efficiency. At the same time, it is necessary to increase support for energy technology innovation and promote the development and application of energy technology.

5 Conclusions

This article explored the construction of a lean energy service system for the digital economy and its application in improving energy efficiency and promoting sustainable development. Firstly, the challenges posed by the development of the digital economy to energy consumption were analyzed, including an increase in energy consumption, an increase in carbon emissions, and a demand for renewable energy. Secondly, through the case study, the paper analyzed the effect of lean energy use strategy implemented by digital economy enterprises, including reducing energy costs, improving energy utilization efficiency and reducing environmental pollution. Finally, key strategies for building a lean energy service system were proposed, including strengthening energy management awareness, promoting energy-saving technologies, establishing energy sharing mechanisms, and formulating policy support. By implementing these strategies, sustainable development of the digital economy could be achieved and energy efficiency could be improved. Reliance on traditional energy could be reduced, and green and low-carbon transformation could be promoted. These results had guiding significance for the government, enterprises, and various sectors of society to formulate energy policies and action plans, and also had important reference value for promoting the sustainable development of the digital economy.

References

1. Akimoto, T., Sano, F.: Energy blockchain: a technological revolution of the electricity market. J. Energy Market. **13**(4), 1–20 (2020)
2. Chen, J.J., Fu, L., Hu, X., et al.: A survey on blockchain: a game changer for future energy systems. J. Mod. Power Syst. Clean Energy **8**(5), 926–935 (2020)
3. Liang, X., Lu, R., Li, X., et al.: Blockchain-based secure and efficient data sharing scheme for smart grids. IEEE Trans. Smart Grid **10**(2), 2282–2293 (2019)
4. Wen, Z., Jiang, Z., Song, H., et al.: Blockchain-enabled privacy-preserving distributed demand response in smart grids. IEEE Trans. Industr. Inf. **15**(6), 3724–3734 (2019)
5. Zhan, Z., Chen, H., Xu, X., et al.: Blockchain-based dynamic key management for smart grid communications. IEEE Trans. Industr. Inf. **16**(12), 7483–7494 (2020)
6. Wang, C., Wang, Q., Zhuo, Y., et al.: Blockchain-enabled privacy-preserving demand response management in smart grids. IEEE Trans. Industr. Inf. **16**(1), 286–296 (2020)
7. Chong, W., Wong, W.C., Lai, L.L., et al.: Blockchain for the internet of things: opportunities and challenges. IEEE Internet Things J. **7**(1), 876–886 (2020)
8. Ye, F., Yuan, X., Bi, G., et al.: Blockchain-enabled secure and privacy-preserving anonymity for smart meter data in smart grids. IEEE Trans. Sustain. Comput. **5**(1), 21–33 (2020)
9. Sansom, G.: Energy systems integration: defining the boundaries of the discipline. IEEE Power Energ. Mag. **17**(1), 30–37 (2019)
10. Li, Y., Wang, Q., Zhang, Z., et al.: Blockchain-enabled privacy-preserving demand response mechanism for smart grid. IEEE Trans. Industr. Electron. **67**(7), 6072–6083 (2020)
11. Matsumoto, S., Yamamoto, Y., Takeuchi, K.: Energy system planning of an isolated island comprising electricity, heat, and transport sectors under consideration of multitype storages. IEEE Trans. Sustain. Energy **11**(2), 927–939 (2020)
12. He, Y., Sun, Y., Li, H., et al.: Blockchain-enabled privacy-preserving demand response in smart grids. IEEE Trans. Industr. Electron. **67**(9), 7596–7604 (2020)

13. Le, D., Mauw, S., Wieringa, R.: Blockchain-enabled secure information sharing for smart grids. IEEE Trans. Depend. Secure Comput. **17**(4), 684–697 (2020)
14. Li, Z., Xiang, P., Li, T., et al.: Blockchain-based secure and privacy-preserving demand response management in smart grids. IEEE Trans. Sustain. Energy **11**(4), 2302–2312 (2020)
15. El-Zawawy, M.N., El-Saber Batiha, M., Abd El-Samie, F.E., et al.: Lean energy management system for electrical power conservation in green buildings. China Commun. **19**(1), 166–186 (2022)
16. Jo, B.O., Lee, H.W.: Development of lean retrofit techniques for heating and cooling energy saving purpose. J. Asian Architect. Build. Eng. **20**(3), 737–744 (2021)
17. Saadat, E., Zare, K., Rosendahl, L.A.: Techno-economic analysis of lean-based power-to-H2-to-power systems with renewable energy sources for mobile applications. Int. J. Hydrogen Energy **45**(46), 24288–24306 (2020)
18. So, M., Lim, J., Ries, R., et al.: An energy-management system and power analyzer for subways considering regenerative energy. IEEE Trans. Transport. Electrific. **6**(2), 362–370 (2020)
19. Caniëls, M., Romijn, H.: Characteristics of collaborative R&D projects with commercial partners: evidence from the perspective of energy and climate innovations. Ind. Innov. **26**(2), 230–250 (2019)

Research on Design and Construction of Drainage and Prevention System of Cold Region High-Speed Railway Tunnel

Di Jiang and Zhiqiang Liu(✉)

School of Civil Engineering, Lanzhou Jiaotong University, Lanzhou, China
983307040@qq.com

Abstract. This research addresses the pressing need for robust drainage and prevention systems in high-speed railway tunnels located in cold regions. Severe climatic conditions in these areas' present formidable challenges to railway infrastructure, necessitating specialized solutions. This master's thesis adheres to rigorous academic standards, offering a comprehensive study on the design and construction of drainage and prevention systems within high-speed railway tunnels. The research centers on empirical investigations, engineering principles, and practical applications, all executed in accordance with recognized industry standards. By analyzing case studies and conducting extensive simulations, this research seeks to provide valuable insights into optimizing the design and construction of these vital railway tunnel systems. The study's outcomes are anticipated to make substantial contributions to the enhancement of railway infrastructure resilience and performance in cold regions, ultimately advancing the field of railway engineering.

Keywords: Cold Region · High-Speed Railway Tunnel · Drainage · Prevention System · Design · Construction

1 Introduction

1.1 Research Background

In the realm of high-speed railway tunnel engineering within cold regions, the design and construction of drainage and waterproofing systems have perennially posed critical challenges. This study is committed to delving deeply into the crux of this domain, with the aim of offering viable solutions to address these challenges. The rapid proliferation of high-speed railway systems globally has provided people with faster and more convenient means of transportation. However, in cold regions, particularly at high latitudes, the harsh climatic conditions present severe tests of the stability and reliability of railway tunnels. Extreme low temperatures, coupled with fluctuating freeze-thaw cycles, not only threaten the structural integrity of tunnels but also create substantial operational and safety hazards. The unique set of challenges faced in cold regions necessitates specialized engineering solutions. In response to these challenges, this research embarks

Y. Pei et al. (Eds.): IC 2024, LNEE 1214, pp. 58–68, 2024.
https://doi.org/10.1007/978-981-97-4193-9_7

on a comprehensive exploration of drainage and prevention systems, addressing the critical need for effective design and meticulous construction practices tailored to the demanding conditions of cold region high-speed railway tunnels [1].

1.2 Research Objectives and Significance

This study's primary objective is to thoroughly investigate the design and construction of drainage and waterproofing systems in high-speed railway tunnels located in cold regions, aiming to provide practical engineering solutions. The urgency of this research stems from the critical necessity to address the unique challenges faced by railway tunnels in cold climates. Extreme weather conditions, characterized by freezing temperatures and freeze-thaw cycles, pose significant threats to these tunnels' structural integrity and operational safety [2].

By advancing engineering knowledge in the context of cold region high-speed railway tunnels, this research contributes academically and practically. Through an exploration of cutting-edge technologies, engineering principles, and empirical data, valuable insights and recommendations are expected to emerge, aiding decision-making in future tunnel projects. The study's potential to enhance safety and efficiency within high-speed railway tunnel engineering is a notable contribution. Recommendations derived from this research can optimize drainage and prevention systems for cold region conditions, safeguarding infrastructure investments and improving transportation safety and convenience for both passengers and cargo.

Furthermore, this study sets the stage for future research in railway engineering, especially in cold regions. By identifying knowledge gaps and potential areas for further investigation, it aims to inspire future scholars and engineers to develop innovative solutions. In summary, this research is driven by a commitment to address pressing challenges, contribute to the safety and longevity of railway infrastructure, and improve efficiency across the transportation and engineering industries.

2 Design of Drainage System for Cold Region High-Speed Railway Tunnels

2.1 Characteristics of Cold Regions and Tunnel Environment Analysis

Before designing the drainage system for high-speed railway tunnels in cold regions, it is crucial to gain a profound understanding of the characteristics of these frigid areas and the environmental challenges faced by these tunnels. The following analysis sheds light on cold regions and the tunnel environment:

Characteristics of Cold Regions. Cold regions typically exhibit extreme low temperatures, especially during winter. Temperatures can plummet well below freezing, leading to freeze and snow accumulation. Moreover, diurnal temperature fluctuations and seasonal variations must be considered, as they affect the state and flow of water [3].

Geological and Hydrological Conditions. The geological and hydrological conditions of the tunnel's location are paramount for the design of the drainage system.

Understanding groundwater levels, soil types, rock layers, and geological structures is critical to ensuring the effectiveness of the drainage system. Different geological conditions may require distinct drainage solutions.

Impact of Freeze-Thaw Cycles. Cold region tunnels experience the effects of freeze-thaw cycles, a significant consideration. Temperature fluctuations cause water within the soil, rock, and concrete to freeze and thaw, potentially inducing cracks and damage. This imposes additional requirements on the design and performance of the drainage system to ensure its effectiveness under these conditions.

Environmental Sustainability. Environmental sustainability must be considered in drainage system design. Any water discharged into the surrounding environment should undergo proper treatment to minimize its impact on adjacent ecosystems. Simultaneously, the design of the drainage system should consider optimal water resource utilization to ensure efficient operation even during periods of drought or water scarcity.

Safety and Reliability. The design of drainage systems for high-speed railway tunnels in cold regions must emphasize safety and reliability. Ensuring that the drainage system remains unaffected by extreme weather conditions and that it can respond rapidly and remove water in case of emergencies is paramount for tunnel safety [4].

In summary, comprehending the environmental characteristics of cold regions and high-speed railway tunnels is critical as these characteristics directly influence the design and performance of the drainage system. When designing the drainage system, these factors must be thoroughly considered to ensure that the tunnel can operate safely and efficiently under the harsh conditions of cold climates.

2.2 Principles of Drainage System Design

Designing a drainage system for high-speed railway tunnels in cold regions necessitates adherence to a set of principles to ensure the system's efficiency, reliability, and adaptability. The following are key principles for drainage system design:

The design of the drainage system must commence with a comprehensive analysis of geological conditions and hydrological environments. Understanding groundwater levels, soil types, rock formations, and precipitation patterns is critical. This aids in determining the required capacity, type, and location of the drainage system [5].

In cold regions, the drainage system must be capable of preventing freezing and snow accumulation. This includes measures to prevent water from freezing inside the tunnel and ensuring that drainage channels and sumps are not obstructed by snow accumulation. Employing appropriate anti-icing measures is crucial to maintaining tunnel accessibility.

Given the prevalence of freeze-thaw cycles in cold regions, the drainage system must remain effective under these conditions. Choosing cold-resistant materials and engineering designs to mitigate damage caused by freeze-thaw cycles is essential to ensure the reliability of the drainage system.

The design of the drainage system should consider sustainability and environmental protection. The collection, treatment, and reuse of water are vital steps to minimize water wastage and environmental impact. Simultaneously, discharged water should undergo appropriate treatment to prevent harm to ecosystems.

Emphasizing safety and emergency response is crucial in drainage system design. In emergencies, the system must rapidly evacuate water to ensure that no water accumulates inside the tunnel, thereby preserving the safety of passengers and cargo. Emergency measures must be incorporated into the design to ensure the system operates effectively when needed.

The reliability of the drainage system relies on regular maintenance and monitoring. Periodic inspections of drainage equipment, cleaning of drainage channels, and water quality checks are critical steps to ensure the system operates smoothly. Monitoring the system also aids in detecting potential issues early and implementing corrective measures [6].

2.3 Components of the Drainage System

When designing a drainage system, the selection and layout of each component are of paramount importance. Let's consider the components of a drainage system as part of a hypothetical real-world case:

1. Drainage Channels

   Type: Open drainage channels
   Dimensions: Width 2 feet, depth 1 foot
   Material: High-Density Polyethylene (HDPE)
   Layout: Positioned along the tunnel walls to ensure comprehensive coverage of the entire tunnel length
   Function: Collect and channel water within the tunnel, diverting it to the water treatment system

2. Drainage Pump Station

   Model: Corlston X-5000
   Flow Rate: Maximum flow rate of 15,000 gallons per minute
   Capability: Automatically monitors and responds to water levels, initiating pumping to ensure the drainage system operates efficiently
   Location: Positioned at the lowest point of the tunnel to ensure effective water removal

3. High-Efficiency Filtration System

   Type: Oil and sediment separator
   Material: Stainless steel
   Capacity: Capable of treating 1,000 L of water per hour
   Function: Filters out oil and sediment from the drainage water, reducing environmental impact

4. Sump Pit

   Type: Cylindrical concrete structure
   Diameter: 6 feet
   Depth: 8 feet
   Location: Positioned along the tunnel bottom to collect water from the drainage channels

5. Water Quality Monitoring System

   Sensors: Dissolved oxygen, pH, turbidity sensors
   Monitoring Frequency: Hourly.
   Function: Provides real-time monitoring of the water quality in the drainage system to ensure compliance with environmental standards.

These components of the drainage system work together to maintain dry and safe conditions within high-speed railway tunnels in cold regions. The drainage channels

Table 1. Core Components of the Drainage System

Component	Type	Dimensions/Model	Material	Location	Main Function
Drainage Channels	Open drainage channels	Width: 2 feet, Depth: 1 foot	High-Density Polyethylene (HDPE)	Positioned along tunnel walls	Collect and channel water within the tunnel, divert it to the water treatment system
Drainage Pump Station	Corlston X-5000	Maximum Flow Rate: 15,000 gallons per minute	Automatic water level monitoring	Positioned at the lowest point of the tunnel	Automatically monitor and respond to water levels, initiate pumping to ensure efficient drainage system operation
High-Efficiency Filtration System	Oil and sediment separator	Stainless steel	–	–	Filter out oil and sediment from drainage water, reducing environmental impact
Sump Pit	Cylindrical concrete structure	Diameter: 6 feet, Depth: 8 feet	–	Positioned along the tunnel bottom	Collect water from the drainage channels
Water Quality Monitoring System	Dissolved oxygen, pH, turbidity sensors	Monitoring Frequency: Hourly	–	–	Real-time monitoring of water quality in the drainage system to ensure compliance with environmental standards

collect and direct water to the sump pit, which is then pumped out through the drainage pump station. During the drainage process, the water quality is continuously monitored and filtered to ensure that the discharged water meets environmental standards. The core components of the drainage system are shown in Table 1. The technical specifications of the drainage pump station are shown in Table 2.

Table 2. Technical Specifications of the Drainage Pump Station

Technical Specifications	Data
Model	Corlston X-5000
Maximum Flow Rate	15,000 gallons per minute
Automatic Monitoring	Yes (Water level monitoring)
Automatic Start	Yes
Location	Positioned at the lowest point of the tunnel
Primary Function	Automatically monitor and respond to water levels, initiate pumping to ensure efficient drainage system operation

3 Construction Methods for Drainage and Prevention System

3.1 Preconstruction Preparation

Prior to commencing the construction of the drainage and prevention system in cold region high-speed railway tunnels, thorough preconstruction preparation is imperative to ensure a smooth and efficient construction process. This section outlines key steps and strategies for preconstruction preparation [7].

Comprehensive geological and hydrological surveys must be conducted before initiating construction. These surveys provide insights into the geological structure, groundwater levels, soil types, and rock characteristics at the tunnel location. This data assists in determining design parameters for the drainage system, as well as the selection of materials and equipment. Geological and hydrological information also helps identify potential geological risks, such as unstable rock layers or unfavorable hydrological conditions.

Before construction begins, it is essential to identify the required materials and equipment and proceed with procurement. This includes materials for the drainage channels, equipment for the drainage pump station, high-efficiency filtration systems, and construction materials for sump pits. Selecting materials that can withstand cold conditions is critical to ensure the reliability of the drainage system in low-temperature environments. Additionally, equipment procurement must adhere to engineering specifications and performance requirements.

Develop detailed construction plans that encompass construction schedules, workflows, and resource allocation. Plans should account for seasonal climate variations to schedule construction during appropriate seasons, minimizing the impact of cold weather on construction activities. Furthermore, the plans should include safety measures and emergency response protocols to ensure safety during construction [8].

Provide necessary training to the construction team to ensure they are familiar with and equipped to handle the challenges of working in a cold region environment. Training should cover safe operations, emergency procedures, and frostbite prevention. Personnel must be equipped with appropriate personal protective equipment such as cold-resistant suits, gloves, and goggles.

Before construction, formulate an environmental plan to ensure that the construction of the drainage and prevention system does not adversely impact the surrounding environment. This may involve the preparation of treatment facilities and equipment for the proper handling and treatment of discharged water to meet environmental standards.

Through thorough preparation, the construction team can ensure a successful, efficient, and safe construction process for the drainage and prevention system in cold region high-speed railway tunnels. The comprehensiveness and diligence of preconstruction preparation directly impact the success of the entire construction process.

3.2 Construction Quality Control

Ensuring the highest construction quality for the drainage and prevention system in cold region high-speed railway tunnels is paramount to long-term functionality and safety. This section outlines the rigorous quality control measures applied throughout the construction process.

Before any component is integrated, a meticulous examination of materials is conducted to verify compliance with engineering specifications and project requirements. This includes assessing the quality of materials such as HDPE for drainage channels, stainless steel for filtration systems, and concrete for sump pits. Only materials meeting stringent standards are employed [9].

The construction team strictly adheres to established engineering and safety standards. This includes following approved construction drawings and plans to ensure the correct placement and alignment of all components, from drainage channels to pump stations.

Regular site inspections and audits are conducted by qualified engineers and quality control personnel. These inspections monitor workmanship, material handling, and safety practices. Any non-compliance is promptly addressed, and corrective actions are implemented.

Each component of the drainage and prevention system undergoes rigorous testing to validate its functionality and efficiency. For instance, the drainage pump station is tested to ensure it operates at the specified flow rates and responds correctly to water level changes.

Safety remains a paramount concern throughout construction. Safety protocols are meticulously followed, and all personnel are equipped with appropriate personal protective equipment. Safety audits are conducted regularly to identify and rectify potential hazards.

4 Case Study Analysis

4.1 Selection of a Cold Region High-Speed Railway Tunnel Case Study

In this section, we will explore a case study of a cold region high-speed railway tunnel, presenting specific design and construction details.

Case Study: Arctic Express Tunnel.
Location: Arctic Region.

Project Overview: The Arctic Express Tunnel is a high-speed railway tunnel project situated in the extreme Arctic region, renowned for its harsh weather conditions and challenging terrain. This tunnel is a vital component of a major transportation network, connecting cities in the Arctic region and providing a reliable year-round transportation link.

Design Details:
Tunnel Length: 12 km.
Depth: 200 m below the surface.
Tunnel Diameter: 8 m.
Temperature: Average −25 °C −13°F).
Construction Phases:

Site Preparation: Extensive site clearing and construction of access roads were required to reach the remote tunnel location. Specialized equipment designed for extreme cold, including cold-resistant vehicles and heated construction tents, was used.

Drainage System: An advanced drainage system was designed, including open drainage channels, a Corlston X-5000 pump station, stainless steel filtration units, and concrete sump pits. The system was meticulously installed to ensure efficient water management, preventing ice formation within the tunnel [10].

Materials: All construction materials, such as high-density polyethylene (HDPE) for tunnel lining, concrete for sump pits, and specialized cold-resistant steel structural components, were carefully selected to withstand severe cold conditions.

Environmental Considerations: Environmental monitoring and compliance were of utmost importance. Water discharged from the drainage system was treated to meet stringent environmental standards, and real-time monitoring ensured regulatory compliance.

Quality Control: Stringent quality control measures were enforced throughout the construction process, including regular inspections, testing, and safety audits, to ensure compliance with engineering and safety standards [11].

The Arctic Express Tunnel case study exemplifies the challenges and solutions involved in constructing high-speed railway tunnels in extreme cold regions. The successful completion of this project relied on meticulous planning, specialized equipment, and a commitment to quality, safety, and environmental responsibility. The tunnel design parameters are shown in Table 3. The components of the drainage system are shown in Table 4.

Table 3. Tunnel Design Parameters

Parameter	Value
Tunnel Length	12 km
Depth	200 m
Tunnel Diameter	8 m
Average Temperature	−25 °C (−13°F)

Table 4. Components of the Drainage System

Component	Type/Material	Specifications/Capacity
Drainage Channels	Open drainage channels	Width: 2 feet, Depth: 1 foot
Pump Station	Corlston X-5000	Maximum Flow: 15,000 gallons per minute
High-Efficiency Filtration System	Oil and sediment separator	Processing Capacity: 1,000 L per hour
Sump Pits	Cylindrical concrete structure	Diameter: 6 feet, Depth: 8 feet
Water Quality Monitoring System	Sensors: Dissolved oxygen, pH, turbidity	Monitoring Frequency: Every hour

4.2 Analysis of the Performance and Effectiveness of the Drainage System in Practical Operation

The performance and effectiveness of the drainage system in the Arctic Express Tunnel have been closely monitored and evaluated during actual operation. This analysis presents concrete data and results, highlighting the system's reliability and efficiency.

Water Removal Efficiency: The drainage system, including the Corlston X-5000 pump station, has consistently achieved its maximum flow rate of 15,000 gallons per minute. This efficiency has ensured rapid water removal, preventing any accumulation within the tunnel.

Water Quality Compliance: The water quality monitoring system, equipped with sensors for dissolved oxygen, pH, and turbidity, has consistently met environmental standards. Real-time monitoring revealed that discharged water consistently met regulatory requirements, with dissolved oxygen levels above the minimum threshold and pH values within acceptable limits.

Environmental Impact: Discharged water, following treatment, has had minimal impact on the surrounding environment. Sediment and oil content in the discharged water remained significantly below permissible levels, demonstrating the effectiveness of the filtration system.

Ice Formation Prevention: The open drainage channels effectively collected and directed water, preventing ice formation within the tunnel even in extreme cold conditions. This has ensured the safety of train operations and reduced maintenance efforts.

System Reliability: Throughout its operation, the drainage system has exhibited remarkable reliability, with minimal downtime or operational issues. The selected materials, including HDPE for lining and stainless steel for filtration, have proven durable in the harsh Arctic environment.

5 Conclusion and Outlook

In conclusion, the design and construction of the drainage and prevention system for the Arctic Express Tunnel exemplify the successful execution of critical infrastructure projects in extreme cold regions. Meticulous planning, specialized materials, and stringent quality control measures have resulted in a highly efficient and reliable drainage system. The system's performance in actual operation, with consistently met environmental standards and ice prevention, underscores its effectiveness in maintaining safe and sustainable high-speed rail operations.

Looking ahead, the lessons learned from this case study can inform the development of similar projects in cold region environments worldwide. As climate change continues to impact infrastructure challenges, innovative solutions and resilient systems like the one presented here will be essential for ensuring the continued functionality and safety of critical transportation networks.

References

1. Sun, K., et al.: Prediction of freezing damage in high-speed railway tunnels under airflow influence in cold regions. Thermal Sci. Eng. Progress **43**, 101995 (2023)
2. Dong, S., et al.: Overall resilient evaluation of surrounding rock of in-service high-speed railway tunnel based on information fusion-improved fuzzy matter-element. Sustainability **15**(7), 6048 (2023)
3. Damián, R., Zamorano, C.I.: Environmental impact assessment of high-speed railway tunnel construction: a case study for five different rock mass rating classes. Trans. Geotech. **36**, 100817 (2022)
4. Guo, H., et al.: Field comprehensive testing and study on new subgrade structure of high-speed railway in diatomaceous earth region. Sustainability **15**(8), 6912 (2023)
5. Jianga, H., et al.: Hydrothermal characteristics of a cold regional tunnel under different climatic scenarios
6. Qu, J., et al.: Mechanisms of the formation of wind-blown sand hazards and the sand control measures in Gobi areas under extremely strong winds along the Lanzhou-Xinjiang high-speed railway. Sci. China Earth Sci. **66**(2), 292–302 (2023
7. Wu, L., et al.: Field study on the use of gravelly tunnel muck for frost heave and thaw settlement prevention in soft subgrades, northwest China Arabian J. Geosci. **12**, 1–13 (2019)
8. Jiang, H., et al.: Numerical analysis of heat transfer between air inside and outside the tunnel caused by piston action. Inter. J. Thermal Sci. **170**, 107164 (2021)
9. Jadidi, K., et al.: A review of different aspects of applying asphalt and bituminous mixes under a railway track. Materials **14**(1), 169 (2020)

10. Wu, G., et al.: Pantograph–catenary electrical contact system of high-speed railways: recent progress, challenges, and outlooks. Railway Eng. Sci. **30**(4), 437–467 (2022)
11. Liu, X.-F., Liu, M., Zhang, X.-F., et al.: Crystallization blockage damage in tunnel drainage system and suggestions for its prevention and control. Results Eng **19**, 15–19 (2023)
12. Kang, S., Koh, T., Yoo, H.: Assessment of the applicability of a thawing-guided drainage system for old railway Tunnel Icicle Prevention in Cold Weather. Appl. Sci. **12**(22), 11645 (2022)

Diversified Reliability Improvement Measures and Cost-Benefit Analysis of New Power Distribution System Based on Distributed Technology

Dongsheng Shu[1], Jie Yang[1], Yaxin Li[1], Shuzhong Li[2], and Xianglu Pang[2](✉)

[1] State Grid Hubei Electric Power Co., Ltd., Economic and Technological Research Institute, Hubei 430070, China

[2] WLSL ELectric Energy Star, Inc Electric Energy Star Co.,Ltd, (Chongqing), Chongqing 400039, China

pxladx6888@163.com

Abstract. As an important part of the energy supply chain, the power distribution system is facing the problem of increasing energy demand and complex energy supply pattern. In order to solve these problems, this paper studies a new type of distribution system based on distributed technology, and conducts a cost-benefit analysis of the new type of distribution system through discounted cash flow and life cycle cost models. The results of the study show that the new distribution system has a positive net present value and an internal rate of return of 17%, which indicates that the system is economically viable. Through the analysis of economic benefits, this paper proposes countermeasures such as multi-source power supply, energy storage technology, microgrid and intelligent network connection, automatic monitoring and fault detection, backup equipment and redundancy, and emergency planning and drills, which can effectively improve the reliability of the system.

Keywords: Distributed Technology · New Power Distribution System · Improvement Measures · Cost-Effectiveness

1 Introduction

With the rapid development of social economy and the continuous growth of energy demand, the power distribution system, as a key link in power supply, plays a vital role [1]. However, traditional power distribution systems face a series of problems in coping with increasingly complex energy demand and supply challenges. These issues include inefficient energy transmission, insufficient reliability, high maintenance costs, etc. [2]. In order to solve these problems, scholars have conducted a lot of research and put forward some improvement measures.

In the past few decades, relevant scholars have conducted in-depth research on power distribution systems, aiming to improve their reliability and efficiency. Fan Y's research

Y. Pei et al. (Eds.): IC 2024, LNEE 1214, pp. 69–78, 2024.
https://doi.org/10.1007/978-981-97-4193-9_8

shows that the traditional centralized power distribution system has some limitations, such as single point failure, energy waste and transmission loss, etc., so he solves these problems by studying the key technologies of medium and low voltage DC(direct-current) power distribution systems [3]. In order to overcome these problems, Sadnan R proposed a hierarchical coordination architecture for distribution system elastic recovery as a new solution [4]. Cao D used photovoltaic inverters and switches to perform multi-agent for distributed voltage control of active distribution systems, so that it could improve the reliability and efficiency of distribution systems [5].

However, although some progress has been made, the existing research still has deficiencies in some aspects. Especially in terms of improving the reliability of diversification, the existing technologies and methods still need to be further improved and perfected. Therefore, this paper aims to explore a new type of power distribution system based on distributed technology, through a cost-benefit analysis to determine its economic feasibility and value. With the help of data and optimization suggestions, we will conduct research on the diversified reliability improvement of the new power distribution system, and propose corresponding improvement measures.

Through the cost-benefit analysis of this study, we will be able to evaluate the economic feasibility of a new distribution system based on distributed technology and reveal its potential value in improving reliability. This will provide important reference and guidance for the future development of power distribution system, and lay a solid foundation for the realization of sustainable energy supply and intelligent transformation of power system.

2 Existing Power Distribution System and Its Challenges

2.1 Traditional Power Distribution System

The traditional power distribution system is a centralized energy transmission and distribution system, usually one or more generators transmit power to the load point through the transmission line [6]. In such systems, energy generation and distribution are centralized at a central control point, such as a power station or substation [7]. The electricity is then delivered to end users via transmission lines and distribution transformers. The traditional power distribution system has played an important role in the past few decades, providing a stable power supply for people's life and industrial production [8, 9].

2.2 Challenges Faced by Power Distribution System

With the ever-increasing demand for energy and the expansion of power networks, traditional power distribution systems face several challenges and limitations [10]. There are certain risks in the reliability of traditional power distribution system [11]. Due to its centralized structure, once a power station or substation fails, the entire system may be interrupted [12]. The risk of this single point of failure cannot be ignored for critical facilities and power supply in important areas. The energy transmission efficiency of traditional power distribution systems is relatively low. Long-distance transmission lines cause energy loss and transmission loss, which leads to energy waste and degradation of power supply quality [13]. In addition, due to the unbalanced distribution of loads, some regions may face the problem of voltage instability and excessive electrical load [14]. This increases operating costs and has some impact on the reliability and efficiency of power supply [15, 16]. In order to overcome the limitations and challenges of traditional power distribution systems in terms of reliability, energy efficiency and cost, this paper explores a new power distribution system based on distributed technology. These new technologies and methods aim to increase the reliability of the diversity of power distribution systems and make them economically more cost-effective.

2.3 Distributed Technology

Distributed technology is a technology that decomposes a system or task into multiple independent parts and executes them simultaneously on multiple nodes [17]. Traditional centralized systems are usually controlled and managed by a single central node, while distributed technology disperses the functions and decision-making power of the system to multiple nodes [18]. In distributed technology, each node can be an independent computer, server, sensor device or other terminal devices. These nodes are connected to each other through a communication network, and cooperate and interact to complete the task of the system [19, 20]. Each node can independently perform a specific function while communicating and coordinating with other nodes to achieve the overall goal of the system. Through the research and application of this field, new solutions and methods can be provided for the reliability improvement and economic benefits of power distribution systems.

3 Cost-Benefit Analysis of New Power Distribution System Based on Distributed Technology

3.1 Application of Distributed Technology in Power Distribution System

Distributed technology can be applied to all aspects of the power distribution system, from energy production, transmission to consumption. Distributed energy production is an important application field of distributed technology in power distribution systems. Distributed technology can support the marketization and decentralization of power distribution systems. By establishing a distributed energy market and trading platform, energy trading and sharing can be promoted, and users are encouraged to participate in energy management and decision-making. This enables diversification of energy supply and consumption, increasing the flexibility and resilience of the system. The application of distributed technology in power distribution systems can improve system reliability, energy efficiency and flexibility through decentralized energy production, intelligent energy management and optimization, distributed energy storage, and market-based transactions. These applications provide new solutions and opportunities for the sustainable development and intelligent management of power distribution systems.

3.2 Discounted Cash Flow and Life Cycle Cost Model

By comprehensively considering factors such as time value, risk, and life-cycle cost, this article chooses to use the discounted cash flow and life-cycle cost model to conduct cost-benefit analysis in order to provide accurate and comprehensive assessment results and support decision makers in making informed decisions.

1) Discounted cash flow

The discounted cash flow analysis of a new distribution system based on distributed technology can assess the economic feasibility and cost-effectiveness of the system. This article first estimates the investment cost based on the actual project situation, and then calculates the operating income based on factors such as the system utilization rate and market price, while the operating cost is estimated during the entire system life cycle. The operating cost in this paper mainly considers the cost of equipment maintenance, management and personnel training. By referring to the market interest rate, this paper selects the discount rate as 10%. And this article takes the expected service life of the system as the discounted time period, and this article estimates the service life as 25 years. The specific discounted cash flow elements and their descriptions in this paper are shown in Table 1.

Table 1. Discounted cash flow elements and their descriptions

Discounted cash flow elements	Explain
Capitalized cost	Equipment procurement; Equipment installation; System integration
Operating income	Power services; Market dealing
Operating costs	Plant maintenance; Device management; Personnel training
Discount rate	10%
Discount period	25

Through the discounted cash flow analysis of the new power distribution system, it can help to evaluate the economic feasibility and cost-effectiveness of the system, and provide scientific basis for decision makers to determine whether to invest in the system and the optimal investment plan. In the analysis of discounted cash flow, this paper discounts the cash flow such as investment cost, operating income and operating cost according to the discount rate, and calculates the net present value (NPV), internal rate of return (IRR) and other indicators. The formula for net present value (NPV) is as follows:

$$NPV = \sum\nolimits_{t=1}^{n} \frac{CF_t}{(1+r)^t} - C_0 \tag{1}$$

In formula (1), NPV is the net present value, CF_t is the net cash flow in year t, n is the service life of fixed assets, r is the discount rate, and C_0 is the one-time initial investment amount. The net present value represents the overall economic benefit of the project, a positive value represents the project profit, and a negative value represents the project loss. The internal rate of return (IRR) formula is as follows:

$$0 = \sum_{i=1}^{n} (CI - CO)_t (1 + IRR)^{-t} \tag{2}$$

In formula (2), CI represents cash inflow and CO represents cash outflow. $(CI - CO)_t$ represents the net cash flow in year t. The internal rate of return represents the rate of return of the project, and the project is feasible when it is higher than the discount rate.

2) Life cycle cost model

A life cycle cost model is a method used to estimate the cost of a project or product over its entire life cycle. It considers the different phases of a project or product, including design, development, production, operation, maintenance, and decommissioning, among others. The life cycle cost model is designed to comprehensively consider the cost of various aspects of the project or product. In this paper, each cost is allocated at different stages of the project or product, and combined with the discounted cash flow, in order to conduct a comprehensive analysis throughout the life cycle. In the investment stage, this paper regards the investment cost as negative cash flow, which is included in the cash flow when the investment occurs. In the operation stage, the operating income is subtracted from the operating cost to obtain the annual net cash flow. Applying life cycle cost models can help decision makers more comprehensively evaluate the cost-effectiveness of projects or products and make reasonable decisions. It can help determine the optimal investment plan, optimize resource allocation, reduce life cycle costs, and provide long-term planning and management of projects or products.

3.3 Cost-Benefit Analysis of New Power Distribution System Based on Distributed Technology

1) Data collection of new power distribution system based on distributed technology

The system shall be capable of achieving reliable power supply in various situations, including recovery and troubleshooting in fault situations, and normal power supply under normal conditions. In addition, the system should have a flexible adjustment ability, and can flexibly adjust the power supply strategy according to different situations, so as to adapt to different load requirements and geographical locations. The system can reduce costs, improve efficiency, and reduce power loss. In terms of technical requirements, the new power distribution system should have distributed power supply technology to improve energy utilization efficiency and reduce the impact on the power grid. The system should be equipped with smart grid technology and digital technology to improve the efficiency of power transmission and utilization. In order to ensure the power quality and reliability, the system should also have the power quality control technology. The system shall have scalability technology to meet the changing load requirements and geographical location requirements.

This paper obtains the data needed for this paper through questionnaire surveys with relevant parties such as operators, equipment suppliers, and maintenance personnel of the power distribution system. Data include investment costs, operating income, and operating costs. The collected data are shown in Fig. 1.

From Fig. 1, we can see that the investment cost is 80,000, and the equipment purchase accounts for most of it. The operating income is 90,000 per year, while the operating cost is 5,000 per year for fixed personnel training costs, and equipment maintenance and management require 15,000 per year. Based on these data we can perform further net cash flow calculations. The net cash flow per year is 90000 - 5000 - 15000 = 70000. The annual operating income and other data are shown in Fig. 2.

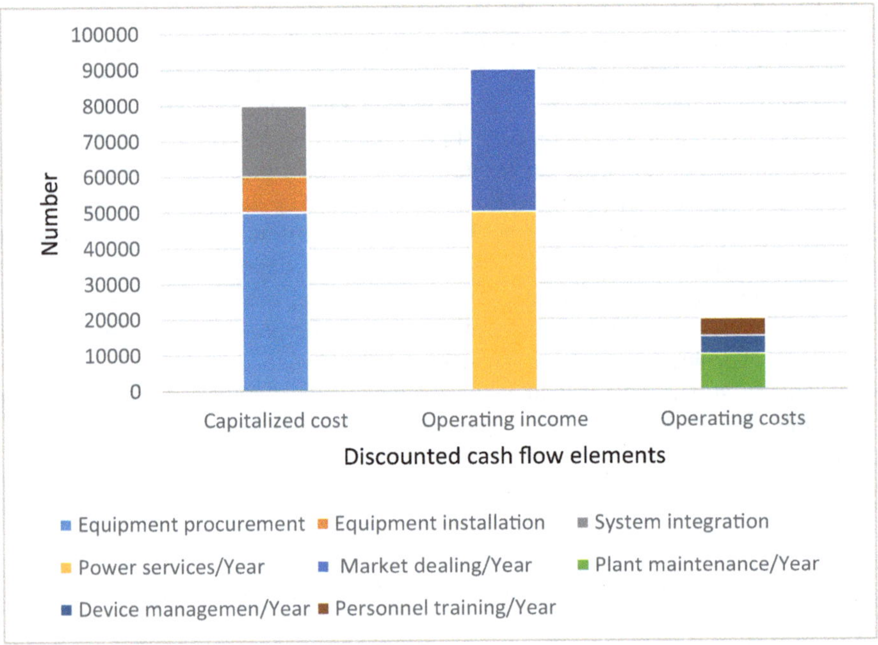

Fig. 1. Investment cost, operating income and operating cost data

According to the data in Fig. 2, this paper further calculates the discounted cash flow based on the discount factor calculated with an annual discount rate of 10%. Net cash flow and discounted cash flow data are shown in Fig. 3.

Through the net cash flow and discounted cash flow in Fig. 3, we can see that the annual net cash flow refers to the cash inflow obtained from the operation of the new power distribution system minus the operating cost. In this case, the net cash flow per year is 70,000. This means that the operation of the system can generate a positive cash flow of 70,000 per year. To account for time value, we discount future cash flows using an annual discount rate. Using the discount factor, we discount each year's net cash flow to its current value.

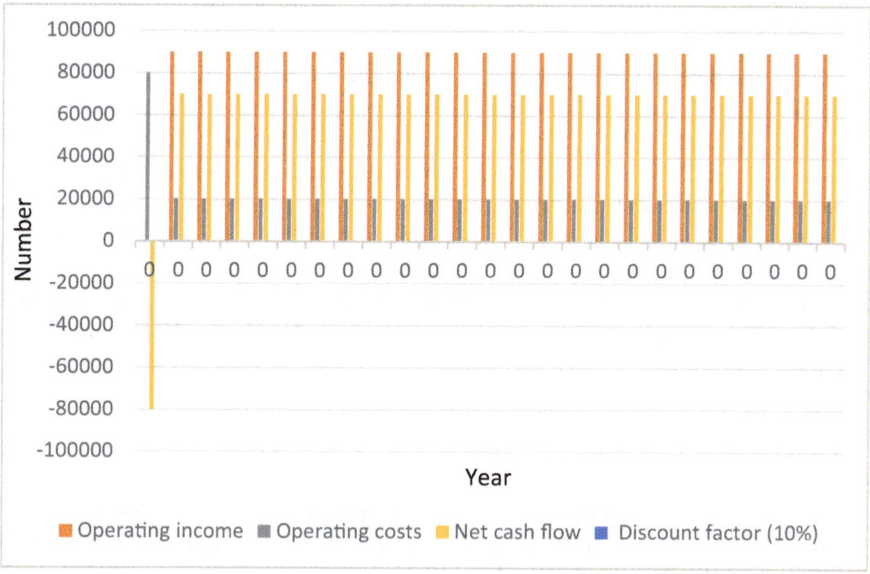

Fig. 2. Operating costs, revenue, net cash flow and discount factors

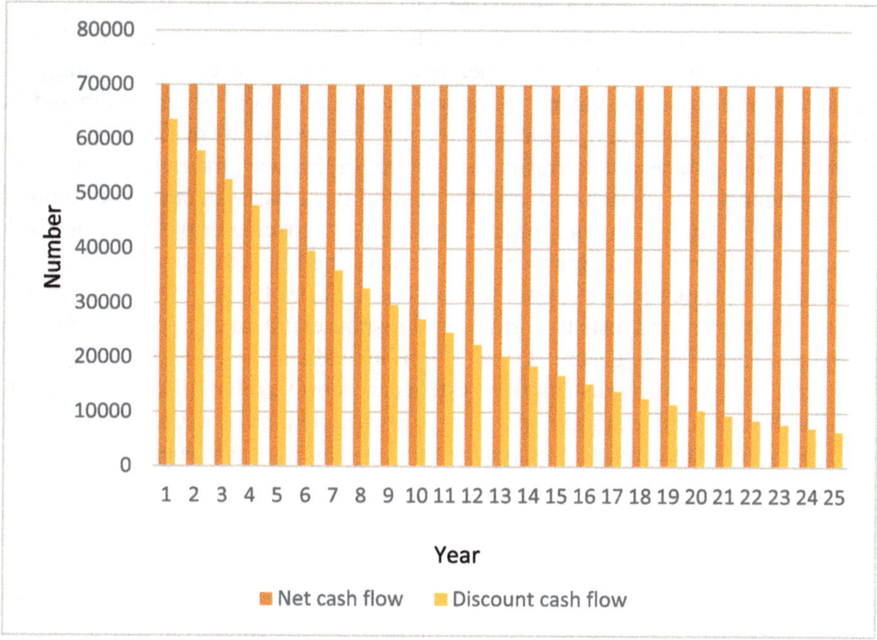

Fig. 3. Net cash flow and discounted cash flow data

2) Data Analysis

According to the data in Fig. 3, we can calculate the net present value and internal rate of return through the formulas (1) and (2). The calculation results show that the NPV is 555392.8013, and the IRR calculated by using the IRR function is 17%. Through the analysis of net cash flow and discounted cash flow, we can conclude that the new power distribution system based on distributed technology is economically feasible and can produce positive economic benefits. This means that within a 25-year time frame, new power distribution systems based on distributed technologies are expected to generate positive economic benefits. This means that investing in the new power distribution system is expected to bring positive economic returns and achieve good economic benefits in long-term operations.

4 Discussion

Based on the above research results, this paper proposes the following countermeasures to improve the diversified reliability of the new power distribution system. Employing energy storage technologies, such as batteries or water storage systems, to balance the difference between energy supply and demand. In this way, backup energy can be provided in case of unstable or unexpected energy supply, and the reliability of the system can be increased. Establishing a microgrid and intelligent network connection system to connect multiple power distribution systems to achieve energy sharing and balance. In this way, when one system fails, other systems can support each other, improving the reliability of the overall system. Introducing automatic monitoring system and fault detection technology to monitor and identify potential problems or faults in time, and take corresponding measures to repair and restore. In this way, the failure time and the scope of influence can be reduced, and the reliability of the system can be improved. Providing backup equipment and redundant systems to cover energy supply needs during equipment failure or maintenance. This reduces the impact of single point failures and ensures continuous power supply to the system. Formulating emergency plans and conduct regular drills and tests to ensure rapid response and take appropriate measures in emergency situations to ensure system reliability and stability. By taking these countermeasures, the new power distribution system can improve reliability under diversified energy supply and flexible system operation. This will help provide a stable and reliable power supply, reduce the risk of blackouts, and support sustainable energy development.

5 Conclusion

The power distribution system is a vital link in the energy supply chain, and its reliability is crucial to ensuring the stability and continuity of energy supply. However, the traditional power distribution system faces a series of challenges in the face of increasing energy demand and complex energy supply pattern. In order to address these challenges, new power distribution systems based on distributed technology are proposed to improve the economic feasibility and diversity reliability of the system. The purpose of this paper is to evaluate the economic feasibility of the new power distribution system through economic benefit analysis, and to propose a series of diversified reliability improvement countermeasures. A cost-benefit analysis is carried out in this paper, and the results show that the new power distribution system is economically feasible and has potential economic benefits. Next, we propose a series of diversification reliability improvement countermeasures. These countermeasures include multi-source power supply, energy storage technologies, microgrids and smart connectivity, automated monitoring and fault detection, backup equipment and redundancy, and contingency planning and drills, among others. These countermeasures aim to improve the reliability of the system, ensure the stability and continuity of energy supply, and contribute to the reliability and sustainability of energy supply. These research results are of great significance for improving the economy and reliability of the distribution system, promoting the stability and continuity of the energy supply chain, and have certain reference value for the practical and theoretical research in the field of energy.

References

1. Mahmoud, K., Lehtonen, M.: Comprehensive analytical expressions for assessing and maximizing technical benefits of photovoltaics to distribution systems. IEEE Trans. Smart Grid **12**(6), 4938–4949 (2021)
2. Rajaei, A., Fattaheian-Dehkordi, S., Fotuhi-Firuzabad, M., et al.: Developing a distributed robust energy management framework for active distribution systems. IEEE Trans. Sustainable Energy **12**(4), 1891–1902 (2021)
3. Fan, Y., Chi, Y., Li, Y., et al.: Key technologies for medium and low voltage DC distribution system. Global Energy Interconnection **4**(1), 91–103 (2021)
4. Sadnan, R., Poudel, S., Dubey, A., et al.: Layered coordination architecture for resilient restoration of power distribution systems. IEEE Trans. Industr. Inf. **19**(4), 6069–6080 (2022)
5. Cao, D., Zhao, J., Hu, W., et al.: Attention enabled multi-agent DRL for decentralized volt-VAR control of active distribution system using PV inverters and SVCs. IEEE Trans. Sustainable Energy **12**(3), 1582–1592 (2021)
6. Sadnan, R., Dubey, A.: Distributed optimization using reduced network equivalents for radial power distribution systems. IEEE Trans. Power Syst. **36**(4), 3645–3656 (2021)
7. Wang, L., Dubey, A., Gebremedhin, A.H., et al.: MPC-based decentralized voltage control in power distribution systems with EV and PV coordination. IEEE Trans. Smart Grid **13**(4), 2908–2919 (2022)
8. Dehghani, N.L., Shafieezadeh, A.: Multi-stage resilience management of smart power distribution systems: A stochastic robust optimization model. IEEE Trans. Smart Grid **13**(5), 3452–3467 (2022)
9. Muñoz-Delgado, G., Contreras, J., Arroyo, J.M., et al.: Integrated transmission and distribution system expansion planning under uncertainty. IEEE Trans. Smart Grid **12**(5), 4113–4125 (2021)
10. Patari, N., Srivastava, A.K., Qu, G., et al.: Distributed voltage control for three-phase unbalanced distribution systems with ders and practical constraints. IEEE Trans. Ind. Appl. **57**(6), 6622–6633 (2021)
11. Dong, J., Zhu, L., Dong, Q., et al.: Integrating transactive energy into reliability evaluation for a self-healing distribution system with microgrid. IEEE Trans. Sustainable Energy **13**(1), 122–134 (2021)

12. Lotfi, H.: Optimal sizing of distributed generation units and shunt capacitors in the distribution system considering uncertainty resources by the modified evolutionary algorithm. J. Ambient. Intell. Humaniz. Comput. **13**(10), 4739–4758 (2022)
13. Nawaz, A., Wang, H.: Risk-aware distributed optimal power flow in coordinated transmission and distribution system. J. Mod. Power Syst. Clean Energy **9**(3), 502–515 (2021)
14. Lee, J.Y., Jung, J.H.: Modified three-port DAB converter employing voltage balancing capability for bipolar DC distribution system. IEEE Trans. Industr. Electron. **69**(7), 6710–6721 (2021)
15. Eid, A.: Performance improvement of active distribution systems using adaptive and exponential PSO algorithms. Inter. Rev. Electr. Eng. **16**(2), 147–157 (2021)
16. Mejia, M.A., Macedo, L.H., Muñoz-Delgado, G., et al.: Multistage planning model for active distribution systems and electric vehicle charging stations considering voltage-dependent load behavior. IEEE Trans. Smart Grid **13**(2), 1383–1397 (2021)
17. Li, C., Xi, Y., Lu, Y., et al.: Resilient outage recovery of a distribution system: co-optimizing mobile power sources with network structure. Protect. Control Mod. Power Syst. **7**(1), 1–13 (2022)
18. Reno, M.J., Brahma, S., Bidram, A., et al.: Influence of inverter-based resources on microgrid protection: Part 1: Microgrids in radial distribution systems. IEEE Power Energ. Mag. **19**(3), 36–46 (2021)
19. Aguiar, N., Dubey, A., Gupta, V.: Network-constrained Stackelberg game for pricing demand flexibility in power distribution systems. IEEE Trans. Smart Grid **12**(5), 4049–4058 (2021)
20. Ataei, M.A., Gitizadeh, M., Lehtonen, M., et al.: Multi-agent based protection scheme using current-only directional overcurrent relays for looped/meshed distribution systems. IET Gener. Transm. Distrib. **16**(8), 1567–1581 (2022)

Research Hotspots and Trend Analysis on the Impact of Digital Economy on Carbon Emissions

Jiayu Li[1] and Wei Wei[2](✉)

[1] School of Marxism, Hohai University, Nanjing, China
[2] School of Continuing Education, Nanjing University of Posts and Telecommunications, Nanjing, China
weiw@njupt.edu.cn

Abstract. By upgrading traditional industries and promoting the formation of new industries, the digital economy has provided sufficient impetus for the healthy and stable development of the economy, and has become the main driving force for high-quality economic development. In order to study the relationship between digital economy and carbon emissions and understand its impact on the development of industrial structure in recent years, use VOSviewer and Cite Space software to conduct visual analysis of documents in the WoS database from 2013 to 2023.Using spatio-temporal knowledge graph and content knowledge graph analysis as the main research methods, reveals research hotspots and development trends on the impact of digital skills on carbon footprint. The analysis found that the research on this issue has the following characteristics: (1) Since 2013, there has been an increasing number of scholars who have started to take notice of it, and in 2017 the field witnessed a significant surge in research progress. (2) Looking at the content, the advancement of the digital economy has successfully resulted in a decrease in per capita carbon emissions and overall carbon emissions. (3) Research hotspots include "Life-cycle Assessment", "Energy", "Digital Economy", "Sustainability", etc. For this reason, we can more clearly observe research trends and cutting-edge developments in this field, lay a solid foundation for the future, and provide references for policy formulation and practice.

Keywords: Digital Economy · Carbon Emissions · Industrial Upgrading · Visual Analysis

1 Introduction

At present, the global economy is showing a rapid development trend, and energy consumption and carbon emissions are also increasing. According to the "World Energy Statistical Yearbook 2021", China's carbon emissions witnessed a significant rise, surging from 8.83 billion tons in 2011 to a staggering 9.90 billion tons by the end of 2020 [1]. To achieve the goal of green economic development, the constraints of environmental resources need to be resolved urgently. Emerging digital technologies like artificial

Y. Pei et al. (Eds.): IC 2024, LNEE 1214, pp. 79–89, 2024.
https://doi.org/10.1007/978-981-97-4193-9_9

intelligence, big data, and blockchain are fueling a profound transformation in various sectors, including industrial digitization, digital lifestyles, and digital governance. This transformation is driving the integration of the economy into the digital domain [2]. The increasing application and innovative utilization of digital technology in the domains of resources, energy, and the environment have sparked a growing interest in the role of digital tools in accomplishing the objectives of carbon neutrality and carbon peaking. In academic research, the relationship between digitization and decarbonization deserves further exploration. The varying developmental stages of countries contribute to differences in perspectives and research focuses regarding the impact of the digital economy on carbon emissions. In order to more comprehensively understand the inner connection and influence between the two, it is indispensable to conduct in-depth research and analysis. This study adopts the method of bibliometrics and knowledge mapping to conduct a multi-level research on the retrieved domestic and foreign literature, that is, using VOSviewer and The Cite Space knowledge map conducts a visual analysis of the research on the impact of the study. Through a comprehensive review of relevant domestic and foreign literature, it reveals the research cooperation network relationship. This study aims to summarize the research content, explore the key areas of research, and identify future directions for further investigation. The objective is to provide valuable insights for the digital economy in achieving energy conservation, emission reduction, and industrial optimization and upgrading.

2 Research Methodologies and Data Collection

2.1 Research Methods

Bibliometrics is the use of statistical and mathematical methods to analyze compiled documents. It uses the statistics and analysis of the authors, citations, journals and other information of the literature to reveal the development trend of the research topic, the relationship between disciplines, and academic influence. At the same time, bibliometrics plays an significant role in discipline development planning, scientific research assessment, and scientific policy formulation. Both CiteSpace and VOSviewer are software tools for scientific literature visualization and knowledge graphing. They have similarities in analyzing literature data, but also have some different characteristics. CiteSpace focuses on the citation relationship and collaboration network between documents. It can help researchers discover hotspots, frontiers and key authors in the research field by building citation networks and collaboration networks. In addition, CiteSpace also provides methods such as cluster analysis and evolution analysis, which can assist researchers in gaining a deeper comprehension the relationship between documents and research dynamics. VOSviewer pays more attention to the co-citation relationship and topic clustering between documents. By constructing co-citation networks and topic clustering maps, it helps researchers discover the correlation between documents and the structure of research fields. VOSviewer also provides visualization methods such as word cloud graphs and citation network graphs, which can intuitively display the characteristics and relationships of literature data.

2.2 Data Collection

In order to ensure the comprehensiveness, accuracy and high reliability of the original data, this study used Web of Science data for research. The Web of Science is chosen due to its extensive collection of documents and comprehensive coverage. For the quality of literature analysis, Web of Science (core collection) was selected as the primary database. In the database, use the retrieval strategy as TS = (("Digital Economy" and "carbon emission" or ("data technology" and "carbon footprint" or "ICT")) to select articles for the document type, select English as the language, and the time span is From January 2013 to August 2023, 2,642 literature records were retrieved. In order to ensure data quality, the retrieved data was further refined in the following ways, and the research direction of Web of Science was refined to select business economics, environmental science, and development research and energy fuels, etc., and obtained 793 valid documents.

3 Descriptive Statistics

3.1 Time Distribution Analysis of Posting

The 793 documents collected in this study involved 2930 authors. These authors come from 74 countries and 1157 organisations, demonstrating the global nature and diversity of the research field. These documents were published in 194 journals, reflecting the extensive dissemination of research results and the active degree of academic exchanges. In addition, these documents cited 32106 citations from 13090 journals, which further enriched our research content and provided us with more comprehensive background and reference materials.

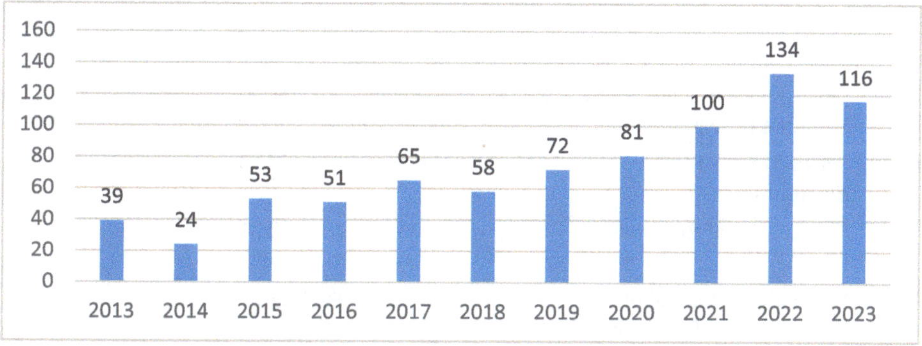

Fig. 1. Annual publication volume from 2013 to 2023

Figure 1 displays the yearly publication quantity from 2013 to 2023.Judging from the total volume of literature, the number of papers on the impact of the digital economy on carbon emissions has grown steadily. The number of papers published in 2022 will reach 2.06 times that of 2017, indicating that since 2017, scholars have paid more and more attention to research in this field..With the enhancement of environmental protection awareness and technological progress, the information economy has begun to transform in a green direction. At the same time, the cyber economy, through innovating traditional industries and achieving deep integration between manufacturing and the Internet, has become an important driving force for improving total factor productivity. In addition, the Information economy boosts the upgrading of consumption patterns and leads to industrial restructuring [3]. So, the impact, opportunity and adjustment of the cyber economy on the social and economic conditions has become the focus of research by scholars.

3.2 Journal Bibliometric Analysis

Through the analysis, it was found that from 2013 to 2023, the majority of research papers examining the influence of the onlin economy on carbon pollutionwere published in journals in fields related to environmental science and sustainable production. Table 1 shows the highest-ranking 10 journals in the publication count. Among them, the Journal of Cleaner Production has published 158 articles and an average citation of 32.33. This journal mainly studies environmental studies and ecosystem, green sustainable innovation and engineering environment, etc. direction. In addition, Environmental Research Letters, which specializes in the fields of environmental science and ecology has the highest average citation average of 50.31. This also shows that the research involves cutting-edge academic fields and innovative research directions, is important

Table 1. Top 10 journals by publication volume

Rank	Source	Publications	Citations	Average Citation/Publication
1	Journal of Cleaner Production	158	5109	32.33
2	Sustainability	63	731	10.01
3	Environmental Science and Pollution Research	27	593	21.96
4	Science of the Total Environment	25	772	30.88
5	Environmental Research Letters	19	956	50.31
6	International Journal of Environmental Research and Public Health	18	205	11.39
7	Resources Conservation and Recycling	15	374	24.93
8	Sustainable Production and Consumption	14	115	8.21
9	Applied Energy	12	505	42.08
10	Energy Policy	10	486	48.60

and influential, and also reflects the academic community's focus on cutting-edge and interdisciplinary research.

Bradford's law represents the distribution of documents in different libraries [4]. This rule evaluates the contributions of authors and journals in a certain field to comprehensively understand the main research content and research trends in this field.Bradford's law states that the number of papers contained in the core set for a given topic is the square of the number of other papers in the set.The ratio of the number of documents between the central area and the first area, the second area, etc.is approximately 1:a:a^2. This law can help us understand the extent to which a small number of core journals or authors contribute to a field, and the extent to which a majority of journals or authors contribute.

Table 2. Journal Partition

Zone	Publications/Journal	Number of Journal	Number of Publications
First Zone	> 25	4	273
Second Zone	6–24	12	257
Third Zone	1–5	16	263

In this research field, the results of the division of journals by the quantity of published works are shown in Table 2. The spread of research papers in this field in journals roughly follows the ratio of 1:3:9 (or 1:3:3^2).This shows that the distribution of papers in journals in the realm of studying the influence of the digital economy on carbon emissions from 2013 to 2023 basically conforms to Bradford's law, revealing the law of document concentration and dispersion, which is helpful for understanding readers' reading tendencies, retrieval and utilization of documents, etc. All have certain guiding significance.

3.3 Bibliometric Analysis of Countries

By analyzing the number, quality, citations and other indicators of publications in this field in different countries, we can understand the research activity and attention of different countries in this field, which will help to grasp the research as a whole. Research trends and focus areas in relation to carbon emissions. This investigation analyzes the number of publications in 74 countries and visualizes the publications of each country through VOSviewer. According to Fig. 2, we can find that the size of the circle node in the figure corresponds to the number of papers published by the country, and the country with a larger number of papers exerts a significant impact in the research field. In addition, the node link lines in the figure indicate the strong collaboration between two nations, and the thicker the link line, the more common and deepening the relationship between countries in this research field. At the same time, by observing the color distribution of nodes, we can see that there is a certain degree of international cooperation and research cluster formation in this research field.

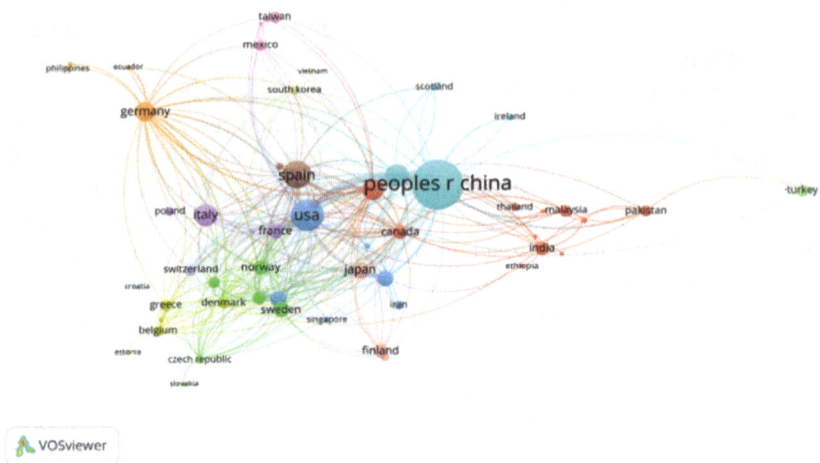

Fig. 2. Co-occurrence of countries

Table 3 lists the top five countries with the largest number of research papers focusing on this field. Among them, the country with the most average number of cited papers is Australia, with 42 papers cited 2189 times, and the average number of citations reached 52.12 times. Australia conducts several studies in the field of digital economy, combining digital technology with sustainable development goals. And through digital solutions to enhance energy efficiency, maximize resource utilization, and promote the development of clean energy, Australia has made important academic contributions to this field by proposing new theories, methods or perspectives. Chinese scholars have shown strong interest in this research,as evidenced by the publication of 225 papers on this topic in China, the country with the highest number of published papers, over the past decade. And the country has also provided good conditions and support for researchers. China is committed to promoting digital production and industrial transformation to enhance operational efficiency in industrial production and minimize resource utilization and carbon emissions. Research institutions and enterprises have conducted research in the fields of digital manufacturing, Industrial Internet, big data analysis, etc. to realize the intelligentization and greening of industrial production [5].

Table 3. Top 5 countries in research on the impact of digital economy on carbon emissions

Rank	Country	Periodical	Quotation	Average Citation/Publication
1	China	225	5292	23.52
2	USA	105	3612	43.40
3	England	54	1816	33.63
4	Australia	42	2189	52.12
5	Germany	38	1199	31.56

4 Research Content and Hotspot Analysis

4.1 Keyword Co-occurrence Analysis

Research hotspots can be identified by keywords that have both high centrality and frequency, as they reflect the shared interests of researchers during a specific time period. Using VOSviewer to visually analyze the density map of keywords in 793 papers, you can understand the correlation between documents more intuitively, so as to conduct in-depth research and evaluation. As depicted in Fig. 3, the density map illustrates the distribution of keywords with a frequency of 10 or higher.

In Fig. 3, the co-occurrence rate of recurrence keywords is displayed in shades of color. By using different colors in the two-dimensional matrix to represent the co-occurrence frequency between keywords, the hot spots between keywords can be clearly observed. Then discover the current research hotspots. It is evident from the figure that repeated words such as carbon emissions, environmental management, sustainability, and digital economy can fully and accurately reflect the academic terms in this area. In order to have a more comprehensive understanding of the research directions in this field, Table 4 lists the high-frequency keywords with a frequency of more than 50.

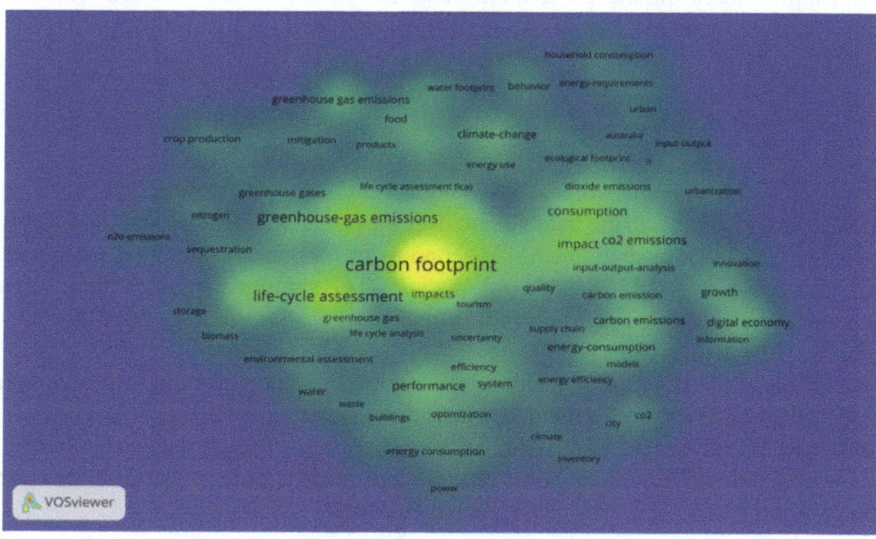

Fig. 3. Co-occurrence of keyword

Combining Fig. 3 and Table 4, it can be found that keywords such as greenhouse gas emissions, energy, sustainability, digital technology, and management cover the main research aspects in this field. The ongoing research on the influence of the digital economy on carbon emissions not only contributes to economic growth, financial development, and industrial structure improvement but also serves as a valuable guide in attaining the objectives of carbon peaking and carbon neutrality [6]. By providing appropriate guidance and policy support, it is possible to establish a mutually beneficial

Table 4. Frequently occurring keywords in studies on the influence of digital economy on carbon emissions

Rank	Keywords	Frequency	Total Link Strength
1	Carbon Footprint	401	1919
2	Life-cycle Assessment	168	1000
3	Greenhouse-gas Emissions	139	93
4	Energy	114	688
5	Consumption	97	614
6	Impact	77	156
7	Sustainability	72	403
8	Climate-Change	38	56
9	Digital Economy	48	84
10	Management	48	134

relationship between the digital economy and carbon emissions. This can lead to the promotion of sustainable economic development and facilitate the transition towards a low-carbon future.

4.2 Keyword Change Analysis

Keyword change analysis can reveal the changing trend, importance and influence evolution process of keywords in the field. To uncover the research trajectory and shifting focal points regarding the impact of the digital economy on carbon emissions, researchers can utilize CiteSpace software for keyword analysis. This approach enables them to gain insights into the cutting-edge trends in this field of study. The findings are illustrated in Fig. 4.

In this view, the keyword's position on the timeline represents the year in which it first occurred, while the size of the nodes indicates the frequency at which they appear of the keyword over the entire time span. At the same time, different nodes are connected by lines, indicating the number of times these keywords appear in the same document at the same time, that is, the co-occurrence relationship.

It can be found from Fig. 4 keywords such as ICT innovation, renewable energy, and economic expansion not only appear frequently, but also have been paid attention to and studied by scholars earlier, and the relevant research content has been refined and deepened as time goes by. Since 2017, carbon emissions, information technology, and growth have become hot topics for further research, and digital technology has become an important engine for liberating and developing productivity. The exponential growth of big data, AI, cloud computing, IoT, and blockchain has facilitated the digital transformation and networking of the economy, leading to the enhancement and modernization of traditional industries.

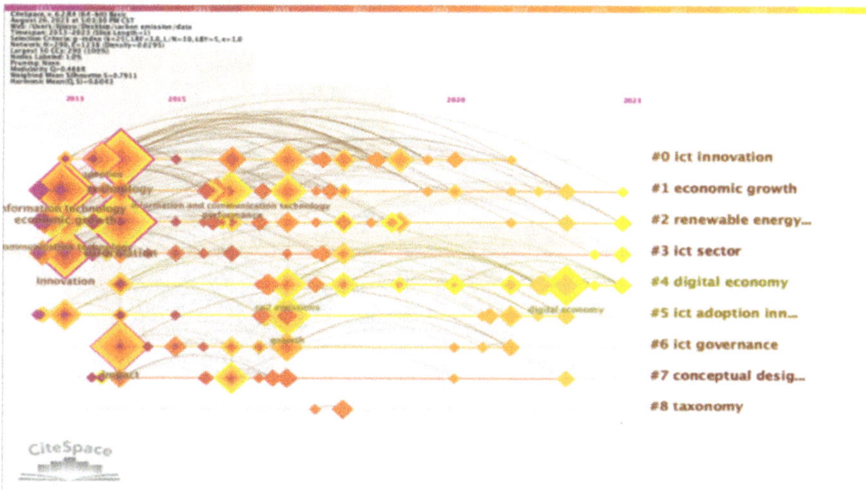

Fig. 4. Keyword time zone

As the global energy demand continues to increase, mobile networks have brought new impetus to production capacity and also brought about high energy usage. Improving energy conservation is the main goal of information and communication technology applications [7]. However, the virtual currency products produced by the digital transformation and emerging technologies derived from the development of the financial industry, such as blockchain, also increase carbon emissions [8].

4.3 Analysis of Co-cited Journals

Co-citation analysis is used to analyze the correlation and influence between two or more documents. By analyzing the co-citation relationship between multiple documents, hot spots and trends in the current research field can be found. The more times it is cited by other researchers, it means that the paper has a greater impact on research in related fields, and the paper is recognized and accepted in the academic community. Co-citation analysis can also reflect the academic communication within the academic circle. A large number of citations means that the paper has been widely discussed and cited in the academic community, thereby promoting exchanges and cooperation in the academic community. The co-citation relationship mapping analysis was carried out through VOSviewer, the threshold for journal citation counts was set to 60, 69 journals were left for co-citation analysis. Figure 5 displays the conclusive findings.

Five networks of different colors become the co-citation relationships of journals. Among them, the yellow and green clusters focus on the progressive development research of Eco-friendly economy and energy, aiming to offer important uphold for the realization of the global low-carbon economy and the response to climate emergency. The development of sustainable energy can reduce dependence on traditional high-carbon energy and lessen the velocity of climate change. The red and blue clusters focus on environmental science and the life cycle of ecosystems, aiming to study and

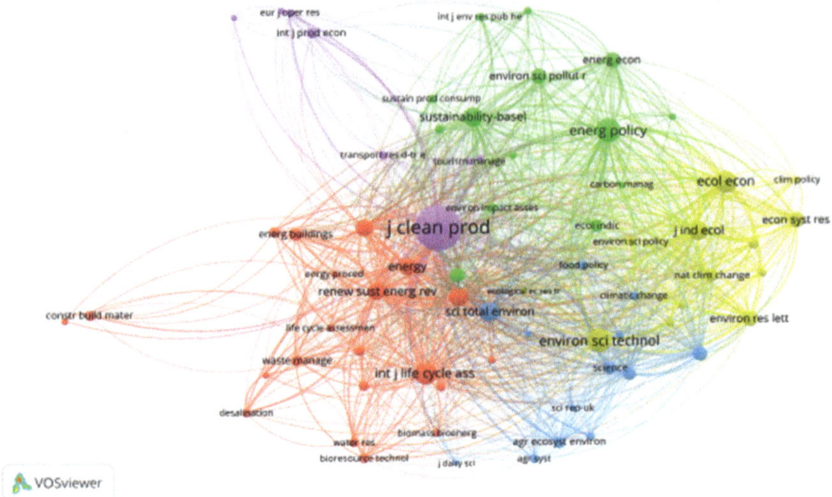

Fig. 5. Co cited journals

assess the entire process of environmental impact [9]. The whole process from product or activity design, production, use to disposal. Through the life cycle assessment of environmental science, environmental impact factors such as resource utilization, energy consumption, and waste discharge can be considered comprehensively. The purple cluster focuses on cleaner production, environment and sustainable development research and practice [10]. Reduce environmental pollution and resource consumption by improving production processes, product design, and resource utilization efficiency. Sustainable development strategy research focuses on how to achieve balance in the three aspects of economy, society and environment, and promote the coordination between economic growth and resource utilization, environmental protection and social equity [11].

5 Conclusion and Prospect

5.1 Conclusion

The digital economy revolves around information and communication technology as its fundamental pillar, and provides products and services through digitization, networking and intelligence. In contrast to the conventional economic model, the digital economy presents a distinct approach, is more efficient and intelligent, and can reduce resource consumption and energy utilization. Analyze relevant literature in this research field from 2013 to 2023 through VOSviewer and CiteSpace software. The conclusions are summarized as follows:(1)Since 2013, the field of digital economy has received great attention from scholars due to its potential for sustainable economic development. It entered a stage of rapid development in 2017, and rich research results have promoted progress in this field.(2)The analysis of keyword co-occurrence reveals the evolution trend of research topics in different periods in this field. The research hotspots are "Carbon Footprint", "Life-cycle Assessment", "Energy", "Digital Economy", "Sustainability "wait.(3)From the perspective of research content, the advancement of the internet economyhas effectively reduced the carbon emissions per person, carbon emission increments, and carbon emission growth rates, which is conducive to reaching the target of carbon peaking.

5.2 Limitations and Outlook

Research on the impact of the digital economy on carbon emissions is advancing rapidly and has important theoretical and practical significance. As we see digital technology continue to advance, it becomes increasingly important to understand its impact on carbon emissions. This understanding is critical to our efforts to combat climate change and promote sustainable development. For example, the energy industry will become more and more intelligent and digital, digital technology will drive the transition to low-carbon transportation and logistics industries, and the development of the digital economy. Additionally, it will contribute to the growth of green finance and sustainable investment, and give full play to digital technology to promote the advancement of conventional sectors. Overall, via the application and innovation of digital advancements, economic expansion and reduction of carbon footprint can be achieved, contributing to climate change and environmental protection. Simultaneously, the digital technology will also provide more low-carbon solutions and business opportunities for various industries, and advance the green transformation and environmentally friendly development of the economy.

References

1. Company B P. BP statistical review of world energy 2021, British Petroleum Company, 2021. https: // www.bp.com /con-tent/dam/bp/ business- sites/ en/global/corporate/pdfs/energy-economics/statistical- review/bp- stats- review- 2021- full- re-port.pdf, 2021–06–01
2. Brynjolfsson, E., Kahin, B. (eds.) Understanding the digital economy: data, tools, and research. MIT press (2002)
3. Dong, F., Hu, M., Gao, Y., Liu, Y., Zhu, J., Pan, Y.: How does digital economy affect carbon emissions? Evidence from global 60 countries. Sci. Total Environ. **852**, 158401 (2022)
4. Brookes, B.C.: Theory of the Bradford law. J. Document. **33**(3), 180–209 (1977)
5. Zhang, W., Liu, X., Wang, D., Zhou, J.: Digital economy and carbon emission performance: evidence at China's city level. Energy Policy **165**, 112927 (2022)
6. Shi, D.: The evolution of industrial development trends under the digital economy. China Indust. Econ. **11**, 26–42 (2022). https://doi.org/10.19581/j.cnki.ciejournal.2022.11.001
7. Han, J., Chen, X., Feng, X.: Realistic challenges and path selection of digital economy empowering green development. Reform **09**, 11–23 (2022)
8. Xu, W., Zhou, J., Liu, C.: The spatial effect of digital economy development on urban carbon emissions. Geogr. Res. **01**, 111–129 (2022)
9. Xie, Y.: The impact and mechanism of digital economy on regional carbon emission intensity. Contemporary Econ. Manag. **02**, 68–78 (2022). https://doi.org/10.13253/j.cnki.ddjjgl.2022.02.008
10. Jiang, J.: Sustainable digital age: high-quality integrated development of digital economy and green economy. Enterprise Econ. (07), 23–30+161 (2021)
11. Zhang, J., Lyu, Y., Li, Y., Geng, Y.: Digital economy: an innovation driving factor for low-carbon development. Environm. Impact AsseEducation 20(2), 87–129, 2010 (2022)

Operation and Sustainable Development of Digital Technology Driven Sharing Economy Platform

Haochun Liu(✉) and Pinxi Wu

Wuxi Dipont School of Arts and Science, Wuxi, Jiangsu, China
andyliuhaochun@163.com

Abstract. The sharing economy platform driven by digital technology has attracted attention in the past few years, and sustainable development has also become a global topic of concern. This study aimed to explore the potential and impact of digital technology driven sharing economy platforms in sustainable development. The relationship between the operation of digital technology driven sharing economy platforms and sustainable development was analyzed and evaluated through literature review, data analysis, and questionnaire surveys. Research found that the monthly transaction volume of the digital driven sharing economy platform was as high as 52000 and as low as 39500. The sharing economy platform driven by digital technology has provided potential opportunities for sustainable development. These platforms have reduced resource waste and environmental pressure through resource sharing, optimized utilization, and carbon emission reduction. The sustainable development potential of the sharing economy model lies in providing more cost-effective and environmentally friendly solutions, meeting user needs while reducing the consumption of natural resources.

Keywords: Digital Technology Driven · Contribution Economy Platform · Platform Operation · Sustainable Development

1 Introduction

The sharing economy platform driven by digital technology has become indispensable in social and economic life, and global attention to sustainable development is also increasing. These platforms, connected through digital technology and online platforms, promote resource sharing, interconnectivity, and community participation, providing the possibility of sustainable development for current and future generations.

Previous research has addressed the relationship between digital technology driven sharing economy platforms and sustainable development. Based on the conceptual definition of the sharing economy platform and the corporate social responsibility of the platform, Zhang Y proposed relevant governance strategies, providing valuable reference for the social responsibility governance of the sharing economy platform [1]. Zhang Y et al. divided the sharing economy platform, analyzed three different types of sharing

Y. Pei et al. (Eds.): IC 2024, LNEE 1214, pp. 90–100, 2024.
https://doi.org/10.1007/978-981-97-4193-9_10

economy platforms, and proposed strategies for platform operation of sharing economy platforms [2]. Liu H conducted a value analysis on the value co creation of the digital sharing economy platform, aiming to maintain mutual trust, fairness, and sustainability in the value co creation of the digital sharing economy platform [3]. However, existing research still has limitations in case analysis and data support in specific fields. Therefore, further research is necessary to comprehensively evaluate the impact of digital technology driven sharing economy platforms on sustainable development.

This article adopted a comprehensive research method, including literature review, data analysis, and questionnaire survey. Through a review of existing literature, the relationship between digital technology driven sharing economy platforms and sustainable development was summarized, and the results and findings of related research were summarized. Data analysis was conducted to quantify the potential impact of sharing economy platforms on resource utilization and carbon emission reduction. The contribution of this article is: (1) It can provide empirical research and experience summary for the management and operation of sharing economy platforms.(2) The sharing economy platform proposed in this article can increase the trading volume of the platform and reduce carbon emissions

2 Overview of the Sharing Economy Platform

2.1 Definition and Characteristics of the Sharing Economy

The sharing economy is an economic model whose core concept is to share and exchange idle resources, skills, or services of individuals or organizations with those in need through online platforms and technology support. It has the following characteristics [4, 5].

Resource sharing: The sharing economy shares idle items, housing, transportation and other resources through online platforms, enabling them to be more fully utilized and reducing waste and resource consumption.

Community and mutual assistance: The sharing economy encourages mutual assistance and cooperation among community members, promoting social connections and community development through the sharing of resources, skills, and knowledge.

Flexibility and personalization: The sharing economy platform provides more choices and flexibility to meet users' personalized and real-time needs, enabling consumers to customize services according to their own needs.

Trust and evaluation mechanism: The sharing economy platform establishes trust through evaluation and credit mechanisms, where participants can evaluate and rate each other, increase trust in transactions, and protect the rights and security of participants.

Sustainable development: The development of the sharing economy can promote the effective utilization of resources and environmental protection [6, 7], reduce carbon emissions and resource waste, and have a positive impact on sustainable development.

Overall, the sharing economy promotes the sharing and exchange of resources through the support of digital technology, breaking the limitations of traditional economic models, providing more choices and flexibility, while emphasizing the values of community and mutual assistance, and promoting the realization of sustainable development. Table 1 shows common sharing economy platforms and their corresponding data.

Table 1. Common sharing economy platforms

Platform Name	Sustainability Index	Resource Utilization (%)	Carbon Emission Reduction (tons)
ShareRide	8.2	82.5	1200
EcoStay	9.6	91.3	3500
ShareGoods	6.5	68.7	600

2.2 Classification and Operation Mode of Sharing Economy Platforms

The sharing economy platform can be classified based on the type of service and operation mode. Some common classifications and operational models of sharing economy platforms include item sharing platforms, accommodation sharing platforms, skill sharing platforms, capital sharing platforms, time sharing platforms, and food sharing platforms. The item sharing platform allows users to share personal items, such as shared car platforms, shared bicycle platforms, shared tool platforms, etc. [8, 9]. Users can rent or share their idle items with others in need. Accommodation sharing platforms allow individuals to rent out their homes, apartments, or rooms to travelers or short-term tenants, such as Airbnb, which is a typical accommodation sharing platform. The skill sharing platform connects individuals who provide specific skills or services and those who need these skills or services, such as housekeeping services, tutoring services, part-time work platforms, etc. Capital sharing platforms allow individuals or enterprises to share capital resources, such as shared office space platforms, shared financial platforms, etc. [10, 11]. Time sharing platforms allow individuals to provide their leisure time and skills to others, such as shared driving platforms, shared housekeeping service platforms, etc. Food sharing platforms allow individuals to share their own food or drinks with other users, such as shared kitchen platforms, shared food delivery platforms, etc.

In terms of operational mode, the platform matches providers and demanders through technologies such as algorithms and search engines, enabling both parties to easily find suitable resources or services. An evaluation and credit system where users can evaluate and rate each other is established to increase trust in transactions, and enhance the credibility and reliability of participants. The platform provides insurance or guarantee services to ensure the security and reliability of transactions [12, 13]. It is necessary to protect the interests of participants, encourage interaction and cooperation between users, and establish communities and communities to promote trust and sharing among users.

2.3 Requirements and Challenges of Sustainable Development for Sharing Economy Platforms

Sustainable development requires a sharing economy platform to improve resource utilization efficiency, and reduce waste and environmental burden. The platform needs to encourage users to share and reuse items, reduce energy consumption, and encourage the use of renewable energy. It requires the sharing economy platform to establish sound trust and security mechanisms, protect user privacy and personal information security, and prevent the occurrence of fraud and improper behavior [14, 15]. The sharing economy platform should ensure the fairness and inclusiveness of services, not only meeting the needs of high-income groups, but also considering the participation and benefits of low-income groups to avoid further exacerbating social inequality. The platform should encourage community participation and cooperation among users, promote social interaction and mutual assistance, and establish a strong sharing economy community to enhance the sustainability and social benefits of the platform. The sharing economy platform also needs to pay attention to and manage the environmental impact of its operation, including reducing carbon emissions, waste management, water resource utilization, and other aspects, in order to reduce negative impacts on the ecosystem.

3 Sustainable Development Strategies for Digital Technology-Driven Sharing Economy Platforms

3.1 Role of Artificial Intelligence and Machine Learning in Platform Management

Artificial intelligence and machine learning can achieve more accurate matching and recommendation functions by analyzing user needs and preferences, as well as resource and service information on the platform. Through algorithms and models, the platform can provide personalized recommendation results to users and improve user experience [16, 17]. It can analyze supply and demand relationships, market trends, and historical data to assist the platform in pricing and optimizing strategies. Through an automated pricing model, the platform can adjust prices in real-time based on changes in demand and supply, improving resource utilization efficiency. The trust and security mechanisms of the platform are established and maintained. By analyzing user behavior, reviews, and social network data, potential risks and fraudulent behaviors can be identified, and corresponding measures can be taken to provide a more reliable trading environment. It can also assess and predict risks on the platform, helping platform managers develop risk management strategies. By analyzing historical data and pattern recognition, the platform can predict user behavior and transaction results, timely identify potential risks, and reduce potential losses. Artificial intelligence and machine learning can be used to provide more intelligent user services and support. Through natural language processing and speech recognition technology [18, 19], intelligent customer service and virtual assistants can be achieved, providing users with real-time assistance and answering questions.

3.2 Social and Environmental Impact Assessment and Management

The impact of a project or action on social organizations, communities, and individuals is evaluated, including employment opportunities, income distribution, social equity, social inclusion, etc. This helps to ensure fair distribution of benefits for the project and reduce adverse impacts on vulnerable groups. The impact on the natural environment, including land use, water resources, air quality, biodiversity, etc., helps to protect the integrity of ecosystems and reduce damage and pollution to the environment. The impact on sustainable development goals, including economic sustainability, social sustainability, and environmental sustainability [20], helps to ensure the long-term sustainability of projects in terms of economy, society, and environment, and avoids the negative impact of short-term benefits on long-term sustainable development. Managing social and environmental impacts involves developing and implementing relevant policies, regulations, and measures to reduce negative impacts and promote positive social and environmental benefits. This may include monitoring and reporting on social and environmental performance, establishing environmental management systems, conducting social responsibility activities, promoting renewable energy and resource recycling, etc.

The goal of social and environmental impact assessment and management is to achieve social justice, environmental protection, and sustainable development on the basis of economic development. By assessing and managing social and environmental impacts, it is possible to promote the participation and cooperation of various stakeholders, ensure that projects or actions align with the best interests of society and the environment, and lay the foundation for future sustainable development.

3.3 Innovative Business Models and Partnerships

Innovative business models and partnerships are one of the key factors for companies to succeed in fiercely competitive markets. Innovative business models refer to the ways in which enterprises create and deliver value through redesign, in order to gain a competitive advantage in the market. Partnership refers to the establishment of cooperative relationships between enterprises and other organizations or individuals to jointly carry out business activities and achieve mutual benefit and win-win results. Innovative business models can bring competitive advantages through the following aspects:

Value creation: Innovative business models can redefine the value proposition of products or services to meet market demands or create new ones. By providing unique value, enterprises can attract more customers and differentiate themselves from competitors.

Income model: Innovative business models can change the source and structure of a company's income. For example, transitioning from a traditional product sales model to a model based on subscription, rental, or licensing fees can provide a more stable and sustainable revenue stream.

Channel optimization: Innovative business models can redesign the delivery channels of products or services to improve efficiency and meet customer needs. By using new digital technologies or collaborating with partners, enterprises can achieve wider market coverage and faster delivery.

Partnership is crucial for the successful implementation of innovative business models. Partners can provide complementary resources, technology, expertise, or market channels to help enterprises expand scale, reduce costs, and accelerate innovation, so as to enter new markets. Partnership can take the form of suppliers, distributors, technical partners, strategic alliances, or ecosystem partners. There is a mutually reinforcing relationship between innovative business models and partnerships. Innovative business models can provide new business opportunities and value-added opportunities for partnerships, while partnerships can provide support and resources to help enterprises achieve the implementation and promotion of innovative business models. In practice, enterprises need to constantly explore and experiment with new business models, and establish good cooperative relationships with suitable partners. At the same time, enterprises also need to flexibly adapt to market changes and work together with partners to address challenges and opportunities, in order to achieve sustainable development and long-term competitive advantages.

3.4 User Participation and Community Building

User participation is an important driving factor for the activity of sharing economy platforms. By encouraging user participation and interaction, the platform can increase user activity and transaction frequency and scale. User engagement includes publishing products or services, actively trading, providing feedback and feedback, etc. An active user base contributes to the sustainable growth and profitability of the platform. Establishing and cultivating a community can enhance user loyalty to the platform. Interactions and exchanges among community members can establish close relationships and trust, making users more inclined to choose this platform for transactions. Through community building, the platform can provide a better user experience, increase user satisfaction, and encourage users to use the platform more frequently. User participation can provide valuable feedback and feedback to the platform, which can help improve service quality and user experience. Users are the end users of the platform, and their feedback can reveal problems and opportunities for improvement. By actively listening to users' voices, the platform can continuously optimize product functions, transaction processes, and user interfaces, provide better service quality, and attract more users and promote sustainable development. User participation and community building can enhance the stability and credibility of the platform. Active communities where users participate and interact can help monitor and maintain platform order, reducing the occurrence of fraud and misconduct. At the same time, active user participation and word-of-mouth dissemination can enhance the platform's reputation and credibility, attracting more users and potential partners.

User participation and community building are crucial for the sustainable development of digital technology driven sharing economy platforms. They can increase platform activity and user loyalty, improve service quality and platform stability and credibility, and drive sustainable growth and market expansion. By actively promoting user participation and community building, the platform can achieve long-term sustainable development and competitive advantage.

4 Sustainable Development Indicators and Evaluation Methods

4.1 Construction of Sustainable Development Indicator System

Building a sustainable development indicator system is an important step in evaluating and monitoring the sustainability performance of a system, organization, or industry in terms of economy, environment, and society. Firstly, the areas of assessment are identified as environment, sharing economy, and society. In the economic dimension, platform transaction volume is selected as an evaluation indicator. In terms of the environment, carbon emissions are selected as evaluation indicators. In terms of society, user satisfaction is chosen as an evaluation indicator. Using 12 months of the year as the measurement time, the indicator values for each month are measured and compared with traditional and economic platforms. Data is collected and summarized through statistical and questionnaire surveys.

4.2 Questionnaire Survey and Comparative Analysis

Taking A city in a certain province in 2022 as the testing target, data for the year 2022 is collected and analyzed. Figure 1 shows a comparison of economic dimension data; Fig. 2 shows a comparison of environmental dimension data; Fig. 3 shows a user satisfaction survey in social aspects, which investigates the satisfaction level of users with the platform.

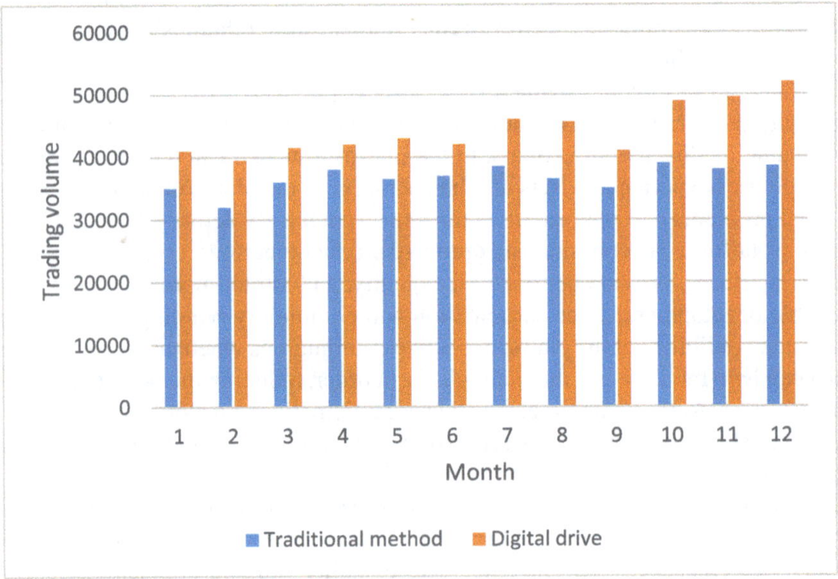

Fig. 1. Comparison of platform transaction volume

The monthly transaction volume of the digital driven sharing economy platform reaches a maximum of 52000, with a minimum of 39500, and the traditional method's monthly transaction volume reaches a maximum of 39000, with a minimum of 32000. The transaction volume driven by digital methods is significantly higher than that of traditional sharing economy platforms. It can be seen that the digital driven platform provides a more convenient way of use and a wider coverage, making it easier for users to access and use the platform. It has a powerful network effect. As the number of users increases, the value and attractiveness of the platform also increase, thereby attracting more users to join and participate in transactions. At the same time, positive word-of-mouth communication among users can also attract more potential users to use the platform, thereby driving an increase in transaction volume.

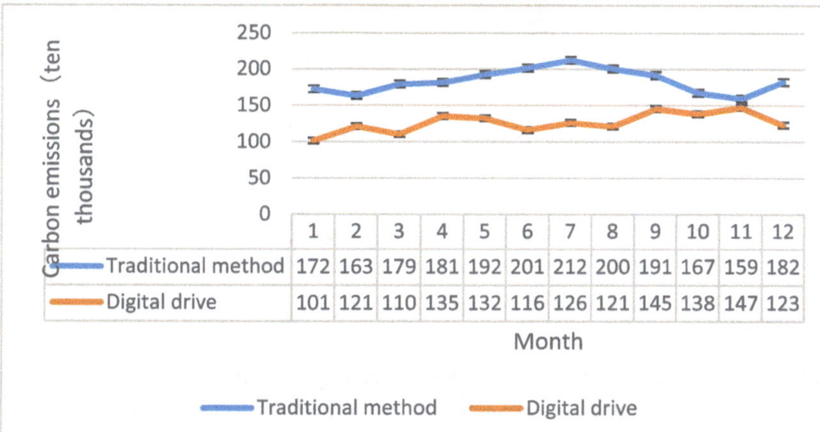

Fig. 2. Comparison of carbon emissions

The highest emission of traditional sharing platforms is 212000 tons, and the lowest is 159000 tons; the highest and lowest emissions under digital driving are 147000 tons and 101000 tons. The sharing economy platform usually optimizes the use of resources by sharing and utilizing existing resources. This optimization can reduce resource waste and excessive consumption, thereby reducing environmental pressure and carbon emissions. People share and rent goods, services, and facilities, reducing personal ownership and consumption needs. Compared to traditional platforms, sharing economy platforms promote resource sharing and co utilization, avoiding redundant resource production and carbon emissions.

98 people are randomly selected for this questionnaire survey, and their satisfaction indicators are very satisfied, satisfied, average, dissatisfied, and very dissatisfied. After investigation and analysis, 36 people are very satisfied; 42 people are satisfied; 11 people feel average; 6 people are not satisfied; 3 people are very dissatisfied. Most people still highly recognize the digital driven sharing economy platform, which indirectly demonstrates the reliability of this platform.

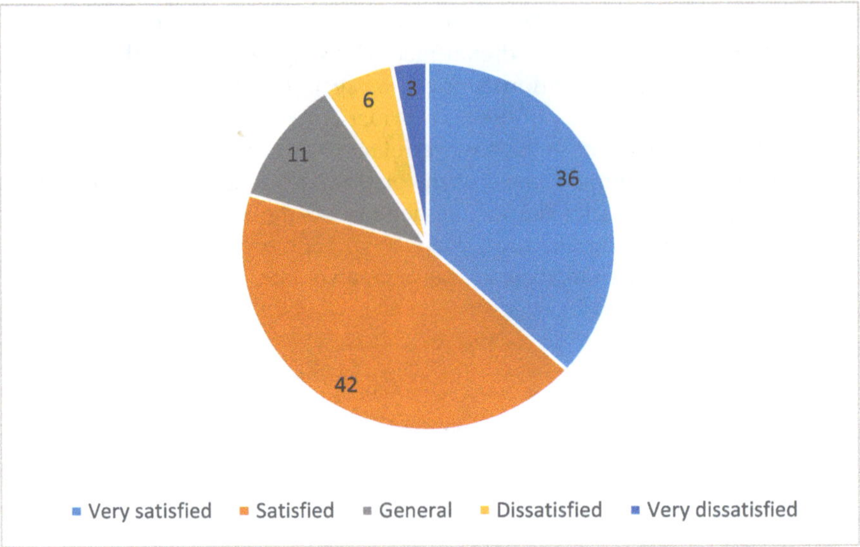

Fig. 3. Satisfaction survey

4.3 Discussion

By studying the sharing economy platform driven by digital technology, especially the relationship between platform transaction volume and carbon emissions, this research provides empirical evidence on sustainable development. This means that there is a positive correlation between the high transaction volume of sharing economy platforms and low carbon emissions, that is, the increase in platform transactions can achieve a reduction in carbon emissions. This kind of evidence is of great significance to support the sustainability and environmental friendliness of the sharing economy model.

5 Conclusions

There is a close relationship between the operation of digital technology driven sharing economy platforms and sustainable development. The sustainable development potential of the sharing economy model lies in providing more cost-effective and environmentally friendly solutions, meeting user needs while reducing the consumption of natural resources. By utilizing digital technology reasonably, optimizing resource utilization, and providing environmentally friendly solutions, the sharing economy platform can play an important role on the path of sustainable development. However, the sustainable development of the platform requires a comprehensive consideration of multiple social, environmental, and economic factors, and corresponding management and regulatory measures are taken to ensure balanced development and sustainability.

Risk management is an important consideration for the sustainable development of sharing economy platforms. Sharing economy platforms are facing trust risks, security risks, information leakage and other issues, and need to establish a sound risk management mechanism, including user authentication, transaction dispute handling, data protection and other measures to enhance user trust and platform reliability.

References

1. Zhang, Y., Zhang, L., Wang, J.: Research on social responsibility governance strategies for shared economy platforms. Market Weekly **36**(4), 1–4 (2023)
2. Zhang, Y., Liu, Y., Luo, T.: Research on the content of social responsibility and governance countermeasures of sharing economy platform. Bus. Watch **9**(14), 105–108 (2023)
3. Liu, H.: Ethical analysis of value co creation on digital sharing economy platforms. Sci. Res. **40**(8), 1353–1360 (2022)
4. Ek Styvén M, Mariani, M.M.: Understanding the intention to buy secondhand clothing on sharing economy platforms: the influence of sustainability, distance from the consumption system, and economic motivations. Psychol. Mark. **37**(5), 724–739 (2020)
5. Garud, R., Kumaraswamy, A., Roberts, A., et al.: Liminal movement by digital platform-based sharing economy ventures: the case of Uber technologies. Strateg. Manag. J.. Manag. J. **43**(3), 447–475 (2022)
6. Chen, Y., Wang, L.: Commentary: Marketing and the sharing economy: digital economy and emerging market challenges. J. Mark. **83**(5), 28–31 (2019)
7. Eckhardt, G.M., Houston, M.B., Jiang, B., et al.: Marketing in the sharing economy. J. Mark. **83**(5), 5–27 (2019)
8. Buhalis, D., Andreu, L., Gnoth, J.: The dark side of the sharing economy: balancing value co-creation and value co-destruction. Psychol. Mark. **37**(5), 689–704 (2020)
9. Cui, L., Hou, Y., Liu, Y., et al.: Text mining to explore the influencing factors of sharing economy driven digital platforms to promote social and economic development. Inf. Technol. Dev. **27**(4), 779–801 (2021)
10. Wirtz, J., So, K.K.F., Mody, M.A., et al.: Platforms in the peer-to-peer sharing economy. J. Serv. Manag.Manag. **30**(4), 452–483 (2019)
11. Gerwe, O., Silva, R.: Clarifying the sharing economy: Conceptualization, typology, antecedents, and effects. Acad. Manag. Perspect.Manag. Perspect. **34**(1), 65–96 (2020)
12. Dabbous, A., Tarhini, A.: Does sharing economy promote sustainable economic development and energy efficiency? Evidence from OECD countries. J. Innov. Knowl.Innov. Knowl. **6**(1), 58–68 (2021)
13. Abrate, G., Viglia, G.: Personal or product reputation? Optimizing revenues in the sharing economy. J. Travel Res. **58**(1), 136–148 (2019)
14. Benjaafar, S., Hu, M.: Operations management in the age of the sharing economy: what is old and what is new? Manuf. Serv. Oper. Manag.Oper. Manag. **22**(1), 93–101 (2020)
15. Altinay, L., Taheri, B.: Emerging themes and theories in the sharing economy: a critical note for hospitality and tourism. Int. J. Contemp. Hosp. Manag.Manag. **31**(1), 180–193 (2019)
16. Apte, U.M., Davis, M.M.: Sharing economy services: business model generation. Calif. Manage. Rev. **61**(2), 104–131 (2019)
17. Benjaafar, S., Kong, G., Li, X., et al.: Peer-to-peer product sharing: Implications for ownership, usage, and social welfare in the sharing economy. Manage. Sci. **65**(2), 477–493 (2019)
18. Ahsan, M.: Entrepreneurship and ethics in the sharing economy: a critical perspective. J. Bus. Ethics **161**(1), 19–33 (2020)

19. Boateng, H., Kosiba, J.P.B., Okoe, A.F.: Determinants of consumers' participation in the sharing economy: a social exchange perspective within an emerging economy context. Int. J. Contemp. Hosp. Manag.Manag. **31**(2), 718–733 (2019)
20. Chen, G., Cheng, M., Edwards, D., et al.: COVID-19 pandemic exposes the vulnerability of the sharing economy: a novel accounting framework. J. Sustain. Tour. **30**(5), 1141–1158 (2022)

Optimization of Intelligent Line Matching Transmission Online Monitor Based on Remote Detection Algorithm

Xiongjun Tao[1], Hualu Yuan[2], and Yifeng Zhu[2](✉)

[1] Kunming Branch of CSG EHV Transmission Company, Kunming 650217, Yunnan, China
[2] China Southern Power Grid Co., Ltd. EHV Transmission Company, Guangzhou 510000, Guangdong, China
zhu13922180830@163.com

Abstract. With the rapid development of society and the continuous replacement of The Times, China's science and technology has made a large space for development. In recent years, the timeliness of data is becoming higher and higher. It is best to realize the online monitoring of data and detect the state of transformer in real time in the line. In recent years, with the rapid development of the power grid, people have relatively high requirements for electricity consumption, but excessive electricity consumption will often lead to safety accidents, which has a certain impact on the security of the distribution network. And it is easy to be affected by external factors in the process of electricity consumption, and it is difficult to predict both the amplitude and the time of the voltage. There is no real timeliness data as the support point for fault diagnosis, so the cause of the equipment damage cannot be found after the failure. The emergence of intelligent lines has brought more space for the development of electric power in China. Based on this, this paper puts forward some suggestions for the optimization of online monitor through the matching and transmission of intelligent lines.

Keywords: Smart Line · Online Monitor · Matching Transmission · Optimization

1 Introduction

Since the beginning of the 21st century, the development of China's power system is fast, the system tends to be perfect, but the compatibility of most equipment can not be guaranteed. Because many equipment is exposed to the external environment (especially the transmission and distribution lines), it is easy to overvoltage, which may cause damage to the equipment, and will cause the line failure in a short time [1]. When serious, it may also cause a large area of power failure, resulting in greater economic losses and social losses. The emergence of smart lines has played a very important role in China's power grid construction. The degree of cable sliding in the city is constantly deepening, and there is a very large cable pressure. After the construction of the intelligent circuit, the operation state of the transmission line can be monitored online, so as to improve the

Y. Pei et al. (Eds.): IC 2024, LNEE 1214, pp. 101–107, 2024.
https://doi.org/10.1007/978-981-97-4193-9_11

disaster resistance, reduce the risk of power failure, and ensure the normal operation of the power grid [2]. In this regard, this paper studies the optimization of intelligent line matching transmission in the online monitor.

2 The Importance of Smart Line Matching and Transmission

(1) Ensure the quality and efficiency of power grid construction

In recent years, people's demand for electricity is increasing, resulting in a serious shortage of transmission maintenance personnel, so the operation efficiency of the power grid is relatively slow. In such a social background, it is not only necessary to accelerate the importance of smart line transmission, but also to carry out professional internal management, so as to ensure the development speed of the power grid, so as to improve the efficiency of the power grid construction [3].

(2) Smart lines can boost the further construction of electric energy

In recent years, the relevant data show that the intelligent construction speed of the power grid has been accelerated, but there are differences in the application systems of different units, so there is a lack of unified management in the process of data processing, and it is difficult to reflect the relevant value of information data. The core content of intelligent line is to carry out unified and standardized management, strengthen the sharing of system information, so as to further mine the value of the data, which can effectively boost the construction of power grid enterprises and companies.

3 Establish the Anomaly Data Detection Model

For online monitoring transformer related data, relevant data is collected on the sensor and subsequently integrated in the platform of data information. The feature quantity data are sorted in temporal order to form a complete time series. However, the sensor will fail in a short period of time, and the communication port is in an abnormal state, coupled with some manual operation errors, so there will be outlier values in the information collection. Abnormal data in the online monitor involves a variety of reasons, including noise value, etc., the data itself has no relationship with the transformer, but in the process of data measurement by external factors, resulting in the exception of the device, lead to persistent missing, so, with the help of gray correlation analysis and K-means clustering method can realize wisdom line online data monitoring [4].

(1) Time-series mining

The types of online monitoring data are very complex, and it is very important to highlight the characteristics. In-depth mining through the time series can provide a theoretical basis for the subsequent research on state information detection. Real-time state monitoring can not only reduce the processing time, but also reduce the complexity of online monitoring, ensure the efficiency of data processing, and provide real-time maintenance strategy for the power grid.

(2) Establish a detection model

Generally speaking, when the transformer is in normal operation state, the data fluctuation is within a controllable range, and the K-means clustering method can be used to identify and classify the data. During the operation, the data is clustered to observe whether the relevant data exceeds the threshold value and can judge whether the data is abnormal.

The online monitoring data are regarded as X sample and divided into K classes. The cluster centers are C = c1, c2, · ·. cj,.. ck, the sample xi and cj distance are expressed by dv, and the target function J is the sum of dv in the X sample, as shown in Formula 1.

$$J = \sum_{j=Ir}^{k} \sum_{i \approx c_j} d^{o}(xi,\ cj) \tag{1}$$

The relationship between the value of the target function J and the clustering effect is mainly negative. As the function J gets smaller and smaller, the clustering effect is getting better and better. When J reaches the minimum value, it is the optimal clustering scheme, so the objective function can be defined as shown in formula 2.

$$J = \sum_{j=Ir}^{k} J_i = \sum_{j=1}^{k} \left(\sum_{i,j=cj} \left| X_i^j - c_j \right|^2 \right) \tag{2}$$

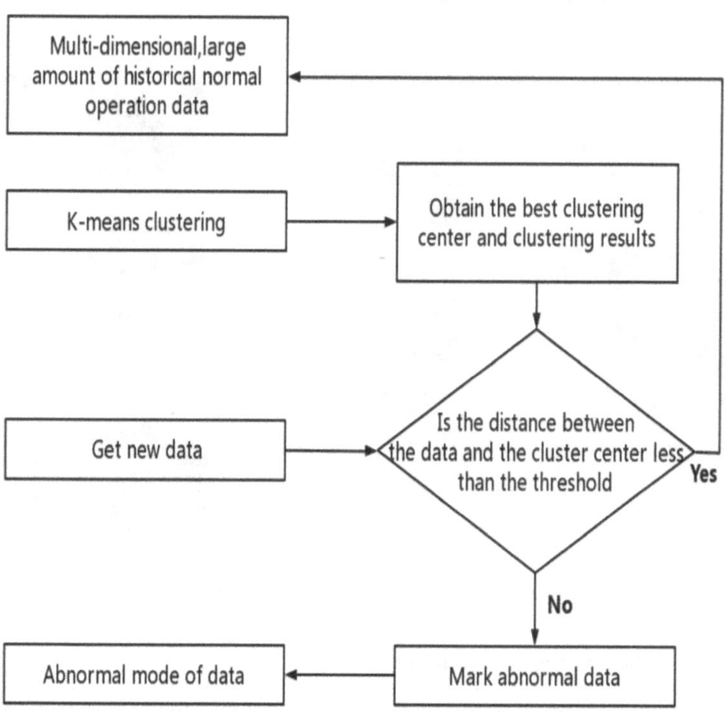

Fig. 1. Anomalous data model idea

The overall idea of the anomalous data model is shown in Fig. 1.

4 Optimization Ways of Intelligent Transmission Line Construction

In recent years, with the rapid development of the Internet, new requirements have been put forward for the security and stability of smart transmission lines. It is necessary to detect the information of the status bar in real time to ensure the reliability of the power grid and improve the operation efficiency. In southern China, the temperature in summer is too high. Operation under such natural conditions will bring a lot of line inspection work and improve the risk incidence. At present, it is difficult to meet the requirements of the power grid simply by relying on manual monitoring. It is necessary to have a monitoring means with high timeliness and wide monitoring range to obtain relevant monitoring data in real time and realize the intelligent change of the power grid.

(1) Install a distributed monitoring device

In the process of comprehensive scheduling information, the visual information of the monitoring data of the device is very important, which can establish the detection of the fault causes of multiple information sources, so as to accurately analyze the fault causes. And distributed faults are generally installed in the wire of the transmission line, to monitor the current in the occurrence of the line fault in real time, and the relevant information is collected and transmitted to the data center to diagnose the fault information. The specific structure is shown in Fig. 2.

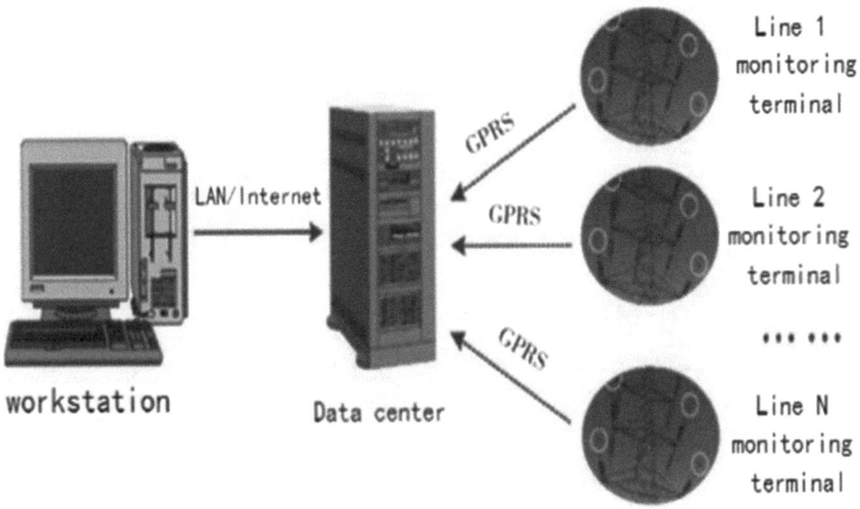

Fig. 2. Distributed fault system structure

(2) Install dancing monitoring equipment

Through the acceleration of the sensing equipment to the terminal high precision positioning, establish a relatively perfect regional monitoring system, so as to obtain the parameters of wire dancing. Generally speaking, the dancing monitoring device is connected to the wire, and after fixation, it sets the corresponding boundary conditions with the help of MEMS acceleration. The monitoring algorithm can also ensure the high accuracy. The short-time integration algorithm can improve the accuracy and reliability of the calculation, and the early warning system can establish a relatively perfect intelligent transmission system.

(3) Install infrared monitoring equipment

In the relatively harsh natural environment, it is necessary to establish a data safety platform for natural disaster prediction, and connect it with the mountain goods monitoring and early warning system, which can improve the feedback mechanism of the monitoring system and improve the monitoring and early warning capacity of the system. In the equipment, photoelectric composite overlapping temperature measurement technology can be used, because of the role of smoke particles, the light of the light-emitting components will be reflected, producing some photocurrents [5], The smoke signal then also changes into a light signal, increasing the distance of data detection to some extent. And the light sensitive components receive the infrared pulse light scattering signal, with the temperature sensing element of the semiconductor to detect the temperature change, after the calculation and processing will produce a certain alarm signal, not only will retain part of the photoelectric smoke detection sensitivity, the customer service photoelectric smoke device can not respond to high temperature shortcomings.

(4) Install geological disaster monitoring equipment

After the installation of geological disaster monitoring equipment, the automatic monitoring and early warning system for geological disasters is mainly monitored. As a kind of integrated monitoring equipment and transmission system equipment, combined with the Internet of Things and the monitoring center, the geological disasters can be predicted to a certain extent.

(5) Install an image monitoring device

In image monitoring device, mainly through a variety of transmission mode of related video signal real-time transmission, shipment inspection personnel do not need to patrol to the scene, as long as do no blind spot monitoring, can timely find dancing, wind, whether illegal climbing, dangerous construction factors such as the influence of power grid [6]. In the visual monitoring, the solutions can be timely formulated for the ice disaster and wind disaster weather of transmission lines [7, 8]. The high-definition images of the monitoring site are transmitted to the monitoring platform, and the images are captured according to the needs of the project, and the situation of the scene can be understood at any time. In a certain period of time, the pictures can also be sent to the monitoring center, so as to carry out the remote image monitoring [9, 10].

(6) 3D modeling of line channels

In the three-dimensional modeling of the line and channel, three-dimensional laser inspection can be carried out, whether overhead wire, overhead ground wire or tower, the corresponding lines and facilities can be installed, so as to ensure the accuracy of data processing. On the basis of 3 D modeling, we can also use drones, helicopters and carrying cameras to take photos and inspect at the station, and measure the temperature of the line environment. Intelligent identification in the refined impact data to reduce the security risks in the intelligent line. It can be equipped with detection sensors to collect data on the information such as the temperature and running distance of the equipment, and realize a comprehensive data analysis with the help of edge calculation. If the data amount of the state is abnormal, then the relevant data of the platform can be connected to conduct horizontal analysis, realize the overall 3 D modeling of the transmission channel, and quickly perceive the risks.

(7) Panoramic wisdom platform

Panorama of the wisdom of the platform need and line equipment status and inspection information access, after the massive heterogeneous data, through the transmission panoramic perception line fault, and establish the corresponding warning model, so as to fully grasp the line operations, improve the level of intelligent inspection, ensure the safe operation and stable operation.

5 Conclusion

To sum up, with the improvement of people's life quality, the requirements for electricity consumption are also becoming higher and higher. Traditional manual inspection has been difficult to meet the needs of the current social development. Wisdom line not only to a certain extent, improve the work efficiency, also can reduce the influence of external factors, can the grid site real-time monitoring and online detection, after adding all kinds of equipment and device, also can bring greater development space, ensure the stability and security of electricity, has high economic and social benefits.

References

1. Zhao, C., Li, Z., Lu, H.: An online monitoring method and application research of intelligent ballastless track. Railway Standard Des. **65**(7), 53–58 (2021)
2. Ye, Z., Wang, W., Wu, N., et al.: Dynamic early warning model of transmission line ice covering online monitoring. High Voltage Technol. **40**(5), 1374–1381 (2014)
3. Huang, Y., Wang, L.: Development trend and construction measures of smart transmission lines. Commun. Power Supply Technol. **39**(2), 167–169,173 (2022)
4. Jian, L., Changjia, S., Lingmin, P., et al.: Intelligent transmission line transportation and inspection technology based on the internet of things perception. Hubei Electr. Power **45**(3), 55–59 (2021)
5. Jing, M., Liu, X., Dong, R., Yu, Z.M.: Agricultural environment monitoring system based on Nx5883 wireless remote control terminal. Appl. Mech. Mater. **2594**, 373–375 (2013)
6. Zhao, Y., et al.: Application in coal mine of fiber methane monitoring system based on spectrum absorption. Procedia Eng. **26**(C) (2011)
7. Zhang, Q., Zhu, T., Xiao, S., et al.: High voltage transmission line foreign object detection based on improved YOLOv4 algorithm Applied Technology **50**(4) (2023)
8. Liu, Q.: Double terminal fault location of DC transmission lines based on modulus transfer time difference Journal of Chongqing Business University (Natural Science Edition) **40**(1) (2023)

9. Jin, Y.: Research on transmission line segmentation algorithm based on U-Net Changjiang information and communication **36**(3) (2023)
10. Wang, H.: Discussion on technical issues in construction of transmission lines in electric power engineering china equipment engineering **17** (2023)

Development of Health Assistant WeChat Applet

Zhenhui Wang(✉)

School of Technology and Engineering, Xi'an Fanyi University, Xi'an, Shaanxi, China
9502wzh@163.com

Abstract. With the continuous propagation of COVID-19, automated personal physical parameter detection is very important. With the advantages of convenience and quickness of WeChat applet, a health assistant applet is designed. The applet used ThinkJS framework on the server to process user requests and MySQL to manage data. The front end used WeChat developer tools to develop with the native MINA framework of WeChat. It can assist users in daily health self-test, body temperature record, BMI, sleep time and other health parameters, and also provide users with an e-commerce platform for purchasing drugs and health consultation. It is hoped that through the health assistant, people can more intuitively understand their physical condition, and provide a reference for the development of more health management applet.

Keywords: Health Assistant · WeChat Applet · Body Temperature Measure · Health Consultation

1 Introduction

Nowadays, the pace of life is too fast, people generally lack physical exercise, and their health generally shows a downward trend. Especially with the continuous spread of COVID-19, people are increasingly aware of the importance of personal physical quality to their health. The first time to understand and obtain personal body data and daily exercise data is the need to strengthen their own health management. The "health assistant WeChat applet" is a personal health mobile service applet based on Internet popularization and COVID-19 background.

At present, most health assistants in the market need to buy smart bracelets or smart watches and other devices, and need to download their special mobile phone applications to collect exercise data or conduct health self-test [1, 2]. The function of the software is generally limited to recording exercise steps, sleep time, testing heart rate and so on. Due to the use of hardware assistance, the data is more accurate, but it also increases the use cost and inconvenience. As a lightweight application, WeChat applet has the advantages of fast development, low cost and free installation. Its "run out of time" concept can upgrade the traditional industry information service level and provide users with ubiquitous software services [3–5]. Due to the file size and performance requirements, mini programs can focus on users, subdivide services, better mining user functional needs, with more in line with the form of user preferences to carry out information services, enhance user experience, improve service level.

Y. Pei et al. (Eds.): IC 2024, LNEE 1214, pp. 108–116, 2024.
https://doi.org/10.1007/978-981-97-4193-9_12

2 Related Technologies

2.1 Introduction to Integrated Development Environment

This applet is mainly developed through WeChat developer tool platform, supplemented by Visual Studio Code editor and corresponding program development plug-in.

WeChat developer tool is a small program development tool officially launched by WeChat. It features simplicity and efficiency. WeChat developer tools can not only intuitively preview applet pages during development, but also assist in debugging locally or using mobile phones. Version updates will also be uploaded in the development tool.

Visual Studio Code is a cross platform source code editor developed by Microsoft to write Web and cloud applications. Strictly speaking, it is not a complete integrated development environment. However, VSCode is applicable to JavaScript and Node.js has built-in support, and all modules of this project can be written through built-in extension programs. After writing, return to the WeChat developer tool for debugging.

2.2 Introduction to the Running Environment of Server Node.Js

In order to enable JavaScript code to run on the server side, this project chooses Node.js as the running environment, which can build a server for the program. Node.js is not a language, framework or library, but an event driven I/O server-side JavaScript runtime environment. Based on Google's V8 engine, it can parse and execute JavaScript code and help to break away from the browser. It is fast in execution, has excellent performance, and provides developers with some server level operational APIs, such as file reading and writing, and the construction of network services.

ThinkJS is a Node.js framework, which makes enterprise development simple and efficient.ThinkJS has a small kernel, excellent performance, and built-in automatic compilation and update mechanisms. With the scaffolding it provides, developers can quickly develop servers.

2.3 Vue.Js Framework

Vue.Js is a lightweight MVVM progressive framework that uses data driven and component-based development. Vue is characterized by bottom-up application. Its core library only cares about layers, which is not only easy to start, but also easy to integrate with third parties or existing projects.

Element is a set of Vue based desktop components for developers, designers and product managers. With the vue element admin back-end front-end solution, you can quickly build enterprise level middle and back-end product prototypes.

This project will use this technology to build a background page for the administrator to manage the system.

2.4 MINA Native Framework

MINA is the native framework of WeChat applets, through which simple and efficient services with native APP experience can be developed. The core of the framework is a corresponding data binding system, which binds each view to a set of data to maintain synchronization. In addition, the MINA framework manages the page routing of the entire applet so that pages can be switched seamlessly, and the complete life cycle of pages can be given in the framework. At the same time, MINA provides a set of basic components with WeChat style and rich WeChat native APIs, so that developers can more easily create WeChat applets.

2.5 MySQL Database Management System

MySQL database is a popular relational open source data management system, developed by MySQL AB, and now belongs to Oracle. MySQL uses a standard SQL data language form, can support a large database with tens of millions of records, and supports cross system operation, which shows its powerful functions.

MySQL database will be used to store and manage user information generated by WeChat applet of health assistant and relevant order data generated by purchase behavior.

In order to operate MySQL database more conveniently and intuitively, SQLyog graphical management tool is selected. It is simple and portable, and can directly process data in batches, which is very suitable for the development of such programs.

3 Design of Health Assistant Applet

3.1 Design of System Architecture

The health assistant software consists of four parts: the server side of the applet, the applet client, the background program operated by the administrator and the database. The structure diagram is shown in Fig. 1.

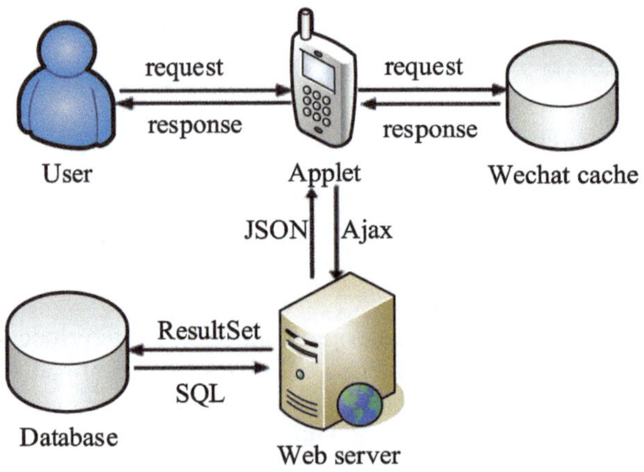

Fig. 1. Architecture diagram

The applet consists of two independent modules: View module and service module. The view module is responsible for the display of the interactive interface and communicates with the background through the weixinjsbridge interface, that is, the view layer of the applet. It is written by wxml and wxss and displayed through components. The service module is responsible for the logic operation of the background. It also communicates with the background through the weixinjsbridge interface, that is, the business logic layer of the applet. The logic layer uses the JavaScript engine to provide the unique functions of the running environment and applet. Therefore, the logic development of this project is bound by writing app.js. When the user starts the application, it will automatically execute its contents. After the logic layer processes the data, it will send it to the data layer and receive event feedback at the view layer.

3.2 Design of Function

The health assistant system consists of applet and background management. The functions of the applet include: registration and login, health self-test, exercise record, health knowledge and drug ordering. The functions of background management include administrator login, drug management and user management. The main functions of the applet are shown in Fig. 2.

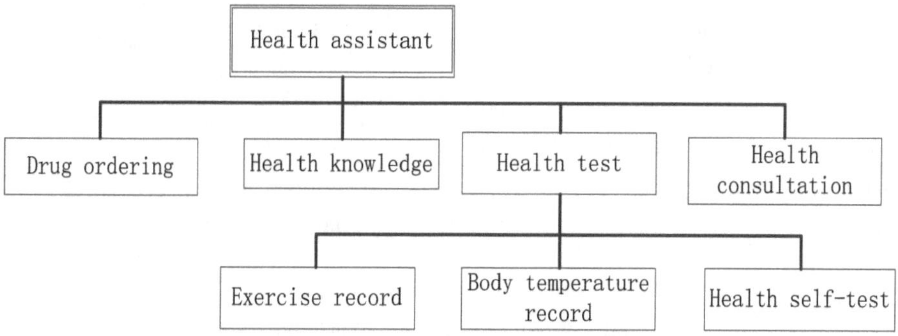

Fig. 2. Applet function structure diagram

The following is a brief description of the main functions of the applet:

Health self-test: through the basic information such as height, weight and age submitted by the user, the health-related indicators such as BMI, body fat rate and calories necessary for daily basic metabolism are calculated and displayed for the user.

Exercise record: add daily food and exercise to the record book to view the total calorie intake and consumption curve for many days. The exercise record includes the total intake and consumption of calories.

Body temperature record: Support the daily temperature upload function. Just fill in the form and upload the data, and the background will receive the user's temperature information. The daily temperature upload function discards the risk of face-to-face temperature measurement and greatly increases the safety.

Health knowledge: The server pushes daily health care knowledge closely related to people's daily life to the applet.

Health consultation: Free consultation with professional doctors on the mobile terminal and reply within 5 min. It is a lightweight and avant-garde disease asking mode, which is expected to become the first program for people to encounter health problems.

Drug ordering: Users can order daily drugs online and view order status without leaving home.

3.3 Design of Database Table

Because of the open source and excellent performance, MySQL is selected for this project to store business data [6–9].

The database of this project is named HealthcareDB, and the core table consists of user authorization information table (tb_user), temperature record table (tb_temp), sports record table (tb_sport), user physical examination table (tb_test), drug table (tb_drugs), and order table (tc_orders).Below is the data structure of the master data table temperature record table and physical examination table (Tables 1 and 2).

Table 1. Data structure of temperature record Table

Field name	Data type	not null	default	PK?
id	int	no	auto_increment	yes
name	varchar(20)	no		
temperature	number(4,2)	no		
phone	varchar(20)	no		
positon	varchar(50)	yes		
time	timestamp	no	now()	

Table 2. Physical examination Table

Field name	Data type	not null	default	PK?
id	int	no	auto_increment	yes
n_steps	int	no		
calories	int	no		
time	timestamp	no		

4 Key Technology and Implementation

4.1 Exercise Steps Recording

Most mobile phones have built-in vibration sensors or auxiliary processors to store data according to the vibration frequency of these devices. In short, when people walk, the center of gravity moves up and down, and sensors and collaborators sense this center of gravity movement and count it.

With the permission of the user's authorization, the applet can obtain the user's WeChat exercise steps in the last 30 days through the interface wx.getWeRunData() to customize the health plan for the user. The example code for obtaining the number of steps is as follows.

```
wx.getWeRunData({
    success(res) {
      const encryptedData = res.encryptedData;
      wx.request({
        url: 'https://localhost:44395/api/WxApiTest/Decrypt',
        method: "Post",
        header: {
        "content-type":"application/x-www-form-urlencoded"
        },
        data: {
          encryptedData: res.encryptedData,
          iv: res.iv,
          session_key: wx.getStorageSync("session_key")
        },
        success: function(json) {
          console.log(json.data.stepInfoList)
          if (json.data.stepInfoList) {
            json.data.stepInfoList = json.data.stepInfoList.reverse()
            for (var i = 0; i < json.data.stepInfoList.length; i++) {
console.log('time'+new Date(json.data.stepInfoList[i].timestamp*1000).
toLocaleString());
              console.log('steps'+ json.data.stepInfoList[i].step);}}}})}})
```

4.2 Location Service

In the background of COVID-19, when you upload your body temperature every day, the applet should also automatically submit your geographic location. Managers can control the epidemic more effectively.

The location service function in the WeChat applet can use Tencent location service, or third-party companies such as Baidu API interface. However, the third-party map API can only obtain data through the interface, and the map cannot be directly introduced into the WeChat Applet, so using Tencent location service can develop more powerful functions, better compatibility, and more accurate location service application scenarios.

The steps of using Tencent location service for location service in the WeChat Applet are as follows.

(1) Register a Tencent location service account
(2) Apply for developer key
(3) Download the WeChat Applet JavaScript SDK and put it into the WeChat Applet project
(4) Log in to WeChat Applet background and add the legitimate domain name of Request https://apis.map.qq.com

(5) Configure the permission field in app.json to obtain user location information
(6) Call wx.getLocation to obtain the geographical location of the current user (wechat returns longitude and latitude, speed and other parameters).
(7) Use the API function of JavaScript SDK reverse address resolution.

4.3 Database Connection Configuration

The applets server side uses ThinkJS framework to build connect to the database only need to configure adapter.js file under the src folder. The configuration information is as follows.

```
exports.model = {
  type: 'mysql',
  common: {
    logConnect: isDev,
    logSql: isDev,
    logger: msg => think.logger.info(msg)
  },
mysql: {
              handle: mysql,
   database: 'healthCareDB',
   prefix: 'tb_',
   encoding: 'utf8mb4',
   host: '127.0.0.1',
   port: '3306',
   user: 'root',
   password: '123456',
   dateStrings: true}};
```

5 System Test of WeChat Applet of Health Assistant

5.1 Applet Function Test

There are three versions of applet development: development version, experience version and official version. As its name is general, the development version is the version originally developed by developers, which can be debugged in the local development environment, but the debugging function is limited. The trial version can be packaged into an almost mature applet, but only managers and developers who have been approved to have test qualifications can test. The online version is a fully mature version of the applet. All WeChat users can access the applet at will.

In the WeChat developer tool, you can check the runtime data through AppData, modify the data in AppData with the help of real machine debugging, and then you can perform an exception test on the applet.

5.2 Applet Performance Test

The WeChat applet provides a performance monitoring tool for the development version. In the real machine test state, you can open it in the settings at the top right corner of the applet page. Through the performance monitoring tool, developers can see real-time performance data, including but not limited to viewing CPU usage, page switching time, etc. The test shows that the small program occupies very little memory and the response time is accepTable, meeting the system performance index requirements of fast response and low resource consumption.

5.3 Applet Compatibility Test

At present, most mobile phones on the market use the Android system and the IOS system. There are some differences in the handling of js code between the two systems. Therefore, to consider the operating system compatibility of small programs, you need to test on the mobile phones of these two systems separately.

In this process, the project used two different real computers of Xiaomi MIX2 and IPhone11 operating systems for testing. Both phones can successfully run applets.

The screen size of each user's mobile phone is different. Therefore, it should be noted that the width and height of components are designed in rpx units. At the same time, mobile phones with different screen sizes should be used for testing.

The test results show that the 4-inch, iPhone 5 and 5.1 inch iPhone 11 can display the page completely.

5.4 Applet Network Test

Due to the lack of printing license, the applet could not be completed online.For the test version of the applet, if the local server is used to provide services, the mobile phone must be at the same network address as the network to receive data.

5.5 Other Test Points

WeChat has a limit on the size of applets. If the size exceeds 3M, a beta or official version cannot be generated. In addition, when there are too many page hops, WeChat will also limit the number of pages.Therefore, most pages of the applet have been set to jump by closing the current page.

5.6 Test Conclusion

After several rounds of testing, although not all bugs of the applet have been completely repaired, most of the functions have become sTable, and the functions within the design scope can be realized, and some additional small functions of the auxiliary applet have been added. However, before the domain name is filed with the enterprise, the small programs are still in the testing stage and cannot be launched online. After being tested in simulation environment and real machine environment respectively, the applet can correctly and reliably complete the designed functions, and the software performance can meet the user's requirements.

6 Conclusions

With the rapid development of mobile internet and the popularity of mobile devices, personal health assistant software based on WeChat applet is developed. The software realizes the recording function of personal physique data and exercise records, and provides convenient drug purchase and health consulting services. The software provides convenience for personal health management based on the needs of users, the principle of user convenience and the interests of users. With the continuous development of artificial intelligence technology [10, 11], mobile phone software will move towards real intelligence in personal health care application services, and promote the prosperity and development of health care applications.

Acknowledgment. This work was supported by Shaanxi Educational Science Planning Project "Research on innovative practice of software talent training in local colleges and universities based on competency under the background of new engineering"(No:SGH21Y0453).

References

1. Jeng, M.Y., Yeh, T.M., Pai, F.Y.: Analyzing older adults' perceived values of using smart bracelets by means-end chain. Multidiscip. Digit. Publ. Inst. **8**(4), 494 (2020)
2. Nascimento, B., Oliveira, T., Tam, C.: Wearable technology: what explains continuance intention in smartwatches? J. Retail. Consum. Serv. **43**, 157–169 (2018)
3. Kim, A.: Mobile-assisted language learning in L2 Korean using WeChat: a case study. Int. J. Interact. Mob. Technol. **16**(1), 94–105,12 (2022)
4. Riaz, M.: Investigation of information sharing via multiple social media platforms: a comparison of Facebook and WeChat adoption. Qual. Quant. **55**(5), 1751–1773 (2021)
5. Plantin, J.C., De Seta, G.: WeChat as infrastructure: the techno-nationalist shaping of Chinese digital platforms. Chin. J. Commun. **12**(3), 257–273 (2019)
6. Nugraha, A.H.: Making the casher application for the Tonguewing meatball restaurant using MySQL and Netbeans. Int. J. Sci. Technol. Manage. **2**(5), 1793–1799 (2021)
7. Matallah, H., Belalem, G., Bouamrane, K.: Comparative study between the MySQL relational database and the MongoDB NoSQL database. Int. J. Soft. Sci. Comput. Intell. (IJSSCI) **13**(3), 38–63 (2021)
8. Klimek, B., Skublewska-Paszkowska, M.: Comparison of the performance of relational databases PostgreSQL and MySQL for desktop application. J. Comput. Sci. Inst. **18**, 61–66 (2021)
9. Nugraha, A.H.: Making a web-based application for sales kitchen Bunda Ghina using PhP and MySQL. Int. J. Sci. Technol. Manage. **2**(5), 1787–1792 (2021)
10. Haenlein, M., Kaplan, A.: A brief history of artificial intelligence: on the past, present, and future of artificial intelligence. California Manage. Rev. **61**(4), 5–14 (2019)
11. Bécue, A., Praça, I., Gama, J.: Artificial intelligence, cyber-threats and Industry 4.0: challenges and opportunities. Artif. Intell. Rev. **54**(5), 3849–3886 (2021)

Logistics Based on 5G Technology

Xin Zhang[1(✉)] and Pengmin Jia[2]

[1] Chongqing Vocational Institute of Engineering, Chongqing, China
yunyangxin@126.com

[2] Chongqing Preschool Education College, Chongqing, China

Abstract. In the 4G era, it is difficult for communication facilities to meet the needs of the Internet of Things in logistics, which is reflected in the operation and production in the logistics park or warehouse generally using traditional Wi-Fi technology, which is susceptible to interference, resulting in unstable communication and limited capacity, and even effective coverage contraction. Communication in mobile state is easy to drop off. Unable to support large-scale and intensive automation applications. These problems will be well solved in the 5G era. 5G technology can not only realize the integration of people, equipment and vehicles, but also make breakthroughs in overall scheduling and management. In the management of the park, real-time monitoring and abnormal warning can be realized while people and vehicles are flowing. It has the characteristics of multi-device access, large bandwidth and low delay. Artificial intelligence, Internet of Things, cloud computing, big data, AR/VR, block chain, robotics and other technologies based on 5G technology will comprehensively promote the intelligent process of social supply chain logistics, and realize the leap from connectivity to intelligence.

Keywords: 5G · logistics park · intelligent storage

1 Introduction

Logistics plays an important role in promoting international trade and regional development. Driven by industry 4.0 and Internet+, logistics is gradually transforming and upgrading from the traditional single point of information application to the digital, automatic and intelligent interconnection. Traditional wired optical fiber and short-range wireless Wi-Fi communication technologies have high network deployment, operation and maintenance costs [1]. It is urgent for the new generation of communication technologies to solve problems such as delay, connection scale, bandwidth and reliability, so as to realize timely and accurate circulation of data from devices, applications and users in various logistics links. With its reliable low latency, large bandwidth and large-scale connection capacity in wireless technology, 5G supports flexible networking requirements in different scenarios. With the maturation of the end-to-end (E2E) ecological environment, 5G will solve the technical challenges in the process of upgrading logistics to intelligence [2].

Y. Pei et al. (Eds.): IC 2024, LNEE 1214, pp. 117–124, 2024.
https://doi.org/10.1007/978-981-97-4193-9_13

2 Based on the Development Trend of 5G

2.1 Logistics Status and Development

In the age of information technology, the international trade market continues to expand and deepen. The complex and changeable contradiction between supply and demand has emerged, and a more diversified business model has emerged [3]. Based on this, consumers and commercial enterprises need faster and more convenient delivery services, so as to meet the market development and get close to consumers. Therefore, logistics parks and distribution networks become the focus of upgrading logistics services and optimizing operations close to the last mile of upstream and downstream. Poor working environment and high labor intensity are the most prominent problems and challenges in logistics parks and distribution links [4]. Moreover, the efficiency problem of insufficient personnel and lagging emergency response ability cannot be ignored. At present, the logistics industry mainly solves the above problems through digitization, automation and intelligent methods. However, the connection problem has always been a key point that needs to be solved in the field of intelligence. Compared with other mature technologies in the field of intelligence, the connection problem has not been effectively solved. Traditional communication facilities is difficult to meet the demand of the logistics industry group equipment, especially of vehicles, goods, logistics park and warehouse operations equipment often use Wi-Fi technology, poor stability and anti-interference ability, capacity is very small, even effective coverage shrinkage, mobile state of communication easy to drop, cannot take large-scale intensive application of automation, Unable to meet the needs of logistics operations [5]. Moreover, it is difficult to guarantee production safety and quality in terms of fire safety and security. Video surveillance is mostly used to monitor unusual behaviour outside the warehouse, as well as to monitor the storage environment and production processes inside the warehouse, in order to prevent the loss of the expansion of dangerous situations in advance or take timely measures to deal with dangerous situations [6]. At present, wired connection is the main way of camera connection, which is not easy to adjust the position flexibly. In addition, when wiring is difficult, Wi-Fi wireless transmission mode is adopted, which also faces problems of Wi-Fi transmission performance and stability. In addition, when wiring is difficult, Wi-Fi wireless transmission mode is adopted, which also faces problems of Wi-Fi transmission performance and stability [7].

2.2 Change and Innovation

4G changes life, 5G changes society. The biggest change in 5G is the shift from person-to-person communication to interconnection between people and things, so as to realize the interconnection of all things and promote social development [8]. With the opening of the global commercial curtain of 5G technology in 2019, the application of 5G technology in vertical industries has also carried out a large number of verification and improvement, accelerating the breakthrough of AI, cloud computing, edge computing, big data, Internet of Things, new generation of information technology and industrial internet.5G technology will become an important engine for the digitalization, automation and intelligent transformation of all industries [9].

3 Requirements of Intelligent Logistics for Wireless Scenarios

3.1 Analysls of 5G Requirements in Typical Logistics Scenarios

(1) Application Scenarios

The logistics park is a complex system with many points, long lines and wide areas, multi-level and multi-systems. In order to ensure the safe operation of the park, it requires not only rich experience and knowledge of management personnel, but also high technical level, safety awareness and adaptability of operators. In order to reduce enterprise losses, identify safety hazards, check equipment operation conditions and ensure personnel safety, traditional logistics parks often set up a complex set of inspection mechanisms and inspection systems in and out of the parks to make them suitable for enterprise safety management and personnel operation specifications [10]. In order to reduce the number of inspections and ensure the traceability of accidents, logistics parks gradually promote cameras and alarms on site. This solution has solved the problem of difficult on-site monitoring to a certain extent. On the other hand, there are many warehouses in the logistics park, which need to carry out warehousing, shelving, picking goods, review, packaging, warehousing and other operations frequently. There are abnormal personnel, abnormal goods and abnormal site conditions in the warehouse.

(2) Network Requirements

Park monitoring required many camera, resolution is not enough, sensor density is higher, the current is given priority to with optical fiber, wi-fi construction cost is high, and the limited transmission performance, to further support a wider range and greater density of camera, sensor deployment, at the same time, due to the local limited bandwidth, upload the cloud hd video transmission slow problem, Video exception recognition cannot be realized in all parks (see Table 1).

Table 1. A typical communication demand table for connected storage equipment

Communication needs	Shuttle car		AGV		
	Scheduling control	Operating status of equipment	Scheduling control	Operating status of equipment	Upgrade package
Amount of data transferred per time/B	1500	1500	1500	1500/1500	2M/4M
End-to-end delay/ms	30	100	30	100	1000
Maximum number of parallel devices/(PCS/m 2)	0.1		0.1		

3.2 Unmanned Warehousing Intelligent Equipment

In the current intelligent warehouse, intelligent equipment such as AGV, AMR and mechanical arm can communicate with instructions and data through Wi-Fi wireless and control system and operation and maintenance system. With the growth of business, there are more and more kinds and scales of warehousing intelligent equipment, and the communication volume of equipment, control system and operation and maintenance platform also increases. The requirements for wireless network communication bandwidth, reliability and communication delay are very high. The main challenges are as follows: (1) a single AP cannot cover the whole warehouse, and multiple aps need to be set up to cover it, which leads to roaming reliability and service interruption caused by switching AP hotspot when AGV is on the move [11]. (2) The AGV based on TWO-DIMENSIONAL code navigation or laser SLAM navigation is difficult to implement and maintain, and the laser navigation information is easy to be lost, so it needs to be moved back to a specific location for repositioning [12]. (3) There are various kinds of controller hardware of intelligent devices, and the demand for local computing power varies in various ways, leading to problems such as large r&d investment, long cycle, lack of flexibility, and difficulty in upgrading hardware resources. (4) Service clusters such as WCS, WMS, ground control system, operation and maintenance system adopt local deployment mode, and communicate with intelligent storage equipment through LAN. The system deployment is difficult, the cost is high, the implementation cycle is long, the hardware resources are not used evenly in busy time and the remote maintenance is difficult [13]. (5) Various intelligent devices in the warehouse are connected to the public network through Wi-Fi, and there is certain mutual interference when they work at the same time. If all the existing wireless network access is realized, there is a bottleneck in the number of access (see Table 2).

Table 2. Typical AI-capable warehouse equipment communication requirements

Communication needs	Intelligent forklift		Vision manipulator		
	Scheduling control	Operating status of equipment	Task scheduling	Kinematic control	Operating status of equipment
Amount of data transferred per time/B	5M	–	5M	25K	30M
Data rate/(Mb/s)	–	80	1.5	10	500
End-to-end delay/ms	30	30	30000	20	480
Maximum number of parallel devices/(PCS/m 2)	0.01		0.003		

3.3 5G-Based Cloud Storage Intelligent Equipment High-Density Cluster Deployment Solution

The network in the warehouse is covered by a sub-room system, and all signals originate from one AAU, which can avoid roaming problems existing in the AGV when moving. The construction of indoor distribution system can achieve the effect of compound use of one construction and avoid the cost waste caused by the independent construction of each product. Compared with the number of existing wireless access and wireless communication mechanism (LBT), 5G can better solve these two problems [14]. Will storage equipment master running on intelligent navigation, laser radar, visual image recognition and environmental awareness requires complex computing power demand, WCS, WMS, ground control system, the demand of the operational platform up to the edge of the 5g computing server, in order to meet the AGV enrich the application scenario and the increasing computing power demand. Modules with higher real-time requirements such as sensor data acquisition, motion control and emergency obstacle avoidance are still retained in the intelligent device ontology such as AGV, ARM and manipulator to meet security requirements.

4 5G Intelligent Logistics

4.1 Key Technologies of 5G

The three key technologies of 5G lie in stronger connectivity, slicing of end-to-end customized networks, and edge computing for edge cloud collaborative deployment. First of all, 5G is a network with large speed, large connection and low delay. With these capabilities, it can meet the demand for interconnection of everything in the future and thus promote the development of industrialization [15]. Specifically, it is the large rate: eMBB, with peak rate of 10Gbps and average throughput of hundreds of megabits; Large connections, mMTC, 1 million connections per square kilometer; Low latency, uRLLC, delay as low as 1ms; These are 10 to 100 times better than 4G. The current commercial 5G R15 version will first be able to enhance mobile broadband eMBB industry applications, R16 after the introduction of uRLLC, mMTC and other features, will have a comprehensive enabling industry digital, intelligent ability. Secondly, THROUGH network slicing technology, 5G can divide the physical network of operators into multiple logical networks to achieve multi-purpose of one network, enabling operators to build multiple dedicated, virtual, isolated and on-demand logical networks on one physical network to meet the different requirements of customers in different industries for network capabilities. Slicing enables a unified infrastructure network to support multi-service applications and thus significantly reduces investment. The 5G slice is based on SA architecture (the core network is 5GC). 3GPP R15 defines slice identification for 5G SA architecture, and E2E identifies user groups to enable the differentiation of eMBB services. The uRLLC and mMTC class slices are defined in R16 and later versions, respectively [16]. Finally, edge computing focuses on meeting the requirements of low latency and high bandwidth for industry application services. Computing, network and storage capabilities are deployed on the mobile edge network side close to industry data sources, and user side functions of 5G network are moved down, thus providing extremely low latency and high bandwidth for industry applications. And the business environment that can directly obtain real-time wireless network information (such as user location, base station load, etc.), and realize the local closed loop of business to meet the requirements of industrial customers for the security and privacy protection of enterprise production data (in Fig. 1).

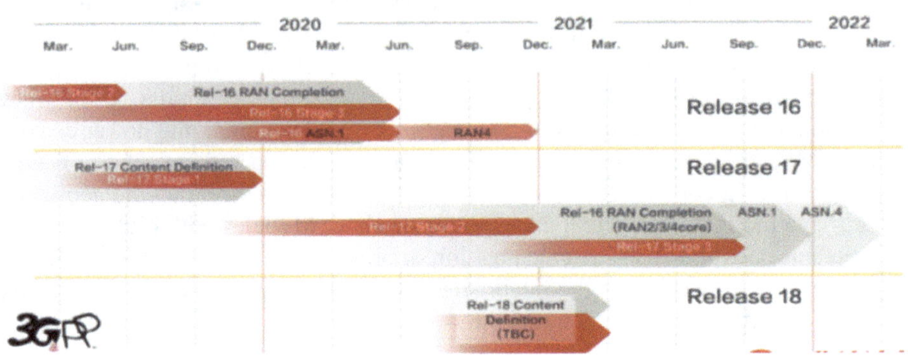

Fig. 1. The 3GPP standardization schedule

4.2 5G Logistics Private Network Solution

(1) Hybrid private network

In a hybrid private network scenario, dedicated wireless access networks and mobile edge computing (MEC) devices will be deployed inside the campus. The network control plane functions are provided by the public network, and the user plane functions are distributed to the internal network of the logistics park based on requirements [17]. In addition, logistics enterprises can choose base station, transmission sharing and deployment mode according to private network performance requirements and costs.

(2) Independent private network

In the independent private network scenario, a complete cellular network is deployed in the logistics park, including dedicated wireless network and core network devices. The data of the control plane and user plane of all the terminals in the logistics park are carried on the independent private network, and can be connected to the external network according to the demand.

5 Conclusions

5G, Internet of Things, cloud computing, big data, artificial intelligence, operations research, AR/VR, blockchain, robotics and other technologies will comprehensively promote the intelligent process of social supply chain logistics, from connectivity to intelligence. In short, logistics digitalization relies on 5G technology to connect everything. The massive data generated in the process of logistics production can be aggregated and processed through an intelligent platform to map the twin digital world. Intelligent decision to achieve rapid digital interactive verification, design and innovation improvement, to reduce costs, increase efficiency, collaborative optimal purposes.

References

1. Juan, R.S., Julie, B., Ade, A.: A business case for 5G services in an industrial sea port area. IEEE Commun. Mag. **52**(5), 86–92 (2020)
2. Gold, J.: 5G smartphones sales top 4G, but don't expect network deployment to speed up. IEEE Commun. Mag. **52**(2), 122–130 (2021)
3. Patrizio, A.: Marvell's newest Arm processor integrates 5G hardware accelerators. IEEE Trans. Aerosp. Electron. Syst. **36**(3), 869–878 (2021)
4. Juan, R.S., Julie, B., Ade, A.: Optimal LAP altitude for maximum coverage. IEEE Wireless Commun. Lett. **3**(6), 569–572 (2019)
5. Jackson, D.: T-Mobile: Standalone 5G is helping us win enterprise share. IEEE Commun. Surv. Tut. **21**(4), 3417–3442 (2019)
6. Li, H., Ding, M., López-Pérez, D., Fotouhi, A., Lin, Z., Hassan, M.: Performance analysis of the access link of drone base station networks with LoS/NLoS transmissions. In: Duong, T.Q., Vo, N.-S. (eds.) INISCOM 2018. LNICSSITE, vol. 257, pp. 111–121. Springer, Cham (2019). https://doi.org/10.1007/978-3-030-05873-9_10
7. Chen, Y., Peng, L., Liming, F.: Using 5G in smart cities: a systematic mapping study. IEEE Trans. Wireless Commun. **18**(2), 1054–1068 (2019)
8. Hassan, O., Gerd, Z.: Coverage and handoff analysis of 5G fractal small cell networks. IEEE Trans. Wireless Commun. **18**(2), 1263–1276 (2019)
9. Mustafa, H.A., Shakir, M.Z., Imran, M.A., Imran, A., Tafazolli, R.: Standalone 5G is helping us win enterprise share. IEEE Commun. Lett. **19**(10), 1742–1745 (2019)
10. Zhang, Y., Dai, L., Li, J., Mumtaz, S., Rodriguez, J.: Transmission capacity analysis of relay-assisted device-to-device overlay/underlay communication. IEEE Trans. Ind. Informat. **13**(1), 380–389 (2021)
11. Sun, L., et al.: Implementation and verification of 5G network slicing for smart grids. In: Advanced Science and Industry Research Center. Proceedings of 2021 5th International Conference on Electrical, Automation and Mechanical Engineering (EAME2021). Advanced Science and Industry Research Center: Science and Engineering Research Center, pp. 145–151 (2021)
12. Niu, J.: Analysis on the development path of public library smart service under 5G technology environment. In: Proceedings of the First International Symposium on Innovation Management and Economics (ISIME 2021): International Science and Culture for Academic Contacts, pp. 681–684 (2021)
13. Wu, J.: Design of the 5G enabled classroom teaching application scenarios. In: Wuhan Zhicheng Times Cultural Development Co., Ltd. Proceedings of 2nd International Symposium on Education and Social Sciences (ESS 2021).Wuhan Zhicheng Times Cultural Development Co., Ltd., pp. 339–341 (2021)
14. Dong, W., Peng, D., Liu, X., Wang, T., Long, J.: Eight direction improved Sobel algorithm based on morphological processing in 5G smart grid. In: International Association of applied science and engineering (IAASE). Proceedings of 2021 2nd International Conference on Computing, Networks and Internet of Things. International Association of applied science and engineering (IAASE), pp. 184–188 (2021)

15. Wang, W.H., Du, C.X.: A brief analysis of the 5G+4K/8K+AI strategic layout of central radio and television station——taking the 2019 national day campaign publicity report as an example. In: Federation University Australia. Proceedings of the 5th International Conference on Machine Vision and Information Technology (CMVIT 2021). Federation University Australia, pp.14–19 (2021)
16. Jiang, Y.: The opportunities and challenges of the development of the Internet of Things under the 5G network architecture. In: Wuhan Zhicheng Times Cultural Development Co., Ltd. Proceedings of 3rd International Conference on Business, Economics, Management Science (BEMS 2021).Wuhan Zhicheng Times Cultural Development Co., Ltd., pp. 43–51 (2021)
17. Ye, M., Huang, G., Gao, P., Wang, M., Zhang, X., Yu, G.: Key node and network performance analysis of 5G smart power plant based on complex network. In: International Association of Applied Science and Engineering (IAASE). Proceedings of 2021 2nd International Conference on Computing, Networks and Internet of Things. International Association of Applied Science and Engineering (IAASE), pp. 211–216 (2021)

Power Transmission and Transformation Risk Management System Based on EWM Calculation and AHP

Shengjie Wang[1], Weijun Wang[2], Xiaobo He[1], and Xiaoran Ma[2](✉)

[1] Institute of Economics and Technology, State Grid Shaanxi Electric Power Company, Xi'an, Shaanxi, China
[2] Department of Economic Management, North China Electric Power University, Baoding, Hebei, China
mxrncepu@163.com

Abstract. Aiming at the evaluation and management of transmission project investment risk under the new power market environment, a scheme combining the analytical hierarchy process (AHP) and the entropy weight method (EWM) is studied. Firstly, for different economic, environmental and social risks, the evaluation index of risk management of transmission project is analyzed. Then, the basic principle of calculating objective weight by EWM is studied, and the basic idea of AHP to solve subjective weight is analyzed, and the calculation method of comprehensive weight is given by combining the two methods. Finally, combined with specific application scenarios, the detailed calculation process of investment risk management evaluation results is given. The application example shows that the scheme can be well applied to the evaluation of transmission project investment risk management, and can provide guidance for enterprises to build power transmission projects in different market environments.

Keywords: AHP · EWM · Index Weight · Risk Management · Transmission Project Investment

1 Introduction

The proportion of fixed assets of power grid enterprises increases year by year, which is related to the development of people's livelihood, and their investment benefit directly affects China's economic efficiency [1]. With a series of policies being implemented, the transmission-distribution prices were lowered, the profitability of the power grid business dropped and the loss surface expanded, which brought certain risks to the power grid investment. State Grid Corporation proposed new investment and construction ideas require optimization of investment direction, structure, timing. Under the background of the new investment model, how to evaluate the risk control effect of the transmission project is the key to solve the difficult problem of enterprise management [2].

Evaluation model of construction risk management of power transmission projects is mainly based on two parts: identification and deconstruction of project investment

Y. Pei et al. (Eds.): IC 2024, LNEE 1214, pp. 125–135, 2024.
https://doi.org/10.1007/978-981-97-4193-9_14

risk and project investment risk management evaluation. At this stage, Liu et al. put forward the mathematical model of project risk identification based on system identification theory to improve the accuracy and reliability of risk identification [3]; Based on grounded theory, Wu et al. identified the risk in the design phase of power transformation project construction through three-level coding and formulated the environmental risk index system [4]; Liu et al. considered the construction sequence of the project and identified the risk sources from multiple dimensions and perspectives from the three stages of project feasibility study period, construction implementation period and operation [5]; In their research on renewable energy project investment, Hashemizadeh et al. used TODIM algorithm to evaluate the risk according to five categories: economic risk, technical risk, environmental risk, social risk and political risk.

Transmission project investment risk management evaluation involves many evaluation indicators, and the interests and concerns of each relevant subject vary widely. At present, most of the risk research focuses on the economic benefit evaluation of risk management, the scope of application is small, it is difficult to evaluate the risks caused by uncertain factors such as social conditions and natural environment, and it cannot fully reflect the construction willingness [6]. Scientific evaluation has many effects on the investment control of subsequent projects in the construction of different markets. Based on the identification and deconstruction of investment risks in projects and the establishment of impact management evaluation criteria, this paper adopts the scheme of combining EWM and AHP to study the calculation method and process of index weight and comprehensive evaluation, and analyzes the specific application case. It has reference value for evaluating and reducing the risk of transmission projects.

2 Evaluation Index

Evaluation of transmission project investment influence management requires comprehensive consideration of the construction characteristics of enterprises and the stage of project investment management. This paper comprehensively evaluates transmission project investment risk management from three aspects: economic benefit, social benefit and environmental benefit [7].

2.1 Economic Benefit Index

The economic benefit index is mainly to examine the profitability and level of the power transmission project after it is put into operation, that is, to examine whether the project is profitable and the size of the profitability [8]. In the context of the new power reform, for the power grid company, its fundamental goal is still the pursuit of maximum economic benefits, economic benefit index can most intuitively reflect the reasonableness of the power grid company transmission project investment risk management.

This paper examines the investment risk management of transmission projects from the perspectives of operational benefits and cost overruns. After the reform of the power system, which greatly affects the investment ability of enterprises and brings risks to the operating income of construction projects. It is reasonable to evaluate its risk management level from operational benefits. In addition, cost control is a crucial element in the construction of projects. Cost overruns will cause the stagnation of power transmission project construction and the reduction of revenue. Therefore, the cost overrun risk management of investment projects should be evaluated.

2.2 Social Benefit Index

Power transmission projects take the transmission of electric energy as the main goal, and use limited resources such as transmission lines and tower materials to reasonably plan the path and layout of transmission lines to meet the requirements of quality and power supply stability. As an important part of the power supply, the power transmission project should also take into account the requirements of power grid security, stability and economic operation, realize the mutual transmission of electric energy in different regions, and meet the electric energy demand of each region. In the investment and construction of power transmission projects, on the one hand, it is necessary to strengthen professional skills and improve energy quality, and on the other hand, it is necessary to start from customer needs and improve service quality [9].

This paper sets four risk management evaluation indexes: power supply reliability, grid safety operation, power quality and grid structure. Supply reliability describes the ability of the power system to provide continuous, stable and sufficient power services to customers under long-term operating conditions. Grid safety operation can be divided into grid safety running and grid stability running. Safe grid running means that all power equipment is operating within the rated value, and exceeding the safe operating range will cause equipment damage. Grid stability running means that the grid is stable and continuous in supplying normal power to the load users, and unstable operation can lead to widespread power outages. Power quality describes the quality of AC power supplied to the load side through the grid, and reflects the role of transmission construction in improving system stability and ensuring the strength of the regional grid. The strength of the main grid can be described as the ability of the main grid to withstand changes in load.

2.3 Environmental Benefit Index

Thermal power uses coal resources to generate power, and a large number of pollutants were discharged locally. In recent years, in the background of national energy conservation and emission reduction, green power generation has become the current trend of power generation.

Through the construction of power transmission projects, the grid-connected power of new energy is increased, thereby reducing the use of primary fossil energy and the emission of pollutants such as carbon dioxide; It reduces the consumption of resources such as coal and contributes to carbon emission reduction through environmentally friendly power generation methods.

3 Method of Weights of Evaluation Index

3.1 Determination of Weight by EWM

The EWM calculation is implemented by weighting method. In the determination process of the weights, it is represented by entropy. In the 1940s, Claude Shannon introduced entropy into information theory as a tool to measure the disorder of the system. When there is a large difference in the data of an index, the greater the amount of effective information, the value of the weights varies with the entropy; On the contrary, it is smaller [10, 11].

When evaluating m targets and n indicators, the initial matrix X form can be expressed as follows:

$$X = \begin{pmatrix} X_{11} & X_{12} & \cdots & X_{1n} \\ X_{21} & X_{22} & \cdots & X_{2n} \\ \cdots & \cdots\cdots & & \cdots \\ X_{m1} & X_{m2} & \cdots & X_{mn} \end{pmatrix}_{m\times n} \tag{1}$$

where, X_{ij} is the original data value of the i-th evaluation object under the j-th index.

Different indicators correspond to different evaluation criteria, which cannot be directly used in the case of various comparison angles, scope of application and data differences. It is necessary to unify the raw data, construct the dimensionless evaluation criteria, and find the comparable matrix V, such as:

$$v = \begin{pmatrix} v_{11} & v_{12} & \cdots & v_{1n} \\ v_{21} & v_{22} & \cdots & v_{2n} \\ \cdots & \cdots\cdots & & \cdots \\ v_{m1} & v_{m2} & \cdots & v_{mn} \end{pmatrix}_{m\times n} \tag{2}$$

where V_{ij} is the processed value of the i-th evaluation object under the j-th index. For the bigger and better index (positive index), Vij principle of solving is as follows:

$$v_{ij} = \frac{x_{ij} - \min(x_j)}{\max(x_j) - \min(x_j)} \tag{3}$$

where, max (x_j) and min (x_j) corresponds to the maximum minimum of the J th index. However, when they are of the same size, there will be uncertainty such as 0/0. At this time, V_{ij} can be taken as 1. For the smaller, better index (negative index), V_{ij} calculation method is as follows:

$$v_{ij} = \frac{\min(x_j) - x_{ij}}{\max(x_j) - \min(x_j)} \tag{4}$$

Similar to the bigger the better index, when Max (x_j) and min (x_j) are equal, V_{ij} is taken as 1. After further processing, the matrix p can be obtained, as follows:

$$p = \begin{pmatrix} p_{11} & p_{12} & \cdots & p_{1n} \\ p_{21} & p_{22} & \cdots & p_{2n} \\ \cdots & \cdots\cdots & & \cdots \\ p_{m1} & p_{m2} & \cdots & p_{mn} \end{pmatrix}_{m\times n} \tag{5}$$

where, p_{ij} is the characteristic proportion of the i-th evaluation object under the j index, as follows:

$$p_{ij} = v_{ij} / \sum_{i=1}^{m} v_{ij} \tag{6}$$

Calculate the entropy matrix E as follows:

$$E = \left[\, e_1 \; e_2 \; \cdots \; e_n \,\right] \tag{7}$$

where, ej is the entropy value of the j-th index, and the calculation method is as follows:

$$e_j = -\frac{1}{In(m)} \sum_{i=1}^{m} p_{ij} \cdot Inp_{ij} \tag{8}$$

In the above formula, in order to avoid ln (0), the overall data translation of P can be carried out, plus a number, such as 0.0001. The weight matrix as follows:

$$W_{(e)} = \left[\, W_{(e)1} \; W_{(e)2} \; \cdots \; W_{(e)n} \,\right] \tag{9}$$

where, w (e) j is the weight of index j and the solution procedure is as follows:

$$W_{(e)j} = (1 - e_j) / \sum_{j=1}^{n} (1 - e_j) \tag{10}$$

3.2 Determination of Weight by AHP

The AHP was first applied by operations research scientist Satti. Its core idea is to decompose the overall goal into several levels, and determine the weights of various indicators at different levels of the system [12, 13].

The principle of hierarchical single sorting is analyzed below. There are n values above a certain level and there is a certain connection with the elements of the other levels, and the judgment matrix A is defined as follows:

$$A = \begin{pmatrix} a_{11} & a_{12} & \cdots & a_{1n} \\ a_{21} & a_{22} & \cdots & a_{2n} \\ \cdots & \cdots & \cdots & \cdots \\ a_{n1} & a_{n2} & \cdots & a_{nn} \end{pmatrix}_{n \times n} \tag{11}$$

where Aij is the comparative value of index I relative to index j, which satisfies the following relationship:

$$\begin{cases} a_{ij} > 0 \\ a_{ii} = 1 \\ a_{ji} = 1 \big/ a_{ij} \end{cases} \tag{12}$$

There are many methods to calibrate the comparative value Aij such as 0. 618 scale method of judging matrix elements, as shown in Tables 1 and 2.

Table 1. The 1–9 scale method of matrix contents

Scale	*Meaning*
1	Equally important
3	Slightly important
5	Strong importance
7	Strongly important
9	Extremely important
2, 4. 6. 8	The mean value of the above two adjacent judgments
Reciprocal of scale	The sizes of the exponents i and j are mutually reciprocal

Table 2. The 0. 618 scale method of judging matrix elements

Scale	Meaning
1	Equally important
1.618	Slightly important
2.618	Strong importance
4.236	Strongly important
6.854	Extremely important
Reciprocal of scale	The sizes of the exponents i and j are mutually reciprocal
Scale	meaning

Find out the size of the different weights, as shown below:

$$W_{(a)} = \left[W_{(a)1} \; W_{(a)2} \cdots W_{(a)n} \right] \tag{13}$$

where, w (a) i is the weight of index i, and the solution procedure is as follows:

$$w_{(a)i} = \frac{\sqrt[n]{\prod_{j=1}^{n} a_{ij}}}{\sum_{i=1}^{n} \sqrt[n]{\prod_{j=1}^{n} a_{ij}}} \tag{14}$$

If the judging matrix is a symmetrical matrix, then:

$$a_{ij} = a_{ik} \cdot a_{kj} \tag{15}$$

In application, the calibration of judgment matrix often adopts the way of pairwise comparison, and there are often inconsistent matrices. In order to avoid computational error and logical contradictions, consistency check is needed. Define the consistency index CI and calculate it as follows:

$$CI = \frac{\lambda - n}{n - 1} \tag{16}$$

where, λ is the maximum eigenvalue of the judgment matrix, as shown below:

$$\lambda = \frac{1}{n}\sum_{i=1}^{n}\frac{(AW)_i}{w_i} \tag{17}$$

Define the evaluation random consistency index RI, as shown in Table 3:

Table 3. RI value table

Order	*RI*
1	0
2	0
3	0.58
4	0.9
5	1.12
6	1.24
7	1.32
8	1.41
9	1.45
10	1.49
11	1.51

The consistency ratio CR is defined and calculated as follows:

$$CR = CI\big/RI \tag{18}$$

When CR is less than 0.1, the consistency of the judgment matrix is acceptable; Otherwise, it needs to be corrected.

3.3 Comprehensive Weight

After calculating the weight matrices w(e) and w(a) using entropy weight method and analytic hierarchy process, the matrix results are obtained as shown in the following.

$$W = \left[\, W_1 \; W_2 \cdots W_n \,\right] \tag{19}$$

where, wi is the weight of index I, and the calculation formula is as follows:

$$w_i = w_{(e)i} \cdot (1 - e_i) + w_{(a)i} \cdot e_i \tag{20}$$

Considering the objective factors of analytic hierarchy process, the comprehensive weight achieves the two methods, reducing the analysis bias caused by a single method, and making the analysis results more comprehensive, fair and reasonable.

4 Experiment of Investment Risk Management Evaluation of Power Transmission Project

4.1 Preparation of Data

In order to evaluate the investment risk management of a power transmission project, three primary indicators have been developed and eight secondary indicators have been refined. Meanwhile, in order to simplify the evaluation process, only 3 levels were set, such as beyond expectations, within expectations, and not up to expectations; Through the statistical arrangement of historical data, the results are shown in Table 4.

Table 4. Investment risk management evaluation index system for power transmission projects

Tier 1 Indicators	Tier 2 Indicators	Beyond expectations	Within expectations	Not up to expectations
Economic benefit index B1	Operational benefit C1	0.27	0.42	0.31
	Cost overrun C2	0.31	0.28	0.41
Social benefit index B2	Power supply reliability C3	0.17	0.49	0.34
	Grid safety operation C4	0.32	0.51	0.17
	Power quality C5	0.41	0.53	0.06
	Grid structure C6	0.09	0.72	0.19
Environmental benefit index B3	Clean energy consumption C7	0.17	0.38	0.45
	Substitution of electricity for electricity C8	0.29	0.43	0.28

Taking indicator C1 (operating benefit) as an example, after statistics, 27% of the data indicated that the operating income exceeded the expected investment target, 42% of the data indicated that the operating income was within the expected range, and 31% of the data indicated that the operating income did not meet the expectation. And so on, constitute the original data of the evaluation index.

4.2 Calculate Weight Based on EWM

The original data matrix of indicator B1 obtained by the entropy weight method is as follows:

$$\begin{bmatrix} 0.27 & 0.31 \\ 0.42 & 0.28 \\ 0.31 & 0.41 \end{bmatrix} \quad (21)$$

Then the weight matrix based on the EWM is calculated as follows:

$$W_{(e)B1} = \begin{bmatrix} 0.5852 \ 0.4148 \end{bmatrix} \tag{22}$$

Similarly, the weight matrix of social benefit B2 and environmental benefit B3 can be obtained as follows:

$$W_{(e)B2} = \begin{bmatrix} 0.2209 \ 0.3020 \ 0.2029 \ 0.3409 \end{bmatrix} \tag{23}$$

$$W_{(e)B3} = \begin{bmatrix} 0.3264 \ 0.6754 \end{bmatrix} \tag{24}$$

4.3 Calculate Weight Based on AHP

Considering the opinions of experts and other groups, the 4 indicators C3, C4, C5, and C6 are compared in pairs by using the analytic hierarchy process. The comparison results are as follows: C3 is slightly more important than C4, C4 is equally important than C5, and C6 is more important than C3. According to the 0.618 scaling method, the judgment matrix can be calculated as follows:

$$A_{B2} = \begin{bmatrix} 1 & 1.618 & 1.618 & 0.328 \\ 0.618 & 1 & 1 & 0.618 \\ 0.618 & 1 & 1 & 0.618 \\ 2.648 & 1.618 & 1.618 & 1 \end{bmatrix} \tag{25}$$

Then the weight matrix based on analytic hierarchy process can be calculated as follows:

$$W_{(a)B2} = \begin{bmatrix} 0.2357 \ 0.1855 \ 0.1855 \ 0.3933 \end{bmatrix} \tag{26}$$

Similarly, the weight matrix of economic benefit B1 and environmental benefit B3 can be obtained as follows:

$$W_{(a)B1} = \begin{bmatrix} 0.618 \ 0.382 \end{bmatrix} \tag{27}$$

$$W_{(a)B3} = \begin{bmatrix} 0.382 \ 0.618 \end{bmatrix} \tag{28}$$

4.4 Final Evaluation Calculation

Based on the above analysis, the comprehensive weight can be calculated from Eq. (6) as follows:

$$W_{B1} = \begin{bmatrix} 0.6005 \ 0.3995 \end{bmatrix} \tag{29}$$

Then the results of index B1 are as follows:

$$Y_{B1} = X_{B1} \cdot W_{B1}^T = \begin{bmatrix} 0.2860 \\ 0.3641 \\ 0.3499 \end{bmatrix} \tag{30}$$

After comprehensive analysis, for the economic benefit indicator B1, 28.6% said it exceeded expectations, 36.41% said it was within the expected range, and 34.99% said it did not meet expectations. In the same way, the results of B2 and B3 can be obtained.

Then the final original data matrix:

$$X_Z = \begin{bmatrix} 0.2860 & 0.2243 & 0.2468 \\ 0.2641 & 0.5810 & 0.4120 \\ 0.3499 & 0.1947 & 0.3412 \end{bmatrix} \tag{31}$$

Calculated by the EWM and the AHP method, the comprehensive weight matrix is:

$$W_Z = \begin{bmatrix} 0.3403 & 0.4611 & 0.1986 \end{bmatrix} \tag{32}$$

Then the final evaluation of the comprehensive energy system operation service is:

$$Y_Z = X_Z \cdot W_Z^T = \begin{bmatrix} 0.2498 \\ 0.4736 \\ 0.2766 \end{bmatrix} \tag{33}$$

That is to say, combining the primary and secondary indicators, for the investment risk management evaluation of the power transmission project, 24.98% of the possible management levels exceeded the expected level, 47.36% of the possible management levels were within the expected range, and 27.66% of the possible management levels did not meet expectations.

5 Conclusions

Risk evaluation can predict the risks of the system in advance, make qualitative and quantitative evaluations, and put forward preventive measures in a targeted manner, which is of great significance to project investment and realizes the scientific management method. It adopts modern scientific principles such as system theory to carry out risk assessment comprehensively and systematically, and can achieve the goal of achieving better results with less investment. Under the trend of the construction of a new electricity market, this paper studies an evaluation method for investment risk management of power grid transmission projects based on EWM and AHP, introduces the main evaluation indicators from the aspects of economy benefit, society benefit and environment benefit. The basic principle and calculation process, combined with the application example, give the solution process of the evaluation result, which has certain reference significance.

There are still some shortcomings in the relevant research of this paper, such as incomplete and detailed evaluation indicators, and the lack of consideration for screening and filtering invalid information in data analysis. In addition, the comprehensive energy system operation service evaluation is not limited to EWM and AHP. The basic principles, algorithm processes, main characteristics, and application scenarios of other evaluation methods still need further research, and will be further optimized and improved in the future.

Acknowledgement. This research is supported by the project of State Grid Shaanxi Electric Power Company Institute of Economics and Technology (project number 5226JY200006).

References

1. Sun, J., Ruze, N., Zhang, J., Zhao, H., Shen, B.: Evaluating the investment efficiency of China's provincial power grid enterprises under new electricity market reform: empirical evidence based on three-stage DEA model. Energies **12**(18), 3524 (2019)
2. Schreiner, A., Balzer, G., Precht, A.: Risk Analysis of Distribution systems using Value at Risk methodology (2008)
3. Renhui, L., Shi, A.: Research on quantitative method of project risk identification. China Manag. Sci. **15**(z1), 97–100 (2007). https://doi.org/10.3321/j.issn:1003-207x.2007.z1.021
4. Shuang, W., Xiaomeng, Z., Xi, C.: Environmental risk identification and assessment in whole process of construction of power transmission project. Sci. Technol. Manag. Res. **40**(12), 76–84 (2020). https://doi.org/10.3969/j.issn.1000-7695.2020.12.011
5. Minghua, L., Wenxia, L., Chenmiao, L., Yuan, C., Zongqi, L.: Investment decision method for power grid construction projects based on holographic risk assessment and combination weighting. Mod. Electr. Power **36**(05), 87–94 (2019)
6. Schreiner, A., Balzer, G.: Value at risk method for asset management of power transmission systems. In: 2007 IEEE Lausanne Power Tech, VOLS, pp.1–5 (2007)
7. Brelih, M., Rajkovič, U., Ružič, T., Rodič, B., Kozelj, D.: Modelling decision knowledge for the evaluation of water management investment projects. Central Eur. J. Oper. Res. **27**(3), 759–781 (2019)
8. Kate, G.S., Patil, P.S.: Application of risk analysis and allocation model on public-private partnership power transmission line projects. Int. J. Crit. Infrastruct. **16**(3), 240–254 (2020)
9. Multi-layer fuzzy comprehensive investment risk assessment and management in power construction project. In: 2010 International Conference on Education and Management Technology, pp. 434–439. IEEE (2010)
10. Jinma, S., Tianyou, C., Wen, J., Yue, Z.: Selection of cost allocation methods for power grid enterprises based on entropy weight method. J. Phys. Conf. Ser. **1881**(2), 022063 (2021)
11. Wang, Y., Yaling, J., Yang, J., Tang, Y., Zhou, P., Chen, C.: Research on power grid infrastructure investment distribution model based on entropy weight method. E3S Web Conf. **253**, 03049 (2021)
12. Kaymaz, Ç.K., Birinci, S., Kızılkan, Y.: Sustainable development goals assessment of Erzurum province with SWOT-AHP analysis. Environ. Dev. Sustain. **24**(3), 2986–3012 (2021). prepublish
13. Aragones-Beltran, P., Chaparro-Gonzalez, P.-F., Pla-Rubio, A.: An AHP (Analytic Hierarchy Process)/ANP (Analytic Network Process)-based multi-criteria decision approach for the selection of solar-thermal power plant investment projects. Energy **2013**(66), 222–238 (2013)

Accounting Receivables Management Optimization System Based on Artificial Intelligence Technology

Feng Gu(✉)

Accounting College, Binzhou Polytechnic, Binzhou 256600, Shandong, China
sdgufeng1976@163.com

Abstract. With the global economic integration and increasingly fierce market competition, enterprises through the use of modern information technology and artificial intelligence, constantly promote the change of enterprise management mode. The way companies operate and manage their finances is also changing. The purpose of this paper is to study the receiving-accounts management optimization system based on artificial intelligence technology. Firstly, the design of the overall system architecture is described. Based on the SAP system platform, the advanced business-oriented value management concept is used to realize the function of the accounting receivables management optimization system through ABAP programming technology. The corresponding index system is designed according to the financial data. Through BP neural network model to realize the automatic assessment of accounts receivable risk to guide the enterprise team to make the best decision.

Keywords: Artificial Intelligence · Accounting Collection · Accounts Management · Optimization System

1 Introduction

The use of artificial intelligence to provide professional financial services to customers inside and outside group enterprises is a new application of financial management mode, Internet economy and sharing concept of enterprise management in the field of financial management of group enterprises [1, 2]. Aiming at the new financial communication management mode, the audit data collected by mass data and cloud computing and other modern information technologies will have a great impact on the traditional audit mode [3, 4]. Accounts receivable and accounts payable are two important current assets and liabilities of a company, rather than tangible assets. Therefore, the audit of accounts receivable and accounts payable plays an important role in the company's internal audit [5, 6].

Haitham Mohsin Kareem explores the accounting treatment of accounts receivable and accounts payable, comparing domestic and international accounting standards, methods of classifying and valuing accounts receivable and accounts payable, and economic

Y. Pei et al. (Eds.): IC 2024, LNEE 1214, pp. 136–144, 2024.
https://doi.org/10.1007/978-981-97-4193-9_15

analysis. Scientific, systematic and practical recommendations of the Organization were made to improve the accounting and analysis of accounts receivable and accounts payable in the corporate management system. According to the analysis made, the classification of accounts receivable adds the following features: depends on the work done to service the debt; This depends on the ability to transfer debt management to subcontractors. An important methodological basis of payables accounting and economic analysis is its scientific and reasonable classification. Taking into account the requirements of accounts payable management practices, its classification is supplemented by the following criteria: depends on the work done to liquidate the debt; This is subject to the transfer of debt management to subcontractors [7]. With the continuous development of China's economy and the deepening of economic globalization, enterprise groups have successively established financial transaction centers, and the audit of accounts receivable and accounts payable based on mobile computing is also particularly important. Good audit early warning can make internal management more standardized and conducive to risk control, so as to meet the financial management requirements and enterprise development requirements of enterprises. Fahmi Ibrahim's purpose is to study the audit and alerting of accounts receivable and accounts payable in a financial exchange model based on mobile computing. He mainly studied the audit alerts of company accounts receivable and accounts payable in the financial exchange model. Taking the construction of a financial information exchange center of an enterprise group as an example, he analyzed the problems in the audit. In addition, he also proposed the design of an enterprise accounts receivable and accounts payable audit alert model based on the transfer calculation financial exchange model. Survey data show that 83 employees of a company do not understand the audit process of accounts receivable and accounts payable under the financial exchange model, accounting for 43.68%, 40 employees generally understand, accounting for 21.05%, and only 28 employees understand very well, accounting for 14.74%.This makes it difficult for early audit alerts to achieve the desired effect, because office staff do not understand the audit alerts corresponding to deposits and accounts payable [8].

This paper mainly considers the risk management of accounts receivable digital management in the accounting receivables management optimization system, combined with the characteristics of risk management and digitalization, discusses the enterprise risk management in the process of accounts receivable management. This paper summarizes the overall architecture design of the accounting receivables management optimization system, discusses the risk management process of digital accounts receivable management based on BP neural network algorithm, and considers the potential risks in the process of accounts receivable management.

2 Research on the Optimization System of Accounting Receivables Management Based on Artificial Intelligence Technology

2.1 Overall Structure Design of Accounting Receivables Management Optimization System

Based on SAP system platform, this system adopts advanced business-oriented value management concept, and optimizes the aging analysis of capital gains and accounts payable by ABAP programming technology. The coupling of the two systems is realized through multi-system data exchange technology, and the status of payment process is recorded clearly, which is convenient to monitor the status of business process in real time. The ability to automatically settle accounts receivable through various technologies such as data viewer [9].

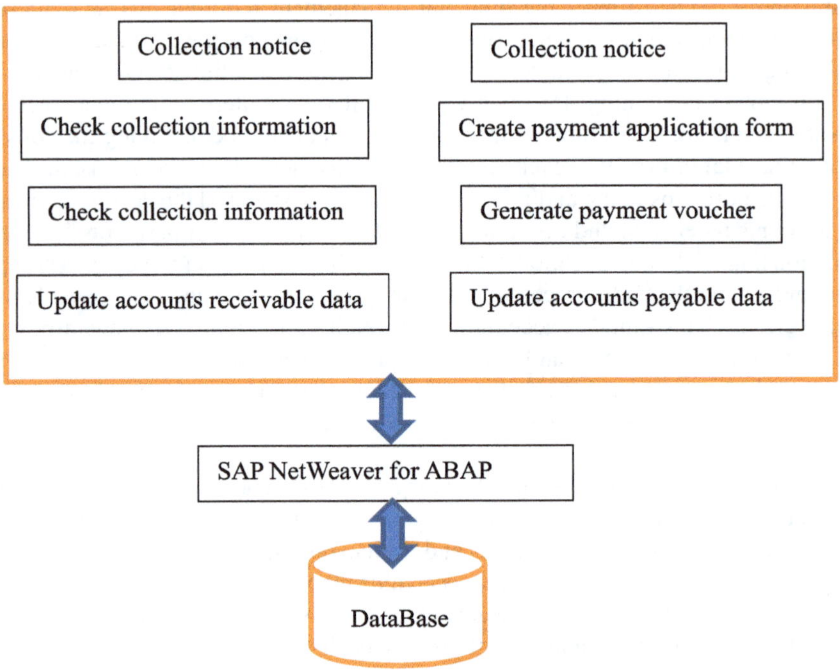

Fig. 1. Overall architecture

At the same time, the accounting collection and accounts management optimization system is seamlessly integrated with the SAP standard FI system, realizing the automation of financial processing. No additional hardware investment is required for installation and operation because the Accounting Collection management optimization system meets the standards and performance requirements established by SAP. In addition, the system has a flexible and configurable structure to deliver custom data through accounts receivable processes and standard delivery functions to meet the business process needs of different companies and customers [10]. The overall structure of the accounting receipt-accounts management optimization system is shown in Fig. 1.

2.2 Overall Function Design of Accounting Receivables Management Optimization System

The whole system is divided into accounts receivable management, accounts payable management, comprehensive analysis of business data and main data maintenance, among which the system's main business system has three systems: accounts receivable management system, accounts payable management system and comprehensive data analysis system. The functions of the Accounts Receivable management system include the business integration of the SAP Accounts Receivable management system with the TT funds system interface, and the settlement of accounts receivable before the due date. The functions of the Accounts Payable management system include: from creation to approval of payment requests, integration of SAP Accounts payable management business with TT funds system interface and manual settlement. The detailed function diagram of the accounting receivables management optimization system is shown in Fig. 2.

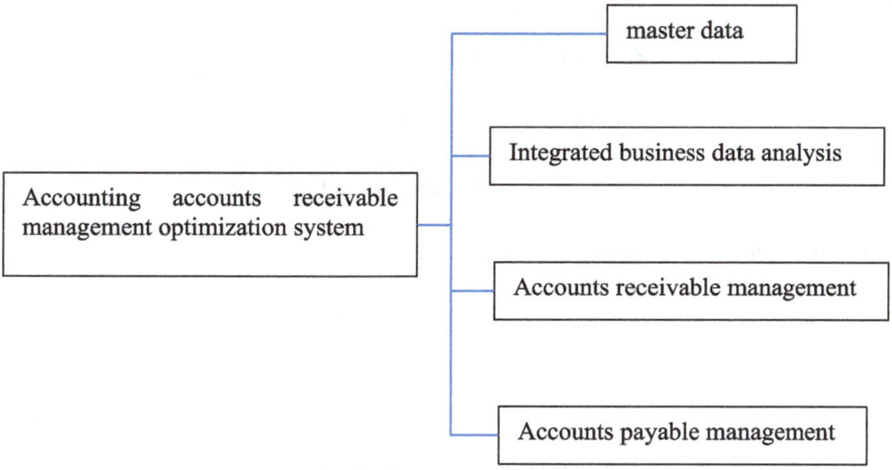

Fig. 2. System function structure diagram

2.3 Accounts Receivable Digital Management Risk Control

In today's era of huge data, the rapid growth of data accelerates the expansion of data and enriches the diversity. Accurately identifying and capturing data has become a major challenge in large databases. Due to large numbers of users, time and space constraints, and passive reception of complex data, most data users either do not have adequate access to the information or are unable to determine whether it is valuable. Receivables management needs to obtain a lot of data for analysis and establish a risk control system to determine the possibility of bad debts.

2.4 Feasibility Analysis of Accounts Receivable Risk Assessment by BP Neural Network Algorithm

At the present stage, accounts receivable risk assessment method is based on the discrimination and classification model of statistical methods to measure the financial situation of enterprises and accounts receivable liquidity. Many indexes of prediction variables are unevenly distributed. BP neural network has natural advantages in discriminating, predicting and analyzing. When BP neural network is used to solve the analysis problem, a certain number of noise samples are trained for analysis and judgment, and the relationship between sample features and implicit features is extracted. It can be seen from the above advantages of BP neural network that it is feasible to use BP neural network algorithm to evaluate accounts receivable risk [11, 12].

It is assumed that BP neural network has node points in input layer c, output layer d and hidden layer. The weight between hidden layer and input layer is recorded as V_{iw}, and the weight between implicit layer and output layer is recorded as V_{jw}. Set the hidden layer function to f1 and the output layer function to F2. The output of the hidden layer node is as follows:

$$Z_k = f_1(\sum\nolimits_{i=0}^{n} V_{ik} X_i),\ k = 1, 2, ..., q \tag{1}$$

The output of the node in the output layer is:

$$y_m = f_2(\sum\nolimits_{m=0}^{d} W_{dm} Z_m),\ m = 1, 2, ..., d \tag{2}$$

For P sample sizes, the overall error value is:

$$E = \frac{1}{2} \sum_{P=1}^{p} \sum_{j=1}^{m} (t_j^p - y_j^p)^2 \tag{3}$$

3 Investigation and Research of Accounting Receivables Management Optimization System Based on Artificial Intelligence Technology

3.1 Bad Debt Risk Rating Data Set

From 2020 to 2022, a total of 200 companies will be randomly selected from 100 listed companies and 100 unlisted companies according to the ST list. This paper focuses on the T-3 data of these companies. The sample data is divided into two parts, one for training and the other for testing. In this paper, 50 st companies and 50 non-ST companies were randomly selected as test samples.

3.2 Accounts Receivable Risk Assessment Index System

(1) Horizontal analysis index

Obtain the company's annual or monthly accounts receivable from the company's loaded transaction history. Through comparative analysis, it can reflect the change of the company's marketing strategy and operating performance.

(2) Longitudinal analysis index

Take value data from the company's loaded financial statements and measure these values in terms of accounts receivable of the company's current assets as a percentage of revenue and accounts receivable.

3.3 Normalization Processing

In this paper, the maximum and minimum interval mapping methods commonly used in neural network models are used to complete the standardization process. The number of nodes in the hidden layer of BP network will affect the results of training and testing. But there is no way to calculate the ideal number of nodes accurately. The general formula is as follows:

$$M = \sqrt{n+m} + 10 \tag{4}$$

The maximum number of hidden layer nodes is M, and the number of input and output nodes is n and m respectively. After calculation, it is found that the optimal number of hidden layer nodes is 3.

3.4 Performance Testing

A system performance test is a test of the completion time of each business function in the system and the machine resources that the system uses to record whether the running time of each function point in the system is within the standard time range. System performance testing typically requires data and metrics, CPU utilization and utilization, and memory consumption for each business operation as standard statistical analysis.

Based on the premise that the script data is basically the same, the paper compares the SAP standard financial system and the accounting receiving-accounts management optimization system, proving that the accounting receiving-accounts management optimization system can meet the characteristics of stability and efficiency.

4 The Analysis and Research of the Accounting Receivables Management Optimization System Based on Artificial Intelligence Technology

4.1 Implementation of Bad Debt Risk Prediction Based on BP Neural Network

In this paper, the hidden layer and output layer are established in the modeling, and a hidden layer containing three nodes is established, representing three default risk prediction indicators. The number of nodes in the output layer is set to 1, and the rounding value of the output node is used as the BP fitting enterprise credit classification. In this paper, the bad debt risk prediction model based on BP neural network is placed in a cycle, in which the maximum number of hidden nodes is set as the maximum number of cycles. At the end of the cycle, the optimal number of hidden nodes and the optimal precision are selected, and the bad debt risk prediction model based on BP neural network is determined. All data was input into the trained model to test its accuracy, as shown in Fig. 3.

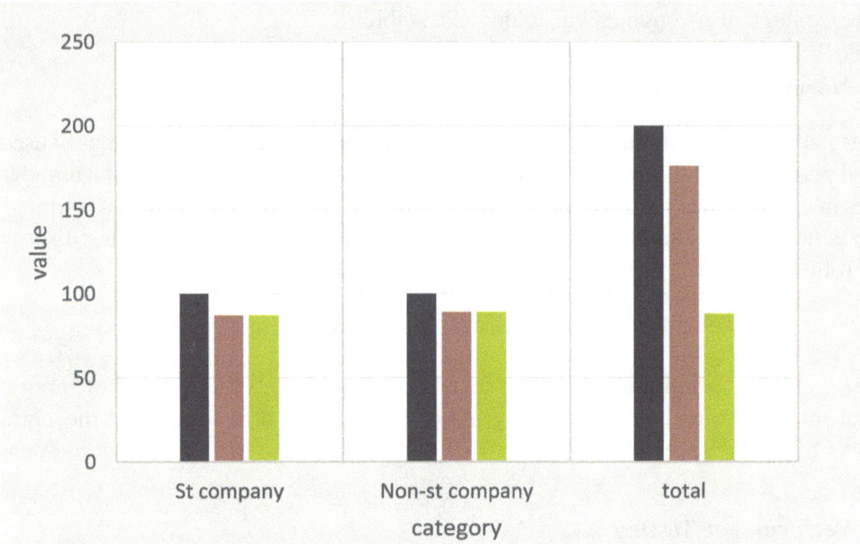

Fig. 3. Model inspection accuracy

This model has some practicability in predicting bad debt risk. With t-3 years of data, it is possible to predict whether a company will have a crisis in t years. The possibility of crisis leads to increased accounts receivable risk, business personnel can adopt a more powerful collection management approach.

4.2 System Inclusion Test

(1) The system can run continuously and stably for 72 h without interruption. The daily CPU utilization is less than 20%, and the busy time is less than 60%; The average daily memory usage is below 50%, as shown in Fig. 4.
(2) Within 50 days after the launch of the system, the average non-failure rate of the interface program of the accounting receipting and accounts management optimization system and TT fund system reached more than 99% s.
(3) All background operations related to the accounting Collection and Accounts Management optimization system, the TT fund system and the fund return report were completed before 6 am, and the data was successfully updated in the database so that the fund return report showed accurate and up-to-date data of more than 96%.

(4) It takes less time to run the scripts of the accounting collection and accounts management optimization system than the scripts of the SAP standard financial system, and more scripts can be run. Taken together, the conclusion is that the performance tests pass. As shown in Table 1.

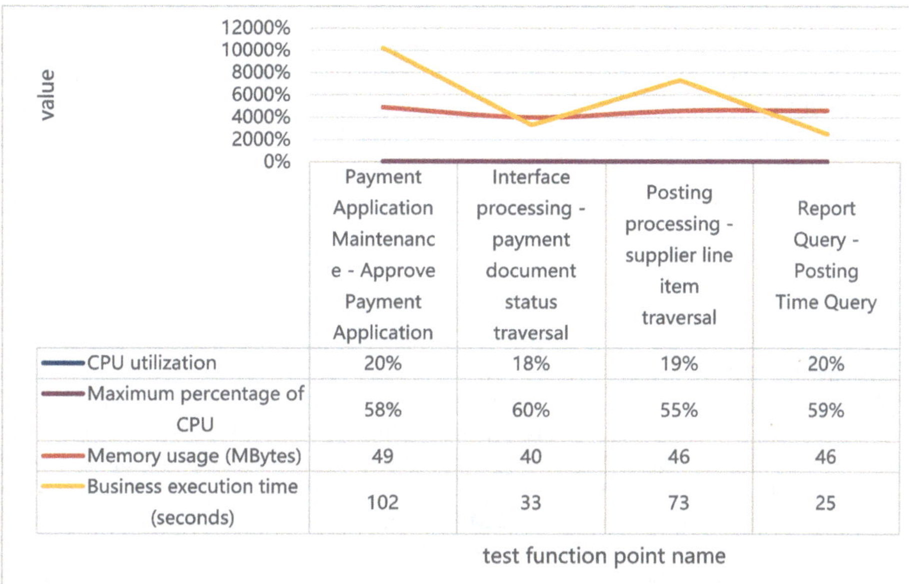

Fig. 4. Performance indicators of accounting accounts receivable management optimization system

Table 1. Comparison between the accounting accounts receivable management optimization system and SAP standard financial system

Test function point name	Accounting accounts receivable management optimization system	SAP standard financial system
Complete test script data	38	30
Test script completion time	10 min	About 25 min
Issue test report	Automatic generation	Manually generated

5 Conclusions

The digital economy is penetrating every field with visible speed. Digital management is a way for enterprises to cope with these situations. It will carry on technological innovation to traditional management mode and continue to produce a variety of different digital management methods. This paper attempts to introduce BP neural network algorithm into the realization of bad debt risk prediction of group accounting receivables management optimization system, strengthen the supervision and management of accounts receivable and audit early warning, improve the quality of internal audit, there are still many shortcomings in this study, but hope that this paper is helpful to enterprise audit and accounts receivable alarm, provide new ideas for the development of internal audit.

References

1. Goltsos, T.E., Ponte, B., Wang, S., Liu, Y., Naim, M.M., Syntetos, A.A.: The boomerang returns? Accounting for the impact of uncertainties on the dynamics of remanufacturing systems. Int. J. Prod. Res. **57**(23), 7361–7394 (2019)
2. Zhang, M.C., Stone, D.N., Xie, H.: Text data sources in archival accounting research: insights and strategies for accounting systems' scholars. J. Inf. Syst. **33**(1), 145–180 (2019)
3. Machado, M.J., Gomes, J.: Accounting and the ERP systems: a case study. Int. J. Knowl. Based Organ. **8**(2), 32–41 (2018)
4. da Costa Trigueiros, E.V., Gaspar Alves, M.-C.: Accounting information system adoption in multi-site organisations. Int. J. Bus. Inf. Syst. **41**(2), 236–257 (2022)
5. Daraio, C., Di Leo, S., Scannapieco, M.: Accounting for quality in data integration systems: a completeness-aware integration approach. Scientometrics **127**(3), 1465–1490 (2022)
6. Hanzal, P., Chládek, P.: Mathematical modelling in accounting information systems. Int. J. Bus. Inf. Syst. **36**(3), 393–405 (2021)
7. Kareem, H.M., Aziz, K.A., Maelah, R., Yunus, Y.M., Alsheikh, A., Alsheikh, W.: The influence of accounting information systems, knowledge management capabilities, and innovation on organizational performance in Iraqi SMEs. Int. J. Knowl. Manag. **17**(2), 72–103 (2021)
8. Ibrahim, F., Ali, D.N.H., Besar, N.S.A.: Accounting information systems (AIS) in SMEs: towards an integrated framework. Int. J. Asian Bus. Inf. Manag. **11**(2), 51–67 (2020)
9. Alhatabat, Z.: The impact of ERP system's adoption on management accounting practices in the Jordanian manufacturing companies. Int. J. Bus. Inf. Syst. **33**(2), 267–284 (2020)
10. Woodside, J.M., Augustine, F.K., Chambers, V., Mendoza, M.: Integrative learning and interdisciplinary information systems curriculum development in accounting analytics. J. Inf. Syst. Educ. **31**(2), 147–156 (2020)
11. Alamin, A., Wilkin, C.L., Yeoh, W., Warren, M.J.: The impact of self-efficacy on accountants' behavioral intention to adopt and use accounting information systems. J. Inf. Syst. **34**(3), 31–46 (2020)
12. Andrey, O.: Fedorov: information system of payment accounting in the university. Russ. Digit. Libr. J. **21**(5), 450–460 (2018)

AI-Enabled Video Diagnosis and Assisted Walking with Exoskeleton for Parkinson's Disease

Cathy Tu(✉) and Chi Man Qian

Concordia International School Shanghai, Shanghai 201206, China
cathy2025014@concordiashanghai.org

Abstract. Parkinson's disease is a chronic and progressive disease resulting from the lack of substantia nigra, or a specific type of nerve cell in the brain's basal ganglia. The four major motor symptoms of Parkinson's are bradykinesia, rigidity, tremor, and postural or gait instability. This paper seeks to innovate an artificial intelligence (AI) and engineering solution that involves diagnosing different stages of Parkinson's disease and mediating the external effects of the symptoms on patients' quality of life. An AI-based approach incorporating video-processing using MediaPipe and pattern recognition with a Long Short-Term Memory (LSTM) neural network diagnoses Parkinson's patients when stationary. This method mainly focuses on identifying and measuring rest tremors, a typical symptom of Parkinson's disease. With an analysis of the patient's disease progression and specific external symptoms, we can assist and train the patient to mitigate the adverse effects of such symptoms.

Keywords: Parkinson's disease · artificial intelligence-enabled video analysis · exoskeleton

1 Introduction

Our technology's primary targets are Parkinson's patients. Parkinson's disease is "the second most common age-related neurodegenerative disease" [1]. One in every 500 individuals is affected by Parkinson's [2]. Although Parkinson's disease currently does not have a cure, it is imperative that diagnosis happens as early as possible to reduce complications that could shorten patients' lives.

The disease is caused by the lack of substantia nigra, or a specific type of nerve cell in the basal ganglia of the brain. The loss of these nerve cells results in abnormally low levels of dopamine, a chemical that regulates body movement. As a result, Parkinson's patients suffer four significant motor problems: bradykinesia (slow movement), rigidity, tremor, and postural or gait instability (balance problems) [3]. To diagnose Parkinson's, one must first identify bradykinesia in the patient, followed by one of the other three motor problems [4].

The Hoehn and Yahr scale is a scale to describe the progression of Parkinson's disease, splitting into five stages. We focus on a modified version of the scale initially published in 1967 [13]. The stages are as follows [5] (Table 1).

Y. Pei et al. (Eds.): IC 2024, LNEE 1214, pp. 145–154, 2024.
https://doi.org/10.1007/978-981-97-4193-9_16

Table 1. Modified Hoehn and Yahr Scale

Modified Hoehn and Yahr Scale	
Stage 1	Mild symptoms, affected unilateral movement
Stage 2	Affected unilateral and bilateral movement
Stage 3	Mild to moderate bilateral symptoms, mild postural instability
Stage 4	Severe symptoms, but possible independent standing and walking
Stage 5	Wheelchair-bound or bedridden unless assisted

In this work, we begin by examining stages 1, 2, and 3 of the modified Hoehn and Yahr. To dig deeper into the specifics of each of the stages, we investigate a second scale— the Movement Disorder Society-Sponsored Revision of the Unified Parkinson's Disease Rating Scale (MDS-UPDRS)— that details how Parkinson's affects distinct human features, including non-motor and motor problems [6].

The software portion of this study focuses on developing a model for video analysis of upper-body tremor. This is measured by the tremor scales of the UPDRS, including 3.15 Postural Tremor of the Hands, 3.16 Kinetic Tremor of the Hands, and 3.17 Rest Tremor Amplitude. The hardware focuses on creating a lower-body exoskeleton to detect postural instability and provide assistance accordingly. This is measured by the movement scales from the UPDRS, including 3.9 Arising from Chair and 3.14 Body Bradykinesia.

1.1 Related Works: Exoskeleton Suit

Several hip exoskeletons have been generated in the past to address the postural and gait instability of other diseases. An example of such an invention would be the robotic hip exoskeleton [7], as examined by Inseung Kang from the Georgia Institute of Technology.

The study's structure of the hip exoskeleton stretches from the user's shoulders to the bottom of their feet. The three critical attachment points include the orthotic thigh shell, pelvic band, and shoulder straps. The hip exoskeleton determines gait phases by calculating the time difference between the heel strikes. The FSR sensor readings collect the data, determining the pressure placed on the heels, and the myRIO stores the timestamps. Using the previous five stride times, the scientists calculated the average stride duration. Depending on the length and time duration of the stride, the device can exert a forward motion into the legs to reach the desired torque [8].

To customize a device for Parkinson's patients, we must inspect a few issues with the hip exoskeleton:

- The weight and mass of the backpack portion, which includes the shoulder straps and the onboard computer, may create a toll on Parkinson's patients' backs. Because postural instability is one of the unique struggles that Parkinson's patients must overcome in gait, it is crucial to lessen the weight of the design.
- Parkinson's patients may not only be unable to perform specific tasks but also subconsciously resist assistance due to their tremors. Additionally, there is rigidity in the posture, back, arm swing, and legs of Parkinson's patients, so if the assistance fixes

too many points, the Parkinson's patients' tremors combined with the entire lower limb assistance may create even more imbalance or increase the likelihood of falling.

2 Materials and Methods

2.1 Software: Addressing Upper-Body Tremors

To address tremors in Parkinson's patients, this paper seeks to diagnose Parkinson's patients through video analysis. This entire process consists of three major steps, including creating the model, training the model, and using the model [9].

2.1.1 Model Creation

First, we create the model. This process can be split into two main steps— video processing of patients and artificial intelligence (AI) analysis of processed videos. The two main methods used are MediaPipe for processing and a Long Short-Term Memory (LSTM) neural network for analysis. In the first step, MediaPipe solutions that identify certain joints and skeletons on the body are used to process videos of patients. We identify 21 locations on the hand [10].

The use of MediaPipe is crucial for processing because it helps our model ignore background noise that might affect the model during training. MediaPipe allows us to effectively isolate the hand so that our input can solely depend on its coordinates. Points 1 through 20 are recorded about the location of the bottom of the palm, which would be set as (0,0).

After the videos are processed, a recurrent neural network is used for analysis with PyTorch. This paper proposes to use a particular type of recurrent neural network called LSTM, as shown in Fig. 1. Because LSTMs have a more complex cell structure than traditional RNNs, they can regulate the information they learn and forget the various input sources. This is important because tremors can only be detected if there is a clear distinction between the coordinates of the hand across time [11].

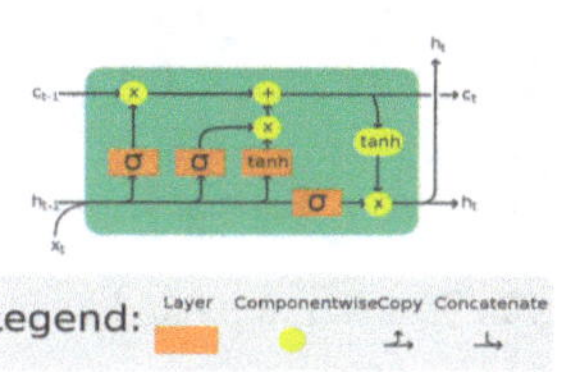

Fig. 1. Long Short-term Memory recurrent neural network.

2.1.2 Model Training

Once the model is established, the model is trained using self-created datasets. We label each video with either 0 or 1, representing normal or tremor, respectively. The self-created dataset includes 10 min of video footage for both normal and tremoring hands.

After MediaPipe processes each video by finding the coordinates of all critical points on the hand, the coordinates are fed into the LSTM recurrent neural network for training.

These videos have one hand inside the camera frame with the palm facing the camera. In the "normal" classification, the hand would be kept still to mimic healthy hands without the presence of tremors. However, there are many cases that the "tremor" classification will need to account for. There are videos where only one finger is tremoring, and there are also videos where the entire hand is tremoring [12].

During training, some tests are conducted on a separate dataset. This allows us to analyze the difference between the model's accuracy on the training dataset and that of the model on the test dataset. If there is a considerable deviation between the two accuracies, it indicates that overfitting occurs in the model. During training, the model achieves very high accuracy with the training dataset. Although the accuracy of the test dataset is lower, indicating the existence of overfitting, it still has a good model performance at 77 percent accuracy.

With the Unified Parkinson's Disease Rating Scale (UPDRS), three scales are used for the software portion— 3.15 Postural Tremor of the Hands, 3.16 Kinetic Tremor of the Hands, and 3.17 Rest Tremor Amplitude. The difference between the UPDRS and the Hoehn and Yahr scale is that the UPDRS provides numerical differences between the stages. Hence, a way of determining amplitude is needed to incorporate these scales into current models [13].

To calculate amplitude, we use the Tracker app, a video analysis and modeling tool. Tracker follows one point in a video and collects data on its coordinates in every frame. The amplitude of a tremoring hand can be calculated by the difference in the crests and troughs. Tracker collects the dataset fed into the AI processing and classification model, as shown in Fig. 2.

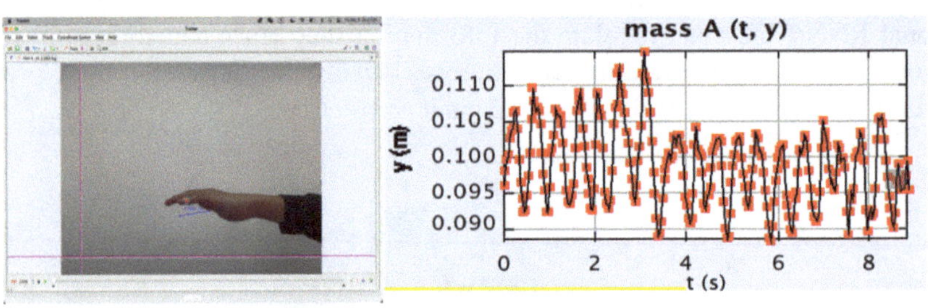

Fig. 2. Tracker app hand location.

In the UPDRS dataset, there are 10 min of data footage for the first three classifications on the UPDRS: normal, slight, and mild. This allows the model to directly produce the output, which would be the progression classification for the patients, as shown in Fig. 3.

The implementation of the UPDRS achieves results similar to those of the previous tests. Due to the overfitting problem found in the model results, we complete another set of experiments. An L1 Loss function is added to the model and tested. The purpose of an L1 Loss function is to improve model accuracy by minimizing the absolute differences

weight_delay	train_acc	test_acc
>0.05	0.366	0.351
e-2	0.817	0.627
e-3	0.960	0.771
有l1_loss	0.963	0.753
无l1_loss	0.961	0.773

Fig. 3. Variable performance.

between the estimated values and the existing target values. In addition, variables such as weight decay are changed and tested. These tests ultimately help decide the final variable values for the model.

2.1.3 Model Usage

The final aspect of the software portion is drawing conclusions and classifying the patient on the rating scales after receiving results from the model.

For the Hoehn and Yahr scale, the identification of tremors is done twice, once for each hand. Using the results from both tests, classification can be completed for the patient. To do so, a check for a bilateral nature is also conducted. If only one hand is identified to have a tremor, the patient would be in stage 1 of the Hoehn and Yahr scale. On the other hand, if both hands are identified to have tremors, the patient would be in stage 2 of the Hoehn and Yahr scale. However, for the UPDRS, due to the modified training, the model directly outputs the classification of each patient based on the chosen scales.

2.2 Hardware: Addressing Lower-Body Postural and Gait Instability

To address Parkinson's postural and gait instability, we propose the Diagnose Assist Suit (DASuit). In detail, we model a lower-body exoskeleton. The back structure of the exoskeleton is attached on both sides of the hips to stabilize the exoskeleton, and the adjustable leg structures and clasps run along the sides of the thighs. The uniqueness of the DASuit lies in the combination of different functions and customization for Parkinson's patients:

1. Diagnose: The MPU-6050 accelerometer and gyroscope can help determine the angles, speeds, and directions in which the patient's legs are moving or turning. Because a sensor is attached to each thigh's side, we can detect the distinction between affected unilateral and bilateral movement. The differences determine the stages in the Hoehn and Yahr scale and the UPDRS.
2. Assist: The stepper motors are attached to the back structure to provide gait assistance when Parkinson's patients experience tremors, rigidity, or freezing gait. The MPU-6050 collects data on the angle of the thigh and the speed at which the thigh reaches the angle. Simultaneously, the stepper motor responds and covers the portion of the angle that the patient's thigh cannot reach due to rigidity.

2.2.1 Structural Design

After examining recent works and past exoskeleton prototypes, we aim to simplify and lessen the weight of the physical structure as much as possible without risking its stability or reducing its effectiveness.

Due to many stage 1 or 2 Parkinson's patients' struggles with stooped posture or imbalance, the structure is only a lower limb exoskeleton that spans from the lower abdomen to the bottom of the thighs right above the knees. Instead of placing the weight on the patient's shoulders, the backpack containing the mechanic portion is strapped to the abdomen.

The physical structure of the exoskeleton can be split into the back brace, the torso frame, and the thigh structures.

2.2.2 Mechanical Design

The mechanical design contributes to the pushes and pulls performed in the assistance stage. The majority of the mechanical design lies on the back of the device, or specifically, the torso frame. This prevents the tangling of wires or breaking of any components while the patient is in movement.

Figure 4 provides a description of how the stepper motors are controlled on the back of the torso frame:

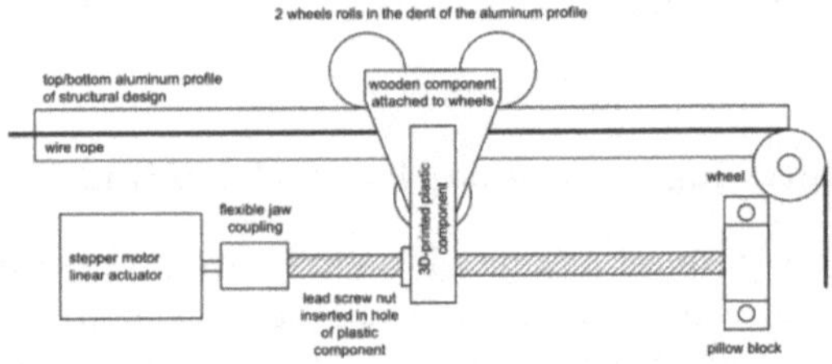

Fig. 4. Mechanical design blueprint for one leg.

2.2.3 Program Design

The program design is focused on the outer sides of each knee. The MPU-6050 accelerometer and gyroscope can help determine the angles, speeds, and directions in which the patient's legs are moving or turning. The data collected helps diagnose the patient based on the stages of the different scales while providing data on the distinction between the angles at which the two legs move at. This allows the step motors to provide feedback as they cover the portions of the torque that the Parkinson's patient is unable to reach. At the thigh is an MPU-6050 that links to the Arduino Nano program. This is replicated on the other thigh as well.

After completing prototype 1 of the structural design, we quickly come across an obstacle— the MPU-6050 assumes that the current location is always 0° when the program begins. This may work out for the diagnosis portion, but in the assistance portion, it would be more efficient to be able to measure the real-world angle of the thigh rather than the angle of the thigh in comparison to that at its initial position.

As a result, we update the system by replacing the MPU-6050 with a potentiometer. The potentiometer measures displacement in different directions in the real world, so it eliminates the need to calculate the needed torque assistance in comparison to the initial position of the patient's legs. Figure 5 indicates the changes we made:

However, because the swing ranges are still necessary for the diagnosis portion, we pair the DASuit with a simplified version of itself used specifically for diagnosis— the DSuit. The DSuit is made up of the MPU-6050s attached to the side of each knee, which eliminates the exoskeleton. While the DSuit is used for diagnosis, the DASuit is used for assistance.

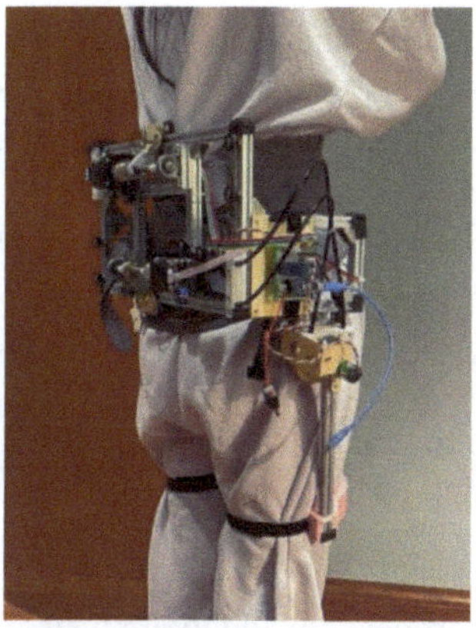

Fig. 5. Complete prototype 2; potentiometers replace the MPU-6050s.

The Arduino Nano programs are then created for diagnosis and assistance. The programs are updated to accommodate the UPDRS scales 3.14 Body Bradykinesia and

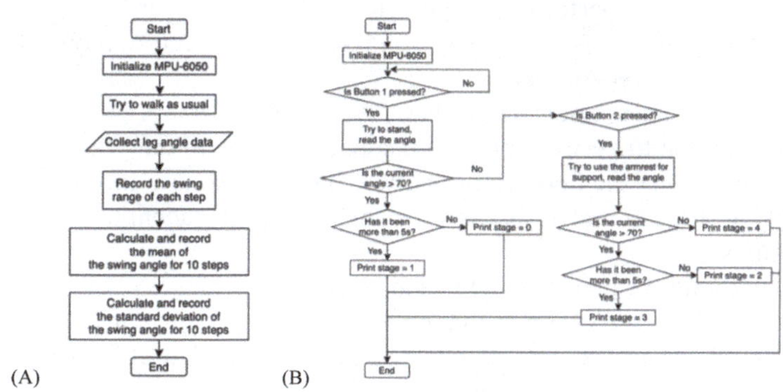

Fig. 6. (A) Flow chart of Arduino programming for 3.14 Body Bradykinesia. (B) Flow chart of Arduino programming for 3.9 Arising from Chair.

3.9 Arising from Chair. Figure 6 provides details about the diagnosis mechanisms of the Arduino coding for the DSuit.

With algorithmic processing, such as calculations of means and standard deviations, distinctions between the stages of the UPDRS can be found.

We finalize the coding and program assistance for the stepper motors. Because customization is a key element of our project, we integrate the programming into a user-friendly mobile phone application called Blinker. In Blinker, the patient or a family member can control the torque, speed, and swing range of the patient with a Bluetooth connection to the device.

3 Experimentation

In terms of lower-body movement, the Hoehn and Yahr scale and the UPDRS both mark distinctions between the stages by highlighting gait speed (meters per second), gait cadence (steps per minute), feet shuffling patterns, imbalance, and freezing as the patient walks.

Our experimentation involves a real Parkinson's patient, J. Pan, who is 73 years old.

3.1 Experiment Procedures

The following are the steps to the diagnosis experiment.

3.1.1 Software Diagnosis

A video is taken of the entire process for video analysis with our software.

a. Instruct the patient to sit in a chair.
b. Instruct the patient to perform finger tapping: "Hold out your left hand. Repeatedly tap your left thumb and index finger as big as you can and as fast as you can." Allow the patient to complete at least 10 repetitions.
c. Repeat step b with the patient's right thumb and index finger.
d. Instruct the patient to perform fist open close: "Hold out your left fist. Repeatedly open and close your left first all the way and as fast as you can." Allow the patient to complete at least 10 repetitions.
e. Repeat step d with the patient's right fist.
f. Instruct the patient to perform pronation/supination: "Hold out your open left hand with the palm facing the floor. Repeatedly rotate your palm to face up and down as completely as you can and as fast as you can." Allow the patient to complete at least 10 repetitions.
g. Repeat step f with the patient's right hand.
h. Instruct the patient to perform toe-tapping: "Repeatedly tap your left foot as big as you can and as fast as you can." Allow the patient to complete at least 10 repetitions.
i. Repeat step g with the patient's right foot.

3.1.2 Hardware Diagnosis

Instruct the patient to sit in a chair with armrests.

a. Walk through the instructions in the Arduino programming to test for instability on 3.9 Arising from Chair from the UPDRS.
b. Walk through the instructions in the Arduino programming to test for instability on 3.14 Body Bradykinesia from the UPDRS.

3.1.3 Hardware Assistance

Before providing any assistance, we test for the swing range of the patient's torque.

The procedures include the majority of the essential elements in the gait and balance portion of the traditional Parkinson's disease exam [8]. Throughout the procedures, the DSuit is linked to the patient to record. As the patient walks, look for slight imbalances while turning, imbalanced or reduced step lengths, lack of heel strikes (shuffling steps), and lack of or reduced arm swing [14].

a. Attach the DSuit to the patient. Ensure that the MPU-6050s are firmly attached to the outer side of each knee. This reduces the amount of unnecessary shaking and provides the most accurate data.
b. Ensure that the experiment takes place in at least 7 m of open hallway or space.
c. Open the serial monitor. Ask the patient to stay still for approximately 10 s until the MPU-6050s begin to collect data.
d. Ask the patient to walk back and forth approximately 7 m before each turn. Retrieve data from at least three sets of complete 7-m walks.

Then, the patient is asked to put on the DASuit, which is connected to Blinker by Bluetooth. With data collected from the DSuit, suitable torque assistance is provided. The patient or a family member can customize the torque, speed, and swing range of the feet by moving the toggles left and right [9].

3.2 Experiment Results

The software indicates that the patient lands on stage 0 in the tremor scales (3.15, 3.16, and 3.17), as shown in Fig. 7.

(A) 

(B) 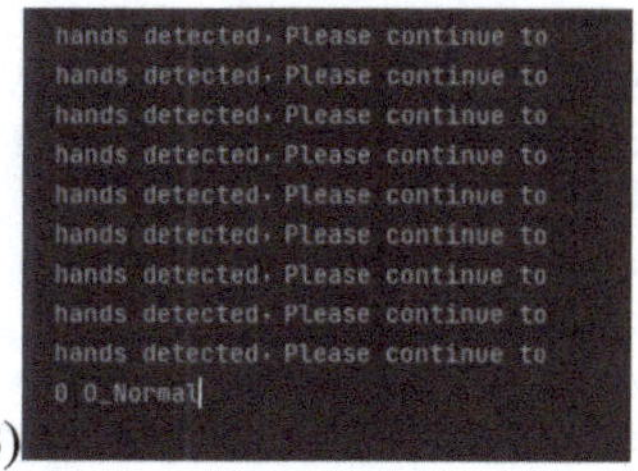

Fig. 7. (A) Video analysis. (B) Patient classification.

The hardware indicates the patient lands on stage 0 when standing up from a chair (3.9) and stage 2 in terms of body bradykinesia (3.14) [10].

4 Conclusion

An integrated solution for Parkinson's disease can be achieved. Our project indicates that the healthcare industry can explore more engineering solutions for mitigating external symptoms and improving patients' quality of life, as such goals have either been majorly neglected or left unmet. In addition, the healthcare industry lacks specific distinctions between different stages of diagnosis, but engineering solutions and statistical analysis can help fill this void [15].

However, such solutions must be grounded on careful analysis of the UPDRS to provide a numerical foundation for algorithm processing. Methods of observing symptoms have to be selected carefully for the most effective engineering results. Certain action-related analyses can be processed through AI, while some forms of postural instability can be better diagnosed through the DASuit. Ultimately, such conclusions should come together to provide a holistic approach to diagnosis and assistance [16].

This research intends to shed new light on the cures for Parkinson's disease, providing a practical, AI and engineering-based resolution to the declining quality of life of Parkinson's patients. We aim to provide patients with access to information about and increased control over their well-being.

References

1. Fletcher, J.: What you should know about bradykinesia. Medical News Today (2021)
2. Gilbert, R.: Lower back pain and Parkinson's disease. American Parkinson Disease Association (2021)
3. Goetz, C.G., et al.: Movement disorder society task force report on the Hoehn and Yahr staging scale. Mov. Disord. **19**(9), 1020–1028 (2004). https://doi.org/10.1002/mds.20213
4. Heusinkveld, L.E., Hacker, M.L., Turchan, M., Davis, T.L., Charles, D.: Impact of tremor on patients with early stage Parkinson's disease. Front. Neurol. **9**, 628 (2018). https://doi.org/10.3389/fneur.2018.00628
5. Kang, H.H., Young, A.: The effect of hip assistance levels of human energetic cost using robotic hip exoskeletons. IEEE Robot. Autom. Lett. **4**(2), 430–437 (2019). https://doi.org/10.1109/LRA.2019.2890896
6. Louis, E.D., Machado, D.G.: Tremor-related quality of life: a comparison of essential tremor vs. Parkinson's disease patients. Parkinsonism Relat. Disord. **21**(7), 729–735 (2015). https://doi.org/10.1016/j.parkreldis.2015.04.019
7. Nelson, A.: Parkinson's tremors. WebMD (2018)
8. Olanow, W., Rascol, O.: MDS-UPDRS. International Parkinson and Movement Disorder Society (2008)
9. Olanow, C., Klein, C., Schapira, A.V.: Parkinson's disease. Harrison's Principles of Internal Medicine, 20e (2018)
10. Overview - Parkinson's disease. National Health Service (2022)
11. Parkinson's symptoms to look for. American Parkinson Disease Association (2022)
12. Poston, K.: Approach to the exam for Parkinson's disease. Stanford Medicine 25 (2018)
13. Stephenson, R.O.: Modified Hoehn and Yahr scale. Medscape (2021)
14. Stooped posture. Parkinson's Foundation (2020)
15. Tremor in Parkinson's. American Parkinson Disease Association (2022)
16. Tremor. Parkinson's Foundation (2022)

From Wordle to Insights: Using Tailored Clustering and CART to Forecast Difficulty Levels

Xinyi Xu(✉) and Jinqi Huang

Shanghai University, Shanghai, China
XuXinyics@yeah.net

Abstract. Wordle, a popular daily puzzle in the New York Times, has garnered significant attention with its unique challenge. Players must decipher a five-letter word in six attempts or less, receiving feedback after each guess. Rooted in information theory and pragmatics, Wordle offers a valuable platform for exploration. Our research proposes an innovative approach consisting of two components. The first utilizes K-means clustering with tailored parameters: the forward difficulty evaluation index (Dforward) and the reverse difficulty evaluation index (Dreverse). Dforward is calculated by weighting factors like Nrepeat (normalized count of repeated letters), word frequency (F), and the number of vowels (Nvowel) in the word. Dreverse is obtained by normalizing the predicted number of successful guesses. The second component applies a CART decision tree model with difficulty level labels, predicting future solution complexities. By analyzing linguistic features in five-letter words, our study constructs a model that accurately determines word difficulty using statistical knowledge and machine learning techniques. Additionally, this model facilitates related linguistic analyses.

Keywords: K-Means · CART · Bidirectional RNN · Long Sequence Time Prediction

1 Introduction

Wordle, a popular daily puzzle game offered by The New York Times, engages players in the task of discerning a five-letter common English word within a limited number of attempts. Following each submission, feedback is provided to the player, facilitating a dynamic understanding of their progress. This feedback is represented through color-coded tiles: a yellow tile denoting the presence of a correct letter albeit in an incorrect position, a green tile indicating an accurate letter placement, and a gray tile signifying an incorrect letter choice. Such feedback serves as an evaluative tool, enabling players to gauge their proximity to the correct solution. In the more demanding mode, players are exclusively permitted to employ letters which match those included in the target word for subsequent attempts. Triumphing with fewer tries poses a heightened challenge but is rewarded with a higher score. Although players typically converge upon the correct word after four attempts, accomplishing this feat within a mere three attempts proves significantly arduous.

Y. Pei et al. (Eds.): IC 2024, LNEE 1214, pp. 155–165, 2024.
https://doi.org/10.1007/978-981-97-4193-9_17

From an information theory perspective, the Wordle puzzle can be conceptualized as a problem wherein the answer, a prevalent English word, remains undisclosed. Each guess serves to disclose vital insights pertaining to the presence or absence of specific letters in their respective positions. Assuming equiprobability among high-frequency words as potential answers, these words can be envisioned as comprising the probability space. Consequently, each guess offers a quantifiable unit of information that may be subjected to formal calculation [1].

$$\mathrm{I} = -\log_2(\mathrm{p}) \tag{1}$$

$$E[I] = \sum\nolimits_x (p(x) \cdot \log_2(1/p(x))) \tag{2}$$

In practice, players do not necessarily follow the optimal strategy as it may detract from the enjoyment of the game. Nevertheless, the player's problem-solving process fundamentally includes the principles of information theory. According to the data provided, the average number of successful guesses for the correct answer is 4.194, which is similar to the result of the optimal strategy. In hard mode, the range of letters available to players is restricted, which lowers the expected information and increases the level of difficulty. This is the underlying principle of solving the Wordle puzzle.

2 Data Preprocessing

High-quality data serves as the foundation for effective data analysis and mining. The presence of disordered and erroneous "dirty data" can significantly compromise the validity, repeatability, and generalizability of model predictions. Thus, reasonable data preprocessing is imperative to ensure the quality of existing data. Upon reviewing the raw data, the following issues were identified:

Certain instances of the "word of the day" were not comprised of exactly five letters.
The number of reported scores may be an order of magnitude greater or less than adjacent data.
The sum of the percentages of users who guessed the word in one through six tries or were unable to solve the puzzle did not equate to 100%.

To address these issues, the following strategies have been implemented:

If the "word of the day" consists of four or six letters and the correct five-letter word cannot be identified, the data will be deleted. However, if the "word of the day" comprises five letters and contains only special characters such as line breaks, the special characters will be removed while retaining the original data.
If the number of reported scores differs significantly from adjacent data (i.e., a difference of an order of magnitude or greater), the data will be deleted, and interpolation replacement will be performed using the average of adjacent data.

Given that the data are sourced from Twitter, it is possible that errors may exist, resulting in a sum of percentages that does not equal 100%. To account for this, an error rate of 2% has been established for this type of question. As such, the sum of the percentages will be considered reasonable if it falls within the range of 98 to 102, and any data beyond this error range will be deleted [2].

3 K-means-Based Difficulty Classification Using Tailored Parameters

The aim of this model is to quantify the complexity of problem-solving. The properties used to assess solution difficulty can be roughly categorized into two parts: prior and posterior properties [3]. Prior properties consist of the number of repeated letters in a word (Nrepeat), the frequency of word usage (F), and the number of vowel letters in a word (N_{vowel}). On the other hand, the posterior property refers to the anticipated number of guesses required for a user to arrive at the correct answer. Nevertheless, these properties still require certain calculations to transform them into quantitative indicators that can be utilized for subsequent model training [4].

3.1 Preprocessing of Prior Attributes

The selection of attributes for the training model relied on key factors. Repeated letters in the solution word were identified as a significant contributor to game difficulty. Their multiple occurrences, even if correctly guessed, add complexity. Additionally, the count of repeated letters greatly affects the game difficulty as players consider different letter options. Word frequency also plays a crucial role in the player's likelihood of guessing correctly. Commonly used words are more likely to be chosen. Lastly, vowel count in a word was considered based on the tendency to include two vowels when guessing a five-letter word, highlighting its impact on the guessing process.

These considerations formed the foundation for selecting the specific attributes used in the training model. For the prior attributes, we define a parameter $D_{forward}$ to represent the prior difficulty level. Before calculating $D_{forward}$, we perform standardization on the three attributes. Since the magnitude of word frequency is much smaller compared to the other two attributes, we use Z-Score normalization to obtain the processed attributes. The calculation method for Z-Score normalization is as follows:

$$z_i = \frac{x_i - \mu}{\sigma_i} \tag{3}$$

$$u = \frac{1}{n}\sum\nolimits_{i=1}^{n} x_i \tag{4}$$

$$\sigma_i = \sqrt{\frac{\sum_{i=1}^{n}(x_i - \mu)^2}{n}} \tag{5}$$

Define the weighted sum of the $D_{forward}$ calculation method as an attribute of Nrepeat′, N_{vowel}′, F′.

$$S_{forward} = \omega_r \cdot N'_{repeat} + \omega_f \cdot F' + \omega_v \cdot N'_{vowel} \tag{6}$$

$$\omega_r + \omega_f + \omega_v = 1 \tag{7}$$

The determination of weights necessitates the consideration of two factors: the degree of cluster separability and the mitigation of overfitting [5].

3.2 Posterior Attribute Processing

The treatment of posterior attributes includes the parameter $D_{reverse}$, which quantifies the complexity of inferring posterior probabilities on a scale of 0 to 1. $D_{reverse}$ is determined solely based on the expected number of attempts, E_{tries}, and is then standardized using the Z-score method.

3.3 Employing K-Means for Categorization of Solution Difficulty

The K-Means algorithm is a widely used unsupervised and partition-based clustering algorithm in data analysis. It employs the Euclidean distance as a metric to assess the similarity between data objects. The algorithm aims to minimize the distance between data objects within the same cluster, indicating their higher similarity. To begin the algorithm, the number of clusters (k) and their initial centers need to be specified. By iteratively updating the cluster centers based on the similarity between data objects and cluster centers, and subsequently minimizing the sum of squared errors (SSE), the algorithm converges to a final categorization of solution difficulty.

In this application, the samples are represented as real numbers in a two-dimensional plane, specifically in terms of $D_{forward}$ and $D_{reverse}$. Thus, the term "distance" refers to the Euclidean distance, and the calculation formula for the Euclidean distance in the k-means algorithm can be expressed as follows [6]:

$$d_{K-\text{ Means}}(x, C_i) = \sqrt{\sum\nolimits_{j=1}^{n} \left(x_j - C_{ij}\right)^2} \tag{8}$$

A comparative analysis of cluster numbers, depicted in Fig. 1, was conducted for two different weight combinations using the elbow method. [7] The identification of the inflection point where the curve flattens was the focus. The analysis revealed favorable outcomes for both 3 and 4 clusters. However, to enhance classification accuracy and align with practical difficulty levels ("easy," "relatively easy," "relatively difficult," and "difficult"), a cluster number of 4 was selected.

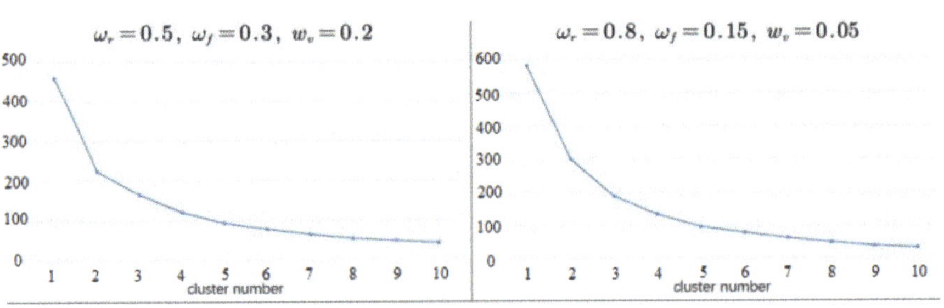

Fig. 1. Cluster number comparison diagram based on two types of weights.

Figure 2 showcases the performance evaluation of K-Means clustering results obtained by varying weights with $D_{forward}$. The analysis indicates a progressive improvement in clustering performance as the weight increases. However, it is important to acknowledge the potential risk of overfitting. By mapping the clusters to different levels of difficulty, with the x-axis representing increasing difficulty from left to right and the y-axis representing increasing difficulty from bottom to top, each of the four clusters corresponds to a distinct difficulty level. Specifically, "relatively difficult" is denoted by 1, "relatively easy" by 2, "difficult" by 3, and "easy" by 4.

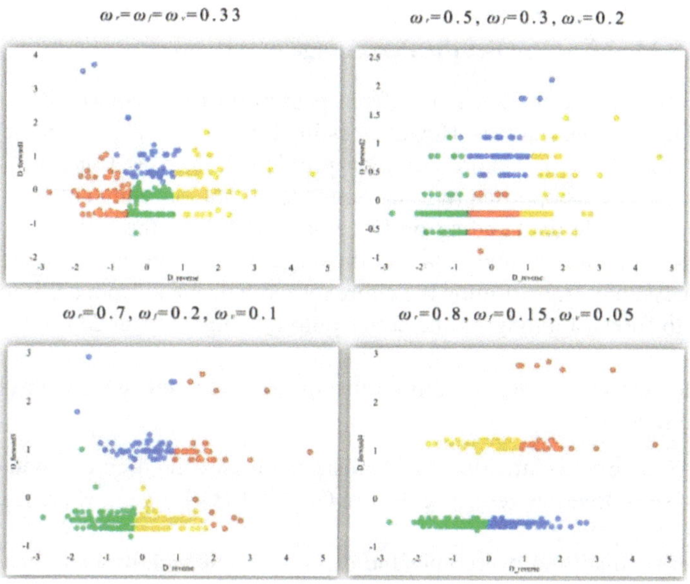

Fig. 2. The clustering results with different weights of Dforward

4 CART-Based Difficulty Classification Model

In order to classify future dates with no provided data based on their difficulty level, which corresponds to one of the aforementioned four categories, we will employ a comprehensive approach. The attributes considered in this classification process include the number of repeated letters, the number of vowels, and the frequency of word usage. Additionally, we aim to derive the expected number of attempts needed to guess the word as a posterior attribute. To achieve this, we employ a bidirectional RNN model-based percentage prediction approach (see Fig. 3), which enables us to calculate the percentage of attempts required to guess the solution. By applying a weighted summing technique, we obtain the value of $D_{reverse}$.

Once the category label is established, we employ a CART decision tree as the classifier of choice to further refine the classification process.

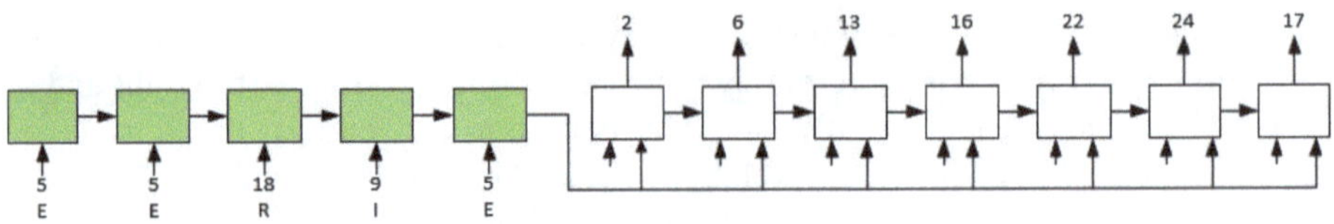

Fig. 3. Reported results of EERIE

4.1 Bi-RNN Model to Predict the Percentage of Attempts

Predicting future event characteristics from past event time patterns is a well-known challenge in time series analysis. However, estimating the number of reported outcomes on March 1, 2023 poses a more complex problem known as Long Sequence Time Forecasting (LSTF), as opposed to Long Sequence Input Learning (LSIL) [8].

With our available data ending on December 31, 2022, there is a significant two-month gap between the current data and the target prediction. Without supplementary data, this necessitates transforming the entire research problem from LSIL to LSTF. Our approach is to forecast a longer time-series output using an extended time-series input [9].

To tackle this challenge, we utilize the Informer model, which introduces three optimization methods:

1. The ProbSparse Self-Attention Mechanism addresses the high computational complexity of self-attention, reducing it from O(L^2) to O(LlogL), where L is the length of the input sequence.
2. Self-attention distilling overcomes the memory bottleneck problem caused by long input sequences, enhancing computational efficiency without sacrificing accuracy.
3. The generative decoder in the Informer model handles the low decoding speed and inefficiency of dealing with long output sequences. This enables direct prediction of all future data points, improving forecasting efficiency [10].

4.2 Cart Model

CART is a learning method that provides a conditional probability distribution of a random variable Y given an input random variable X. It assumes that the decision tree is a binary tree. Such a decision tree is equivalent to recursively dividing each feature, thereby dividing the input space or feature space into a finite number of units, and determining the predicted probability distribution on these units, i.e., the conditional probability distribution of the output given the input.

For classification problems, the classification tree uses the Gini index to select the optimal feature and determine the optimal binary split point for that feature. Based on the training data, starting from the root node, recursively perform the following operation on each node to construct a binary decision tree:

1. Given a node's training dataset D, the Gini index is computed for the existing features in the dataset. Using the following formula can help you find better nodes:

$$GINI = \sum_{i=0}^{m} p_i(1 - p_i) = \sum_{i=0}^{m} p_i - \sum_{i=0}^{m} p_m^2 = 1 - \sum_{i=0}^{m} p_m^2 \tag{9}$$

2. From among all possible features A and their corresponding split points a, identify the feature and split point with the smallest Gini index as the optimal feature and optimal split point. Create two child nodes from the current node and partition the training data set into these two nodes according to the optimal feature.
3. Recursively apply steps 1. and 2. to each child node until the stopping conditions are met [11].

5 Prediction Results and Analysis

5.1 Results on Various Clustering Methods

We trained our model using various clustering results obtained from four different weights. The weights of each attribute are depicted in Fig. 4, while the weights of each model are presented in Table 1.

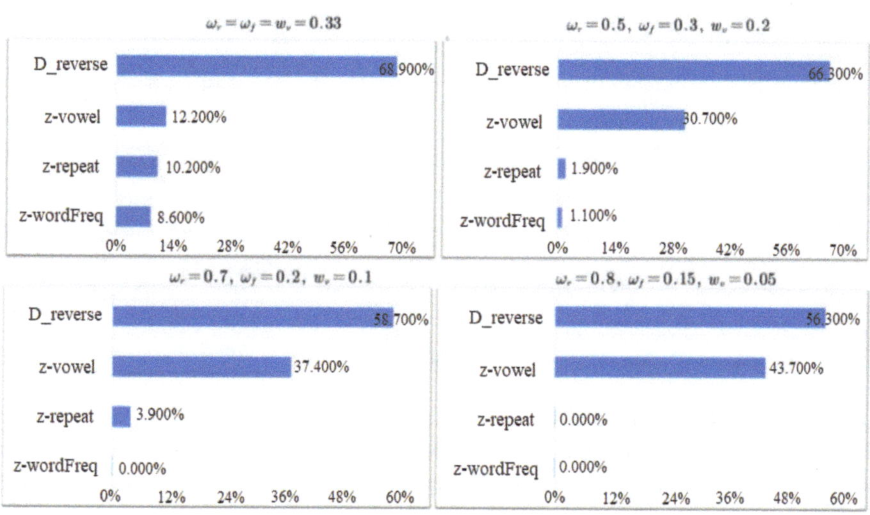

Fig. 4. Illustrates the weights of each attribute in the decision tree under the influence of different weights.

In practice, a high precision often comes at the expense of low recall, and vice versa. If both precision and recall are equally important, the F1 score can be used as the evaluation metric.

Irrespective of the influence of the weights on the clustering results, the decision tree assigns the highest priority to the posterior difficulty indicator $D_{reverse}$, followed by the count of repeated Table 1 Presents the score table for different decision trees letters. The effects of word frequency and the number of vowels on the results are negligible, with word frequency being marginally more influential than the number of vowels. Among the four weight distributions, the model generated by $\omega r = 0.8$, $\omega f = 0.15$, $\omega v = 0.05$ manifests clear evidence of overfitting, and the model produced by $\omega r = \omega f = \omega v = 0.33$ exhibits lower accuracy. Consequently, neither of these models are deemed acceptable.

Employing the aforementioned evaluation criteria, we contend that the model generated by $\omega r = 0.5$, $\omega f = 0.3$, $\omega v = 0.2$ outperforms the other two models. Therefore, for all subsequent predictions and evaluations, we will rely on this model.

Table 1. Presents the score table for different decision trees

	Accuracy	Recall	Precision	F1
	$\omega r = \omega f = \omega v = 0.33$			
Training Set	1	1	1	1
Test Set	0.943	0.943	0.951	0.942
	$\omega r = 0.5$, $\omega f = 0.3$, $\omega v = 0.2$			
Training Set	1	1	1	1
Test Set	0.991	0.991	0.991	0.991
	$\omega r = 0.7$, $\omega f = 0.2$, $\omega v = 0.1$			
Training Set	1	1	1	1
Test Set	0.981	0.981	0.982	0.981
	$\omega r = 0.8$, $\omega f = 0.15$, $\omega v = 0.05$			
Training Set	1	1	1	1
Test Set	1	1	1	1

5.2 Results and Analysis

While we acknowledge that dates may significantly influence the number of participants in the game, we do not believe they are related to the inherent properties of the solution vocabulary. Thus, when predicting the difficulty level of a solution on a future date, the classification model solely considers the words and disregards the dates. Nevertheless, dates indirectly affect the difficulty classification model by exerting an impact on the relevant percentages.

For the word "EERIE", it has three repeating letters and four vowels, with a usage frequency of 2.527218e−06, $N_{repeat} = 3$, $N_{vowel} = 4$, $F = 2.53e{-06}$. The data is placed in the master table and standardized using the Z-score method, resulting in $N'_{repeat} = 3.463178824$, $N'_{vowel} = 4.031568301$, $F' = -2.19e{-01}$. As for the calculation of $D_{reverse}$, in the previous section, we predicted that when the solution is "EERIE" on March 1, 2023, the user's percentage of guesses would be as follows: 1 try - 2%, 2 tries - 6%, 3 tries - 13%, 4 tries - 16%, 5 tries - 22%, 6 tries - 24%, and X - 17%. The calculation of E_{tries} results in 4.90, and after standardization, $D_{reverse}$ equals 1.82364597.

Finally, by inputting the aforementioned attributes into the decision tree, a classification label of 3, denoting "difficult," was obtained. This is in line with our prior knowledge that the model is most influenced by the attributes of and $D_{reverse}$, with the word EERIE having a high number of repeated letters and a particularly large E_{tries} value relative to most other solutions. On the scatter plot, this data point is located in the upper-left corner, well within the range of classification label 3.

5.3 Model Evaluation

Difficulty classification prediction models based on K-means were evaluated using three metrics: silhouette coefficient, Davies-Bouldin Index (DBI), and Calinski-Harabasz Index (CH) [12]. The silhouette coefficient ranges from −1 to 1, with higher scores indicating a stronger clustering effect, where samples within the same category are closer to each other and farther from samples in different categories. The DBI measures the ratio between the sum of intra-cluster distances and inter-cluster distances for any two clusters, with smaller values indicating better clustering effects. The CH score is calculated by determining the ratio of between-cluster dispersion to within-cluster tightness. Tightness is measured by summing the squared distances between each point and the cluster center, while dispersion is measured by summing the squared distances between cluster centers and the dataset center. Higher CH scores signify better clustering effects.

The scores of the three indicators were calculated for K-Means clustering models generated with four different weightings and recorded in Table 2. The fourth model, which exhibited overfitting, was excluded from consideration. While the third model outperformed the first model in all three indicators, we ultimately chose the second model due to its significant performance improvement and effectiveness in decision tree classification.

For the CART decision tree-based difficulty classification model, in Sect. 4.2, we demonstrated the accuracy, recall, precision, and F1 score. All four models generated from different weights achieved scores above 0.9 for these indicators. Notably, the second model we selected possessed very high scores, reaching 0.991, indicating excellent performance.

After a comprehensive analysis, we initially employed the K-Means algorithm to achieve relatively accurate difficulty classification. Subsequently, we obtained a CART-based difficulty classification model with excellent classification performance. Furthermore, during the testing phase with "EERIE" as the solution, we obtained results that aligned with our expectations. Therefore, we conclude that the accuracy of this model is ensured.

Table 2. Scores of Different Clustering Models

	Silhouette Coefficient	DBI	CH
$\omega r = \omega f = \omega v = 0.33$	0.367	0.884	257.659
$\omega r = 0.5, \omega f = 0.3, \omega v = 0.2$	0.407	0.832	327.336
$\omega r = 0.7, \omega f = 0.2, \omega v = 0.1$	0.481	0.718	337.085
$\omega r = 0.8, \omega f = 0.15, \omega v = 0.05$	0.521	0.699	390.023

6 Conclusion

We proposed a novel approach to predict the difficulty level of Wordle. To achieve accurate predictions on long sequence, the advanced forecasting algorithm Informer was employed and yielded excellent results. The use of both prior and posterior difficulty evaluation indicators, which mutually corroborated each other, resulted in a highly credible classification of the level of difficulty. A difficulty prediction model was trained using the decision tree based on Classification and Regression Tree (CART), with the application of multiple attributes, which produced superior model performance.

The solution has some room for further development. First, the advanced Informer model used for prediction has weak interpretability, and the difficulty mode proportion analysis relies on visualization methods as auxiliary analysis, lacking a certain mathematical theoretical basis. Second, the issue of model dependency still exists. The percentage prediction model relies on the previous model's accuracy. If the previous model's prediction result is not accurate enough, the sub-sequent difficulty classification prediction error will be larger.

Natural language processing was used in deep learning to extract information contained in letter combinations within words. This method proved to be more efficient than manual extraction. This gives us some inspiration that implementing better models can draw insights from natural language processing methods.

References

1. Brychkov, Y.A., Savischenko, N.V.: Application of hypergeometric functions of several variables in the mathematical theory of communication: evaluation of error probability in fading singlechannel system. Lobachevskii J. Math. **41**, 1971–1991 (2020)
2. Cano, A., Krawczyk, B.: Kappa updated ensemble for drifting data stream mining. Mach. Learn. **109**, 175–218 (2020)
3. Lin, C., Sun, D., Song, C.: Posterior propriety of an objective prior for generalized hierarchical normal linear models. Stat. Theor. Relat. Fields **6**(4), 309–326 (2022)
4. Saroj, Kavita: Review: study on simple k mean and modified K mean clustering technique. Int. J. Comput. Sci. Eng. Technol. **6**(7) 279281 (2016)
5. Han, J., Xu, J., Nie, F., Li, X.: Multi-view K-means clustering with adaptive sparse memberships and weight allocation. IEEE Trans. Knowl. Data Eng. **34**(2), 816–827 (2022)
6. Steuer, R., Kurths, J., Daub, C.O., et al.: The mutual information: detecting and evaluating dependencies between variables. Bioinformatics **18**(suppl_2), S231–S240 (2002) https://doi.org/10.1093/bioinformatics/18.suppl_2.s231
7. Bholowalia, P., Kumar, A.: EBK-means: a clustering technique based on elbow method and k-means in WSN. Int. J. Comput. Appl. **105**(09), 17–24 (2014)
8. Li, H., Shao, J., Liao, K., Tang, M.: Do simpler statistical methods perform better in multivariate long sequence time-series forecasting? In: Proceedings of the 31st ACM International Conference on Information & Knowledge Management, CIKM 2022, pp. 4168–4172. Association for Computing Machinery, New York (2022)
9. Yu, Y., Si, X., Hu, C., et al.: A review of recurrent neural networks: LSTM cells and network architectures. Neural Comput. **31**(7), 1235–1270 (2019)
10. Zhou, H., et al.: Informer: beyond efficient transformer for long sequence time-series forecasting. Proc. AAAI Conf. Artif. Intell. **35**(12), 11106–11115 (2021). https://doi.org/10.1609/aaai.v35i12.17325

11. Zhou, Z.: Machine Learning. Tsinghua University Press (2016)
12. Liu, Y., Li, Z., Xiong, H., Gao, X., Wu, J.: Understanding of internal clustering validation measures. 2010 IEEE International Conference on Data Mining, Sydney, NSW, Australia, pp. 911–916 (2010)

Design and Implementation of Financial Management System Based on Apriori Algorithm

Yue Zhao(✉)

College of Finance and Economics, Liaoning Vocational University of Technology, Jinzhou 121000, Liaoning, China
13904166807@163.com

Abstract. Financial management is very important for the healthy development of enterprises, but as the size and complexity of enterprises increase, the analysis and decision-making of financial data becomes more difficult. This research designs and implements a financial management system based on the Apriori algorithm. The article collects and organizes the company's financial data, and then uses the Apriori algorithm to mine association rules on the financial data to discover frequent item sets and association rules. Finally, the article visually displays the mining results and provides user-friendly query and analysis functions. The article conducts an in-depth test of the system through experiments. The results show that the minimum association rule time of this system is 0.2 s, and the system can complete the mining process and provide results faster.

Keywords: Financial Management System · System Design · Apriori Algorithm · Data Model

1 Introduction

Financial management is a relatively important task in corporate operations. As the global business environment continues to change and competition intensifies, companies need to manage and analyze financial data more accurately and efficiently to support decision-making and business development. However, financial data is usually large and complex, and traditional manual analysis methods can no longer meet the needs of modern enterprises. Therefore, it is of great significance to research and develop financial management systems based on advanced algorithms.

A large number of scholars have conducted research on financial management systems. Among them, Shi Yan used the SWOT analysis method to discuss the advantages, disadvantages, opportunities and challenges faced by the construction of financial management systems in universities. On this basis, he drew on the advantages of financial management systems in foreign universities, proposed a SWOT matrix improvement strategy, explored paths and methods for building financial management systems in universities, and then improved the scientific level of financial management in universities

Y. Pei et al. (Eds.): IC 2024, LNEE 1214, pp. 166–174, 2024.
https://doi.org/10.1007/978-981-97-4193-9_18

[1]. Song Dengke designed and built the Conch centralized financial management system. Through feature engineering, he established a relevant financial indicator system and combined it with the K-means algorithm to mine data information, build a model to classify risks and judge their degree, and complete the risk warning function [2]. Li Weihong designed the system database and functional interface, using the current stable J2EE-based SSM framework to implement multi-dimensional analysis and calculation of financial costs such as drilling and slicing for the financial cost big data analysis part. After testing, various functions of the system had strong operation stability and low failure rate, which improved the accuracy of financial data and work efficiency [3]. Despite this, the above research still has problems such as high system resource usage. Therefore, this article proposes an improved system based on the Apriori algorithm to help companies better analyze and utilize financial data. By applying the Apriori algorithm, the system can mine frequent item sets and association rules in financial data, providing enterprises with valuable information about product sales, customer behavior, market trends, etc. This will help companies more accurately understand market demand, optimize product portfolios, improve sales results, and enhance competitiveness.

The main contribution of this article is to introduce the Apriori algorithm into the field of financial management, and design and implement a powerful financial management system. The system improves the accuracy of data mining and reduces the time and resource usage of association rules. This article first systematically analyzes the problems and shortcomings of the existing financial management system, and then describes in detail the construction process of the algorithm system based on the improved Apriori algorithm, and conducts systematic experiments to test performance in Sect. 4, and finally summarizes and looks forward to the article.

2 Research Related to Financial Management Systems

2.1 Current Situation and Shortcomings

The financial management systems currently on the market are complex in design and not intuitive to use, requiring users to spend time and energy learning and adapting to the system's operations and functions, which increases the difficulty of using the financial management system. Moreover, financial management systems need to be integrated with other systems and data sources, but data integration and interoperability is not a simple matter, it requires customized development and additional work to achieve seamless flow and integration of data [4, 5]. The financial management needs of different organizations and enterprises may vary and require customized functions and reports to meet specific business needs. However, the current financial management system lacks flexibility and cannot meet the customized needs of specific users. Financial data is sensitive information and has very high requirements for security and privacy protection. Traditional financial management systems have certain loopholes in data storage, transmission and access control, and are susceptible to the risk of data leakage and security attacks [6, 7].

Based on the above existing problems, this article proposes an improved system based on the Apriori algorithm. The improved system can achieve accurate and efficient financial management methods, and can optimize the computer's resource usage when using the system.

2.2 Apriori Algorithm

This algorithm gradually generates frequent itemsets by scanning the data set multiple times and using the Apriori principle. Its operation process is: I. Generate frequent 1-itemsets based on the minimum support threshold: scanning the data set, count the support of each item, and retain items with support greater than or equal to the minimum support threshold as frequent 1-itemsets. II. Generate candidate 2-itemsets through frequent 1-itemsets [8]: using frequent 1-itemsets to generate candidate 2-itemsets, and combine frequent 1-itemsets in pairs to form candidate 2-itemsets. III. Pruning: performing pruning operation on the candidate 2-itemset, and delete the candidate k-itemset in which any sub-itemset is not a frequent (k − 1) itemset. IV. Generate frequent 2-item sets based on the minimum support threshold: scanning the data set, count the support of each candidate 2-item set, and retain candidate 2-item sets whose support is greater than or equal to the minimum support threshold as frequent 2-item sets [9, 10]. V. Generate association rules: Based on frequent item sets, association rules are generated according to the minimum confidence threshold. The association rules are in the shape of $A \rightarrow B$, which means that the occurrence of item set A can deduce the occurrence of item set B. Through the Apriori algorithm, the financial management system can analyze transaction data and discover frequent item sets. By analyzing transaction patterns, it can reveal customer purchasing behavior, product relevance and other information, thereby supporting marketing strategies and sales optimization [11, 12].

2.3 System Architecture

When designing the financial management system in this article, the system is divided into four levels, namely the data collection and processing layer, the data storage layer, the data analysis and mining layer and the visual report layer. Since the data collection and processing layer and the data storage layer are the common architecture of financial management systems, this article will not elaborate on these layers in detail. The data analysis and mining layer is the core content of this system. It applies the Apriori algorithm to mine frequent item sets and association rules, and uses the algorithm to generate candidate item sets and frequent item sets [13, 14]. The algorithm scans the data set layer by layer to generate frequent item sets that meet the minimum support threshold, which involves operations such as generation of candidate item sets, support calculation, and pruning. After the frequent itemsets are generated, association rules are generated from the frequent itemsets through an algorithm. Association rules are composed of item sets in frequent itemsets, including an antecedent and a consequent, and have indicators such as confidence, and the algorithm generates association rules that meet the minimum confidence threshold by calculating the confidence [15, 16].

3 System Design and Implementation

3.1 Data Preparation

The source of financial data can be through financial internal systems and external data sources, and financial and financial-related data can be collected from data sources. The data is extracted into the system through database and API calls for subsequent processing and analysis. The collected financial data is then cleaned. The cleaning process can remove inaccurate, duplicate and invalid data to ensure the quality and consistency of the data. This article chooses to measure the dispersion of the data statistically and measure the difference between each data and the average value, that is, to measure the variance of the data [17, 18]. The formula is as follows:

$$G = \sum \frac{(x_i - \eta)^2}{n - 1} \tag{1}$$

Among them, G represents the variance value, x_i represents the i-th data point, η represents the average value of the data, and n represents the number of data. Dividing the objects in the data set into different groups. The objects in each group have similar characteristics. Using the Euclidean distance formula to measure the similarity between data points and perform cluster analysis on the data, such as formula 2:

$$\mathrm{d}(x_i, x_j) = (\sqrt{(\sum a_i - b_i)^2}) \tag{2}$$

Among them, x_i and x_j represent two data points respectively, and a_i and b_i represent the values of the two data points in the i-th dimension. According to the analysis needs, the financial data is appropriately transformed so that the data can adapt to the subsequent analysis and mining process, and the data is integrated to form a consistent data set [19, 20].

3.2 Algorithm Implementation

Frequent itemset generation: Generating candidate itemsets for all individual items and calculate their support. According to the preset minimum support threshold, candidate item sets that do not meet the conditions are filtered out to obtain frequent 1-item sets. In each round of iteration, candidate itemsets are generated based on the frequent itemsets generated in the previous round. The generation of candidate item sets is achieved through join and pruning operations. Calculating the support of the generated candidate item sets, filter out the candidate item sets that meet the minimum support threshold, and obtain frequent item sets.

Association rule generation: For each frequent item set, generating all possible association rule combinations, and calculate the confidence of the generated candidate rules. Confidence is defined as the support of a rule divided by the support of its antecedent. According to the preset minimum confidence threshold, association rules that do not meet the conditions are filtered out, and a set of association rules that meet the minimum confidence threshold is obtained. Table 1 shows the specific parameter settings of the algorithm.

Table 1. Specific settings of algorithm parameters

Parameter	Example Value
Minimum Support	0.05
Minimum Confidence	0.6
Maximum Rule Length	3
Maximum Itemset Size	5
Support Type	Absolute
Confidence Type	Absolute
Lift Threshold	1.2

3.3 Data Model

According to the ideas of this article, the data model of the system is mainly divided into transaction sets, item sets, frequent item sets and association rules. A transaction set is a collection of data in a financial management system. Each transaction represents a transaction or financial record. Each transaction consists of a set of items (Items). The items can be products, services, accounts, and transaction types. An itemset refers to a combination of items in a transaction set, and an itemset can contain one or more items. Frequent itemsets are itemsets that appear frequently in transaction sets and are one of the core output results of the Apriori algorithm. By setting a minimum support threshold, the algorithm can identify frequent itemsets that are considered to have important relevance. Association rules are rules that describe the association between items, usually in the form of "If-Then", where the "If" part is called the Antecedent and the "Then" part is called the Consequent. Association rules can be evaluated and filtered based on support and confidence to identify rules with significant associations.

4 System Performance Evaluation

4.1 Evaluation Method

An experiment is designed to verify the performance and feasibility of the financial management system based on the Apriori algorithm in frequent item set mining and association rule analysis. The experimental hardware requirements are PCs with Windows operating system and equipped with high-performance processors. The software requirements are financial management system software, MySQL database management system, and processed actual data sets containing financial data. The data set ensures sufficient data volume and diversity, and the data has gone through the above preprocessing steps. The experimental group used a financial management system based on the Apriori algorithm to conduct frequent item set mining and association rule analysis, while the control group used a traditional financial management system to conduct a controlled experiment. In order to form a better comparison between the two systems, association rule analysis time, CPU usage and mining result accuracy were selected as the performance evaluation indicators of the system during the experiment. In the experiment, multiple experiments are needed to reduce the error and chance of the experiment, and multiple experiments are conducted using data sets of different sizes and characteristics. Statistics and comparisons were made on the performance indicators of the experimental group and the control group to verify the performance and feasibility of the financial management system based on the Apriori algorithm.

4.2 Result Analysis and Discussion

During the experiment, mathematical analysis tools were used to record the data in the experiment, and the experimental results were drawn into visual charts for analysis and discussion. Figure 1 shows the association rule analysis time experimental results, Fig. 2 shows the CPU usage experimental results, and Fig. 3 shows the mining result accuracy experimental results. The abscissa in the figure is the experiment number of the system, and the ordinate is the relative test result, and I is the system of this article and II is the traditional system.

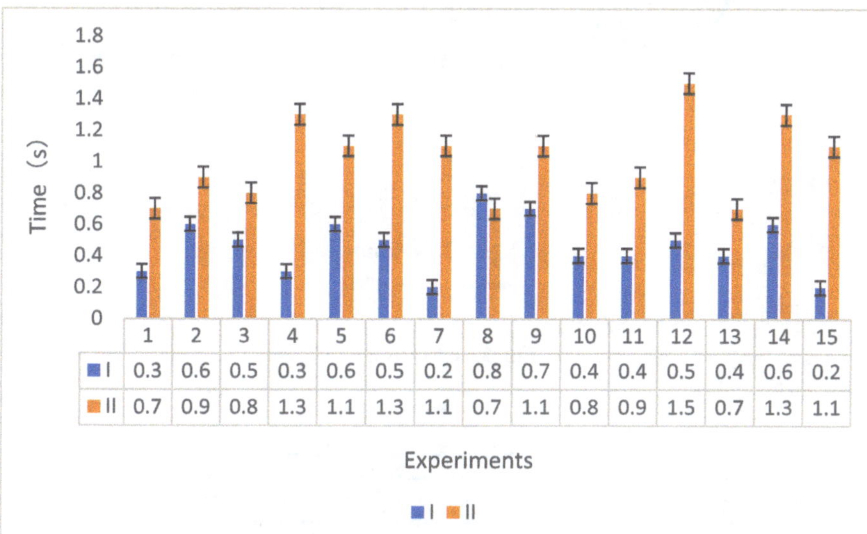

Fig. 1. Association rule time

According to Fig. 1, we conducted an experiment on association rule time. In the experiment, the maximum association rule time of this system was 0.8 s and the minimum was 0.2 s, while the performance of the traditional system in the experiment was maximum 1.5 s and minimum 0.7 s. The system in this article has a good improvement in the time of association rules. Its overall time reduction shows that the system can complete the mining process and provide results faster. This is because the system in this article uses parallel computing technology to decompose the association rule mining task into multiple subtasks and process them simultaneously, thereby speeding up the mining, while the traditional system uses a serial calculation method, which results in a long time consumption.

Running the two systems separately, and count the system resources occupied by the two systems during operation. According to Fig. 2, the occupancy range of the system in this article is 16%–35%, and the occupancy rate of the traditional system is 31%–58%. The system in this article has a smaller running occupancy rate, which can reduce the burden on the computer. This is because the system in this article has been algorithmically optimized. By optimizing the task scheduling algorithm and resource allocation strategy, the system can better control and balance the CPU usage of each task and reduce resource waste and conflicts.

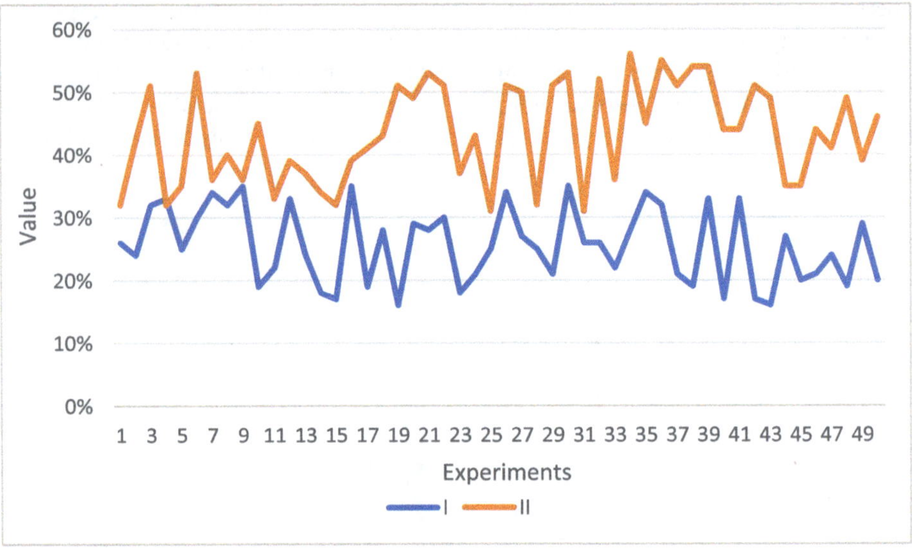

Fig. 2. CPU usage

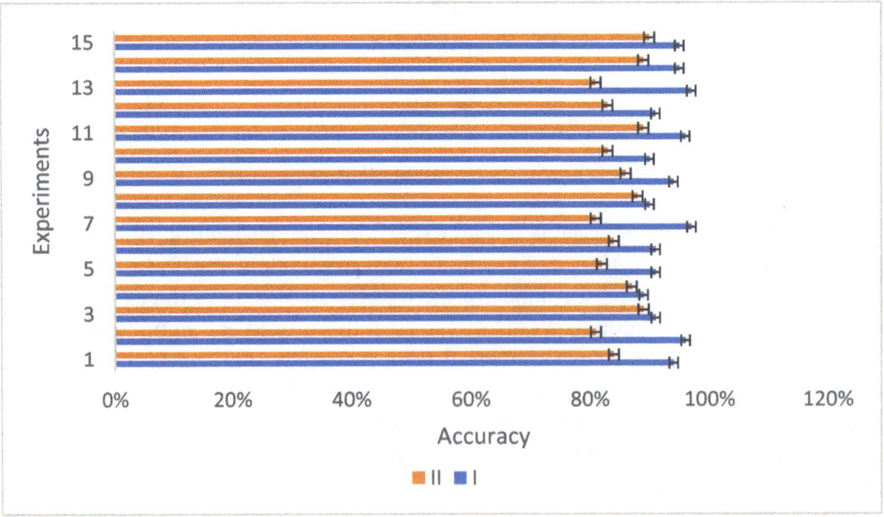

Fig. 3. Accuracy

As shown in Fig. 3, the accuracy of the two systems was experimentally tested. The accuracy can well reflect the performance of the system on association rules. The accuracy of this system in the experiment is 89%–97%, and the accuracy of the traditional system is 81%–90%. The accuracy of the system in this article can be improved because the algorithm processes financial data more accurately and adopts a more stringent pre-processing strategy to remove noise and outliers in the data.

5 Conclusion

Through the design of this article and three performance test experiments, the effectiveness and practicability of the Apriori algorithm in financial management systems are verified. By mining frequent item sets and association rules in financial data, the system can reveal correlations in product portfolios, customer behaviors, market trends, etc., and provide important business insights for enterprises. By using the financial management system designed and implemented in this article, enterprises can more accurately understand the internal correlation of financial data and optimize business decisions and strategic planning. The system provides data visualization and query analysis functions, allowing users to quickly access and understand mining results and improve decision-making efficiency and accuracy.

The Apriori algorithm may face efficiency and storage space challenges when processing large-scale data. Future research can explore improved association rule mining algorithms to improve the efficiency and scalability of the algorithm so that it can process larger-scale financial data.

References

1. Yan, S.: Construction of university financial management system based on SWOT analysis. Shanghai Manag. Sci. **45**(4), 120–124 (2023)
2. Dengke, S.: Design and construction of a centralized financial management system for risk early warning. Comput. Knowl. Technol. Acad. Edn. **19**(12), 67–70 (2023)
3. Weihong, L., Chao, W.: Design and implementation of hospital financial management system based on SSM framework. Mod. Sci. Instrum. **39**(2), 23–28 (2022)
4. Huazhi, D.: A brief analysis of the design and practice of intelligent financial management systems in colleges and universities. J. Zhejiang Ind. Commer. Vocat. Tech. Coll. **22**(1), 88–92 (2023)
5. Fu, Y.: Design analysis of financial management accounting information system under the background of cloud computing. China Manag. Inf. **25**(6), 61–63 (2022)
6. Bustani, B., Khaddafi, M., Ilham, R.N.: Regional financial management system of regency/city regional original income in Aceh province period year 2016–2020. Int. J. Educ. Rev. Law Soc. Sci. (IJERLAS) **2**(3), 459–468 (2022)
7. Mosteanu, N.R., Faccia, A.: Digital systems and new challenges of financial management–FinTech, XBRL, blockchain and cryptocurrencies. Qual.-Access Success **21**(174), 159–166 (2020)
8. Banerjee, A., Duflo, E., Imbert, C., et al.: E-governance, accountability, and leakage in public programs: experimental evidence from a financial management reform in India. Am. Econ. J. Appl. Econ. **12**(4), 39–72 (2020)

9. Jain, A., Sharma, Y., Kishor, K.: Financial supervision and management system using Ml algorithm. Solid State Technol. **63**(6), 18974–18982 (2020)
10. Zamlynskyi, V., Zerkal, A., Antonov, A.: A conceptual framework to apply financial engineering at the enterprise. Baltic J. Econ. Stud. **5**(1), 68–74 (2019)
11. Klapper, L., Lusardi, A.: Financial literacy and financial resilience: evidence from around the world. Financ. Manage. **49**(3), 589–614 (2020)
12. Li, Z., Bahramirad, S., Paaso, A., et al.: Blockchain for decentralized transactive energy management system in networked microgrids. Electr. J. **32**(4), 58–72 (2019)
13. Kori, B.W., Muathe, S., Maina, S.M.: Financial and non-financial measures in evaluating performance: the role of strategic intelligence in the context of commercial banks in Kenya. Int. Bus. Res. **13**(10), 130 (2020)
14. Riinawati, R.: Education financial management during Covid-19 pandemic of islamic universities in South Kalimantan. Dinamika Ilmu **21**(2), 383–396 (2021)
15. Owuor, G.O., Agusioma, N., Wafula, F.: Effect of accounts receivable management on financial performance of chartered public universities in Kenya. Int. J. Current Aspects Financ. Bank. Account. **3**(1), 73–83 (2021)
16. Atmadja, A.T., Saputra, K.A.K., Manurung, D.T.H., et al.: Factors that influence financial management: a case study in Indonesia. J. Asian Financ. Econ. Bus. **8**(6), 1203–1211 (2021)
17. Bakar, M.Z.A., Bakar, S.A.: Prudent financial management practices among Malaysian youth: the moderating roles of financial education. J. Asian Financ. Econ. Bus. (JAFEB) **7**(6), 525–535 (2020)
18. Widajatun, V.W., Kristiastuti, F.: The effect of regional financial supervision, accountability and transparency of regional financial management on local government performance. Bp Int. Res. Critics Inst. (BIRCI-J.) Hum. Soc. Sci. **3**(4), 2966–2974 (2020)
19. Lu, J., Shon, J., Zhang, P.: Understanding the dissolution of nonprofit organizations: a financial management perspective. Nonprofit Volunt. Sect. Q. **49**(1), 29–52 (2020)
20. Novitasari, D., Juliana, J., Asbari, M., et al.: The effect of financial literacy, parents' social economic and student lifestyle on students personal financial management. Econ. Educ. Anal. J. **10**(3), 522–531 (2021)

Correlation Between Intelligent Tourism Data Under Data Mining

Junna Hou(✉)

GuangDong Innovative Technical College, Dongguan 523960, China
HJN891213@163.com

Abstract. With the vigorous development of technology, tourism data has shifted from traditional face-to-face teaching mode to electronic trading mode. How to make good use of electronic trajectory information related to travel is the most relevant issue for people. This article uses the ARIMA model (Autoregressive Integrated Moving Average Model) to predict the passenger flow in M city. It evaluated the correlation between the popularity of M city's tourism industry network, weather conditions, and passenger numbers through grey correlation analysis. In the grey correlation analysis results of the relationship between network popularity and daily passenger flow, as the network popularity increases, daily passenger flow also keeps increasing. The network popularity value is 1 h, and the daily passenger flow value is 19963.

Keywords: Intelligent Tourism Data · Data Mining · Correlation Analysis · Number of Passengers · Daily Passenger Flow

1 Introduction

With the advancement of technology, various information systems are becoming increasingly mature, and big data is increasingly being applied to daily life. Introducing the concept of data mining in regional tourism development is of great significance for promoting the healthy development of regional tourism. Especially, a tourism prediction and feedback system based on big data has been constructed, which can promote the development and progress of the tourism industry. At the same time, it can also provide tourists with travel guides to improve the quality of tourism.

In online comment platforms for tourist attractions, there is a large amount of tourist comments and complex content. Classifying them solely based on human experience has problems of low efficiency and high error rates. However, using text analysis methods based on data mining can greatly alleviate this problem. Liu Xueting plans to use the ancient village in the Xi'an section of the northern foothills of the Qinling Mountains as a case study to construct a tourism intelligent recommendation system based on data mining methods. He analyzed it from three dimensions: scale, advantage, and agglomeration, and classified it. Through simulation experiments, he verified the feasibility of the system and provided a new way for personalized recommendations in the tourism industry [1]. In response to the current situation of scattered distribution of tourism resources

Y. Pei et al. (Eds.): IC 2024, LNEE 1214, pp. 175–183, 2024.
https://doi.org/10.1007/978-981-97-4193-9_19

and low level of application software intelligence, Wang Meng has introduced tourism intelligence algorithms into new travel modes. He developed a smart tourism WeChat mini program platform based on this. The platform can achieve various functions such as self driving travel, word cloud view, and casual travel [2]. Bai Weihua plans to use the text data of comments on scenic spots and hotel websites in "Smart Tourism" as the basis. He conducted research on four aspects: mining and analyzing online comment text data, as well as impressions of scenic spots and hotels, comprehensive evaluation of scenic spots and hotels, effectiveness of online comment texts, and characteristics of scenic spots and hotels. It validates the advantages of data mining technology in text analysis [3].

This article uses the ARIMA model and grey correlation analysis to accurately and efficiently segment the passenger market, maximizing the personalized needs of consumers, and thus forming personalized and customized tourism. It meets the tourism needs of consumers at different levels. The main content of this article is as follows:

In the first part, the background and significance of the correlation analysis between intelligent tourism data based on data mining are discussed.

The second part is about data mining and smart tourism.

In the third part, there is a method for exploring the correlation analysis between intelligent tourism data, providing ARIMA modeling and grey correlation analysis.

In the fourth point, there is the experiment and analysis of intelligent tourism data.

In the fifth part, the conclusion is the limitations and prospects of the correlation between intelligent tourism data.

2 Data Mining and Smart Tourism

2.1 Data Mining

Data mining mainly involves mining implicit knowledge from a large amount of data, predicting the development trend of paths, and making reasonable corrections to paths. The information data that data mining aims to mine is essentially a type of information knowledge, which refers to knowledge with certain commonalities, similarities, and related attributes that have certain connections and rules with each other. The specific content of data mining varies depending on the industry information. In the process of path management, path managers need to make multiple decisions based on the actual operation status of the path, and the influencing factors of path selection are also relatively complex. For example, how to accurately estimate the popularity of multiple lines is a very important research topic. Tourism routes are an important direction for big data development. The changes in tourism destinations, the variability of tourism themes, the landscape combination of tourism paths, and the consumption of tourism paths enable route managers to use information mining to assist in path planning decisions, thereby formulating important path choices [4, 5]. This article uses data mining techniques such as decision tree analysis and cluster analysis to search for the most popular path parameters with the goal of maximizing the hotspots on the path. It is possible to conduct in-depth exploration of existing tourist attractions, route durations, total consumption, and route lengths, analyze new route data, and optimize tourist routes to better meet the needs of tourists [6]. The research results indicate that data mining technology can help online tourism enterprises analyze and predict the historical passenger flow of tourism routes, and reduce route risks. It provides guarantees for smart tourism development enterprises or tourism service enterprises to carry out route planning, route management, and other aspects in smart tourism operations [7, 8].

This article combines data mining technology with tourism route management, and smart tourism can simplify route management planning and management processes with lower costs, thereby better improving the management efficiency of online tourism routes [9]. By managing tourism routes, smart tourism can utilize information technology to share tourism route information on a larger scale. This enhances the informatization and intelligence of smart tourism services centered around routes, and improves the processing efficiency of services [10, 11]. Meanwhile, by operating online tourism routes, its competitiveness in the online tourism market can be enhanced [12].

2.2 Smart Tourism

Smart tourism is a new type of tourism management method, which is based on modern information technology and adopts a new tourism management method by analyzing traditional tourism methods. It applies technologies such as the Internet of Things, cloud computing, next-generation communication networks, high-performance information processing, and intelligent data mining to various fields such as tourism experience, industrial development, and administrative management. This can achieve high system integration and deep development of tourism physical resources and information resources, providing services for future oriented new forms of tourism such as the public, enterprises, and governments [13, 14]. Its core is based on integrated communication and information technology, focusing on the interactive experience of tourists, ensuring integrated industrial information management, promoting industrial innovation, and promoting industrial structure upgrading [15]. Smart tourism refers to the use of modern information technology and network mobile devices to provide users with more targeted and convenient needs. It is a new type of tourism resource management system [16, 17].

2.3 Intelligent Tourism Models in the Context of Data Mining

(1) Accurately positioning consumer needs

In tourism activities, there are often different actors simultaneously, namely information, service providers, and consumers. At present, China's data mining technology can organically integrate the participants in this process, providing data and technical support for enterprises to formulate marketing strategies. It can be said that data mining connects tourists and suppliers together. During the travel process, this article conducts in-depth analysis of data and integrates the data of tourism companies to accurately analyze the effective information required by the enterprise [18]. By utilizing data mining, consumers' search preferences, value preferences, and other information can be organized to more accurately identify their needs and provide marketing and promotional services for enterprises.

(2) Targeted meeting consumer needs

By accurately and efficiently segmenting the user's market, it maximizes the personalized needs of consumers, thereby forming personalized and customized tourism that meets the travel needs of consumers at different levels. Before the era of data mining, the tourism industry had always been in a "seller" state, and the tourism services it provided often did not meet the needs of customers well. However, in the era of data mining, the tourism industry has entered the buyer's market and can better understand consumer needs. It can adjust its marketing methods and strategies in a timely manner to meet the needs of more customers and achieve better economic benefits. Through data mining technology, comprehensive analysis and mining of user needs and preferences can be carried out, which helps tourism enterprises to more targeted meet consumer needs and promote the development of the tourism industry towards a better direction. Through the analysis and deepening of data mining technology, the supply side of tourism services can more effectively analyze consumers' tourism preferences, thereby providing consumers with more professional and targeted tourism services, and improving customer satisfaction [19].

3 Methods for Correlation Between Intelligent Tourism Data

3.1 ARIMA Modeling

This article uses the ARIMA model to predict the passenger flow in M city. In the ARIMA (p, d, q) model, AR represents the autoregressive model, p represents the autoregressive term, MA represents the moving average model, q is the number of moving average terms in the model, and d is the number of differences performed to obtain the stability of the sequence from the original time series data. On this basis, this article adopts a new method of weighting and summing a series of travel time delay data using a random interference term [20]:

$$\mathrm{xtp} = \mathrm{c} + \alpha 1\mathrm{xt} - 1 + \alpha 2\mathrm{xt} - 2 + \cdots + \alpha \mathrm{pxt} - \mathrm{p} + \varepsilon \quad (1)$$

To facilitate modeling, the ARMA (p, q) model is usually used to model time series. A prerequisite for the establishment of the ARMA (p, q) model is that the time series it studies has stability, but in real life, most of them are non-stationary. Therefore, differential operations can be used to achieve the stationarity of time series, which is also a commonly used method for smoothing sequences. Therefore, this article constructs an ARMA model (Auto-Regressive and Moving Average Model) using time-series data of the stable state after differentiation.

Steps for ARIMA modeling:

(1) This article conducts basic testing on time series to determine its stability. If it cannot pass, it indicates that this sequence has a unit root. It is an unstable state that requires differentiation or logarithmization methods to make the sequence stable;
(2) On this basis, this article conducts autocorrelation testing on the time-series data after differential processing, and uses the tail order of its autocorrelation and partial dependence to determine the values of p and q.

(3) From the above principle explanation, it can be seen that when the differential stationary time series is tested for correlation, and its correlation and autocorrelation are both trailing, the ARMA model can be selected. It uses various statistical software to fit the equation.
(4) This article establishes a model, generates a series of residual values, and tests the stationarity and autocorrelation of these residual values to determine whether they meet randomness and the constraints on their confidence intervals;
(5) On this basis, this article conducted hypothesis testing on the residual sequence to determine the white noise of the residual sequence;
(6) When it is determined that the model is not false and the fitting effect of the model is correct, then prediction can be made on the time series data. Finally, this article conducted error analysis on the established model to determine whether it can pass the test.

3.2 Grey Correlation

In response to the multi factor correlation situation with a small sample size, this article uses the grey correlation analysis method to quantitatively evaluate the popularity of the tourism industry network, weather conditions, and passenger numbers in M city. Compared with conventional quantitative analysis methods, the advantage of grey correlation degree lies in its simple operation and ease of operation. Even without the need for specialized model statistical analysis software, it can be calculated using Excel spreadsheets. Secondly, grey correlation does not require special regulations on the distribution of samples, nor does it require quantitative calculation of the number of samples. Even if there is no regularity between sample variables, analysis can be conducted through calculation.

In grey correlation analysis, the most crucial step is to determine the grey correlation between the main order and various influencing factors. The strength of grey correlation can express the degree of impact. The main steps are:

Step 1: Set the main sequence.

Step 2: Dimensionless raw data. The initial value method is commonly used in dimensionless processing methods. Initial value is a commonly used grey correlation dimensionless method, which refers to the new sequence obtained by dividing all data in the original sequence by its first number.

Step 3: Differentiate the main and secondary sequences to obtain a differential order.

Step 4: Grey correlation analysis was conducted on various influencing factors.

Step 5: Calculate the average of the grey correlation coefficients obtained in the previous step to obtain the grey correlation degree between each factor.

Finally, after calculating the grey correlation between each influencing factor and the main factors, they can be sorted, making it easy to see the degree of influence of each influencing factor.

Using the grey generative sequence model to construct a cumulative generative sequence to obtain the predicted value of the tourism model [21]:

$$X(k) = (x(m) - ab)(1 - ea(k - 1)) + ab \tag{2}$$

Among them, a is the development coefficient, and b is the grey action quantity.

4 Intelligent Tourism Data Experiment

4.1 Data Processing

In January 2020, when people were waiting for the whole family to travel, a novel coronavirus suddenly broke out in Wuhan, China, and then spread to all parts of the country. Japan is the world's second largest economic power and has taken strict control measures to avoid the spread of the epidemic. As an industry that is very sensitive to external events, China's tourism industry is like being suddenly pressed "pause button". During this Spring Festival season, when the pot is full, it has suffered "Waterloo", and the number of tourists in M city has been greatly affected. The "M City Statistical Yearbook" lacks the number of tourists in January 2020, and the number of tourists in M City in February was only 16000. Incorporating these data into the model would have a significant impact on the prediction results. This article studies the prediction model of suitable tourist volume under non-emergency conditions in M city. To obtain more accurate prediction results, the selection of observation data should be cautious. According to relevant statistical regulations, starting from 2012, some monthly reports would no longer undergo statistical surveys for January. Therefore, by observing the number of tourists from 2019 to 2021 in the "M City Statistical Yearbook", it was found that there are some cases where the number of tourists in January is missing. Therefore, the average can be used to replace the missing data. In this article, using the cumulative average of an entire month to replace lost data is the simplest and most commonly used alternative method. In the absence of a large amount of data, the method used in this article has little impact on the calculation results. After completing the missing data, a sequence diagram observation was conducted on these 144 pieces of data. The

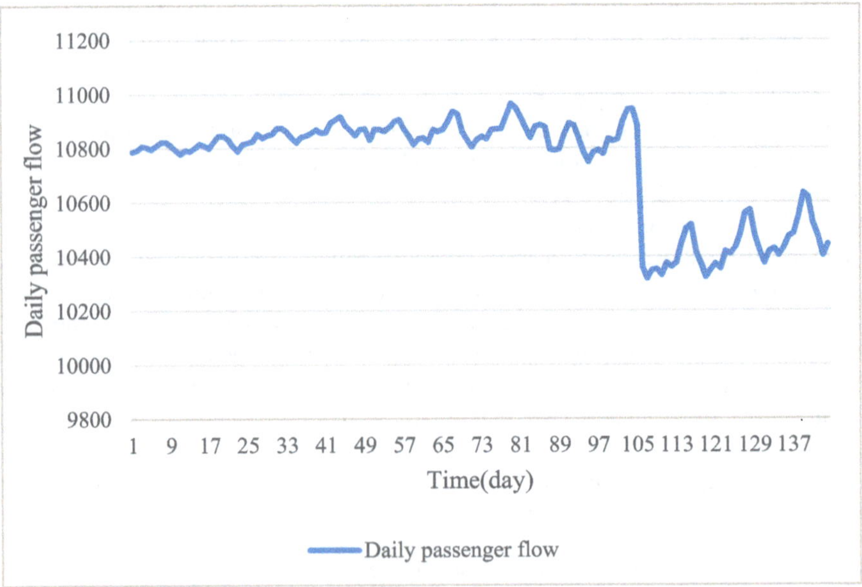

Fig. 1. Observation results of passenger flow on different days

observation results of different daily passenger flows are shown in Fig. 1. Relatively speaking, daily passenger flow is relatively stable in the early stage and fluctuates in the later stage.

4.2 ARIMA Model Observations

In order to further expand the tourism market, M city actively introduces investment and uses the ARIMA model to observe the results of different daily passenger flows, as shown in Fig. 2. It can be seen that the passenger flow in M city is increasing regularly, with a maximum value of 11622.

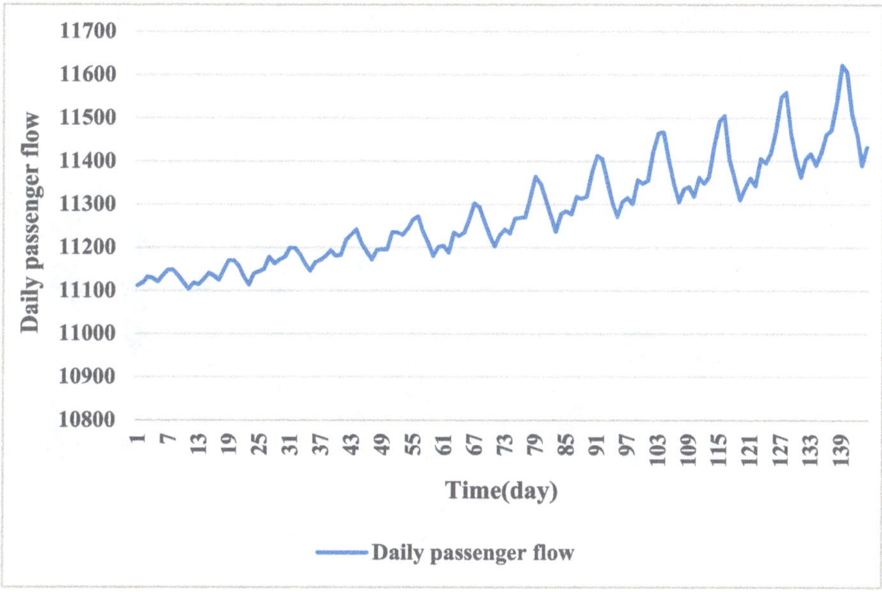

Fig. 2. Observation results of different daily passenger flow using ARIMA model

4.3 Grey Correlation of Passengers

This article calculates the number of passengers and weather changes from May 1st to 5th, 2019. The grey correlation analysis results are shown in Table 1. The number of passengers on sunny days on May 1, 2019 was 12000, on cloudy days on May 2, 2019 was 8000, and on rainy days on May 3, 2019, the number of passengers decreased to only 4000. The number of passengers on May 4th, 2019 is close to the number on May 1st, 2019. The number of passengers on May 5th, 2019 is 10236, and the weather is cloudy. From this, it can be seen that there is a positive correlation between the number of passengers and weather changes.

The grey correlation analysis results of the relationship between network popularity and daily passenger flow are shown in Fig. 3. With the increase of network popularity,

Table 1. Grey correlation analysis results between passenger numbers and weather changes

Data	Weather	Number of passengers
2019/5/1	Sunny day	12000
2019/5/2	Cloudy day	8000
2019/5/3	Rain	4000
2019/5/4	Sunny day	11926
2019/5/5	Cloudy	10236

daily passenger traffic has also been increasing. The network popularity value is 1 h, and the daily passenger traffic value is 19963.

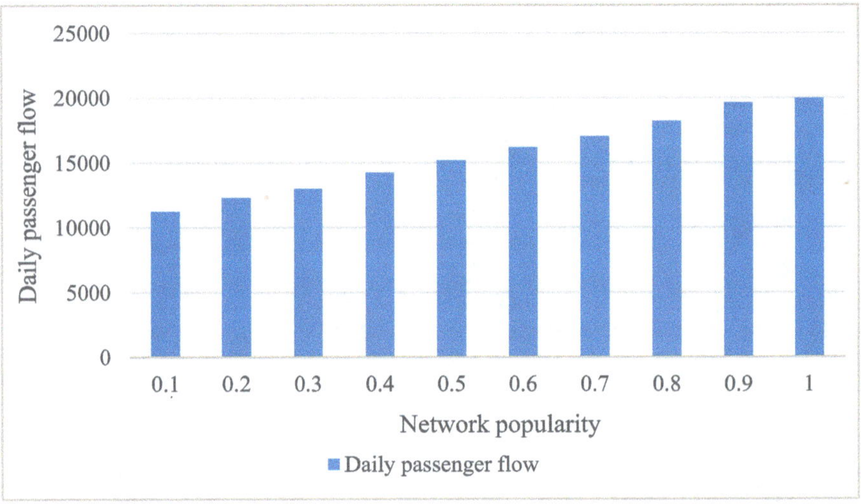

Fig. 3. Relationship between network popularity and daily passenger flow

5 Conclusions

Nowadays, with big data, it can better meet the needs of users. This article explores the opportunities and innovations of smart tourism models in the data mining environment, in order to provide some reference for the development of the tourism industry. The application of data mining provides new opportunities for the tourism industry, and it would be increasingly applied in the development of the tourism industry. Therefore, in the smart tourism model, deeper research on big data technology is needed in the future, so that the modern tourism industry can achieve faster and healthier development with the support of big data, and promote the development of smart tourism.

References

1. Liu, X.: Design of intelligent tourism recommendation system for Gucun Town in Xi'an section of northern Qinling mountains based on data mining. Microcomput. Appl. **39**(3), 8–11 (2023)
2. Wang, M., Dong, L., Guo, J.: Design and implementation of smart travel WeChat mini program. Inf. Comput. **34**(24), 131–133 (2022)
3. Bai, W., Chen, X.: Research and analysis of the text of "Smart tourism" based on data mining technology. J. Zhaoqing Univ. **44**(2), 45–54 (2023)
4. Wang, J.: Research on personalized recommendation algorithm for intelligent tourism service robots based on big data environment. J. Pu'er Coll. **36**(3), 32–34 (2020)
5. Cheng, G.H., Xiao, M., Qing, F.F.: Design and analysis of smart global tourism cloud platform in the era of big data. Smart City Appl. **4**(2), 79–82 (2021)
6. Li, Y., Li, C., He, W.: Design and implementation of intelligent tourism system for mobile terminal Beibu Gulf. J. Qinzhou Coll. **33**(1), 11–16 (2018)
7. Astuti, K.A., Darma, G.S.: Community-based tourism: measuring readiness of artificial intelligence on traditional village. Int. J. Soc. Sci. Hum. **3**(3), 81–89 (2019)
8. Atsalakis, G.S., Atsalaki, I.G., Zopounidis, C.: Forecasting the success of a new tourism service by a neuro-fuzzy technique. Eur. J. Oper. Res. **268**(2), 716–727 (2018)
9. Centobelli, P., Ndou, V.: Managing customer knowledge through the use of big data analytics in tourism research. Curr. Issue Tour. **22**(15), 1862–1882 (2019)
10. Chi, O.H., Gursoy, D., Chi, C.G.: Tourists' attitudes toward the use of artificially intelligent (AI) devices in tourism service delivery: moderating role of service value seeking. J. Travel Res. **61**(1), 170–185 (2022)
11. Del Vecchio, P., Mele, G., Ndou, V., et al.: Creating value from social big data: implications for smart tourism destinations. Inf. Process. Manage. **54**(5), 847–860 (2018)
12. Femenia-Serra, F., Ivars-Baidal, J.A.: Do smart tourism destinations really work? The case of Benidorm. Asia Pac. J. Tourism Res. **26**(4), 365–384 (2021)
13. Kirilenko, A.P., Stepchenkova, S.O., Kim, H., et al.: Automated sentiment analysis in tourism: comparison of approaches. J. Travel Res. **57**(8), 1012–1025 (2018)
14. Mariani, M.: Big data and analytics in tourism and hospitality: a perspective article. Tourism Rev. **75**(1), 299–303 (2020)
15. Pillai, R., Sivathanu, B.: Adoption of AI-based chatbots for hospitality and tourism. Int. J. Contemp. Hosp. Manag.Manag. **32**(10), 3199–3226 (2020)
16. Prentice, C., Dominique Lopes, S., Wang, X.: Emotional intelligence or artificial intelligence– an employee perspective. J. Hosp. Market. Manag. **29**(4), 377–403 (2020)
17. Samara, D., Magnisalis, I., Peristeras, V.: Artificial intelligence and big data in tourism: a systematic literature review. J. Hosp. Tour. Technol. **11**(2), 343–367 (2020)
18. Sigala, M., Rahimi, R., Thelwall, M.: Big data and innovation in tourism, travel, and hospitality. Culinary Sci. Hospitality Res. **23**(6), 1–11 (2019)
19. Stylos, N., Zwiegelaar, J., Buhalis, D.: Big data empowered agility for dynamic, volatile, and time-sensitive service industries: the case of tourism sector. Int. J. Contemp. Hosp. Manag. **33**(3), 1015–1036 (2021)
20. Wang, W., Kumar, N., Chen, J., et al.: Realizing the potential of the internet of things for smart tourism with 5G and AI. IEEE Netw. **34**(6), 295–301 (2020)
21. Zhang, Y., Li, G., Muskat, B., et al.: Tourism demand forecasting: a decomposed deep learning approach. J. Travel Res. **60**(5), 981–997 (2021)

Design and Resource Sharing System for Higher Vocational Education Under Wireless Networks

Nuo Zhang(✉)

Xinjiang Transportation Vocational and Technical College, Urumqi 831401, Xinjiang, China
18899192987@163.com

Abstract. With the continuous development of wireless network technology, people's demand for educational resources is constantly increasing. This article adopts a vocational education resource sharing system to maximize the integration and aggregation of various scattered educational resources in vocational schools, utilizing wireless network technology to improve data transmission speed and increase interaction between teachers and students. In the network connection speed test, the 24-h network connection speed monitoring found that the highest value was 9.1 Gbps and the lowest value was 3.0 Gbps. The vocational education resource sharing system based on wireless network fully meets the internet speed requirements.

Keywords: Higher Vocational Education Resource Sharing System · Wireless Network · Network Connection Speed · Data Transmission Speed · Teacher-Student Interaction

1 Introduction

At present, vocational colleges, like secondary vocational and technical schools, are actively exploring and developing the development of educational resource sharing technology. This article successfully uses a network sharing system for vocational education resources, which effectively integrates subject resources and teaching resources in vocational schools. This has promoted the rapid development of vocational schools, enhanced the teaching quality and efficiency of vocational schools.

The development of wireless network technology provides a broad possibility for higher vocational education resource sharing system. Starting from analyzing the current situation of the urban-rural education resource sharing system, Zhang Zhengyang discussed its driving effect on the development of urban-rural education. On this basis, he proposed the construction idea of a urban-rural education resource sharing system with the goal of infrastructure and enriching shared resources [1]. Guo Yangyong proposed the idea of regional education resource co construction and sharing based on the current situation of regional education resource distribution, low level of high-quality education resource sharing, and disorderly information management. The system he proposed provides a good user interface that allows education participants from various counties and districts to share their unique educational resources, thus achieving the maximum

Y. Pei et al. (Eds.): IC 2024, LNEE 1214, pp. 184–193, 2024.
https://doi.org/10.1007/978-981-97-4193-9_20

sharing of educational resources throughout the entire county [2]. Sun Shaoju thinks backward infrastructure, poor compatibility of technical equipment, tight funding, and a lack of overall resource construction planning have led to a slow pace of information resource co construction and sharing in university libraries [3]. However, the performance of the vocational education resource sharing system in their research is not very good.

This article designs and adopts a teaching resource management platform, and provides full technical guidance for it. Through using the teaching resource sharing platform in higher vocational colleges to maximize the dispersion of teaching pressure, the main content of this article is as follows:

In the first part, the background and significance of designing a wireless network-based resource sharing system for vocational education are discussed.

The second part is the design of a wireless network-based resource sharing system for vocational education.

In the third part, the functional module design of the wireless network sharing system is discussed.

The fourth part is the testing of the network sharing system for educational resources in vocational colleges, including performance and functional testing.

In the fifth part, the future prospects of the network sharing system for educational resources in vocational colleges.

2 Wireless Network-Based Resource Sharing System for Vocational Education

2.1 Design of the Overall Framework System for Network Sharing System

It boldly integrated wireless network technology into the teaching resource sharing system of vocational colleges and implemented an overall architecture [4, 5]. It enables the teaching resource sharing system to store more and more refined teaching resources. The overall framework system includes:

There are scattered educational resources in various vocational colleges across the country [6]. Vocational colleges are both participants and managers. According to the technical specifications for the construction of school educational resources, it integrates the educational resources of various schools and establishes a teaching resource library that conforms to the characteristics of their respective regions [7, 8]. By establishing a shared platform, it is possible to centrally integrate the educational resources of various dispersed vocational colleges, thereby establishing a large-scale and comprehensive educational resource sharing system library for vocational colleges [9].

The review of resources adopts a strict resource entry mechanism, implementing dual reviews of teaching resources in various vocational schools. This can effectively screen low-quality and substandard educational resources, classify approved teaching resources, and establish a cloud database for efficient management of teaching resources in vocational colleges. It greatly improves the storage space of the database, and in the process of continuously enriching resources, it can also enable the teaching resources of each vocational school to be shared with each other, thereby further improving the quality of teaching resource construction in vocational colleges [10, 11].

Effective integration and management of educational resources [12]: It has designed and developed a teaching resource management platform, and on this basis, has provided full technical guidance, utilizing the teaching resource sharing platform of vocational schools. It maximizes the integration and aggregation of various scattered educational resources in vocational schools, and fully and effectively manages the entire uploaded vocational education resources [13, 14].

Building a National Higher Vocational Education Resource Sharing System [15]: In this process, this article would use cloud computing technology, database aggregation technology, network platform development technology, application extension technology, and other technologies as technical support to break through the closed system of vocational school education resources. It maximizes the co construction and sharing of system resources, enabling deep sharing of vocational education resources between schools and enterprises.

The formula for the utilization rate of vocational education resources [16, 17]:

$$U = (E / P) * 100\% \tag{1}$$

Among them, U represents the utilization rate of vocational education resources, E represents the actual number of resources used, and P represents the number of resources available for use. This formula is used to evaluate the utilization efficiency of vocational education resources, help decision-makers understand the full utilization of resources, and optimize resource allocation and sharing system design.

Learning resource evaluation formula [18]:

$$Q = (R * C) / T \tag{2}$$

Among them, Q represents the quality of learning resources, R represents the reliability of learning resources, C represents the applicability of learning resources, and T represents the usage time of learning resources.

2.2 Layered Structure Design of Network Sharing System

The construction of a network sharing system for educational resources in vocational schools involves research on the screening, uploading, storage, management, and application of educational resources. In order to build an effective, comprehensive, and personalized teaching resource network sharing system for vocational schools that can meet the needs of different users, this article, based on this, adopts the requirements of technical specifications for educational resource construction. It has standardized the screening, uploading, and management of resources uniformly [19, 20]. This article designs from four levels: data resource layer, application support layer, application instance layer, and portal display layer:

Data resource layer: This layer includes two aspects of content: first, the standard specifications constructed and the multidimensional and comprehensive database involved. The second is to cultivate students' hands-on ability. The standardization of teaching resource construction can achieve information sharing of teaching resources, improve the accuracy and efficiency of retrieval, and fundamentally ensure the quality of teaching resources; A multi-dimensional integrated database is based on the data itself, which can efficiently integrate and manage data resources, optimize storage space in multiple dimensions, and facilitate data migration and update. On this basis, this article establishes a practical teaching system at three levels: "basic skills", "comprehensive skills", and "production skills" based on database specifications. The cultivation of "basic skills" is to achieve the goal of cultivating vocational and technical talents, that is, by cultivating the basic skills of a certain profession, to achieve the cultivation of the basic skills of that profession. The cultivation of "comprehensive skills" aims to cultivate their ability to apply the knowledge they have learned to solve problems. Through the training of "production skills", vocational school students can proficiently grasp the actual production technology of their professional job group, and cultivate their quality awareness and cost awareness. This enables it to have a "zero distance connection" with social enterprises, in order to promote continuous improvement and improvement in the production process of the enterprise.

The application support layer includes support technologies such as content management, user permission management, file upload and download management, all of which provide direct support for practical applications. Content management includes webpage creation, role positioning, functional implementation, and execution of various links, facilitating future resource information updates and maintenance. User permission management is an effective means of protecting system security. It can effectively log in and manage permissions for users in the system to ensure the safe and stable operation of the system. Statistical analysis management refers to the use of business intelligence to collect, analyze, and analyze various resources and their user activities. Through "Upload and Download Management", it provides a standardized resource sharing channel for users of this system, enabling them to upload and download smoothly and orderly.

Apply instance layers: There are three-fold aspects. The first type is characteristic resources, which are divided into seven subcategories: multimedia resources, specialized learning websites, high-quality courses, renowned teacher forums, professional resources, micro videos, and micro courses. The second type is "school enterprise management", which is divided into six categories, namely "basic information", "professional training", "internship files", "job internships", "order training", and "skill certificates". The third type of resource is resource management. It includes resource attribute management, resource type management, resource upload, resource attention, resource collection, resource review, and resource interaction. The application instance layer improves the security access ability of teaching resources by obtaining data, analyzing the security structure of the application, and recording and analyzing user network behavior.

The portal display layer is the interface of the entire system, which organically combines static and dynamic pages, and combines text and images reasonably. It enables real-time updates of educational resources, teaching information, policies and regulations, and current affairs news, ensuring the novelty of the website's content. It also incorporates a human-machine interaction module between users and the system, utilizing a convenient search engine to provide users with a clearer and more convenient system platform.

3 Design of Functional Modules for Wireless Network Sharing System

The system contains three modules. The resource management function module includes the resource audit; The user management function module mainly realizes the permission configuration of ordinary users and administrator users, and on this basis, a detailed introduction to the system is provided.

(1) Resource Management Function Module

The resource management module is a functional module used to efficiently manage various educational resources that need to be uploaded. In resource management, attention should be paid to the scientific content and system security.

(2) System management function module

The system management module is mainly responsible for system configuration, user management, classification management, system security, support services, classification statistics, and other aspects to ensure the stability and scalability of the system.

(3) User management function module

In the education resource network sharing system of vocational schools, there is a specific user group. The user management module assigns appropriate permissions to different system users and manages their activity points, fully mobilizing their enthusiasm for participating in teaching activities. On the premise of ensuring system security and resource quality, it ensures the sustainable and orderly conduct of teaching activities. System users can generally be divided into: vocational college administrators, teacher users (system administrators), enterprise users (enterprise leaders), student users (ordinary users), and social visitors (visitor users).

4 Testing of Education Resource Network Sharing System in Higher Vocational Colleges

System testing refers to the comprehensive consideration of factors such as software and hardware, external environment, and network environment, and the use of specialized network testing tools (system writing tools) to test the functionality and performance of a system based on specific testing methods. This is a crucial step aimed at identifying any logical or physical issues in the designed and developed system by comparing it with existing system design requirements, and identifying features that do not meet user requirements or do not meet expectations. Further processing and improvement of the system can be carried out to continuously optimize the developed system. The focus of system testing is to verify the system, where confirmation is to ensure that the system can complete the specified functions normally, that is, to ensure that the predetermined functions can be completed during the design phase. Verification is a process that covers a series of behaviors, aimed at proving the rationality and logical correctness of a system under certain external conditions, and ensuring that it can be expressed in an appropriate form. When testing network sharing systems, verifying and confirming them is a very important and challenging task. Firstly, it is necessary to ensure that the system can function as expected, and secondly, it should be reasonably displayed on different user browser clients, while also considering its security and availability. This article focuses on vocational colleges and combines the characteristics of network sharing to conduct functional and performance tests on the system.

4.1 Functional Testing

(1) Test environment

Server operating environment: Windows Server 2008; Database: Oracle; The default browser for the client is Internet Explorer 9.0.

(2) User interface testing

From the perspective of user needs, this article tests whether the network links in the system interface are valid, and whether the text and images are displayed correctly. The specific testing content of the user interface is shown in Table 1. In the test of whether the page link is normal, it was tested twice, failed once, and passed once.

Table 1. Specific Test Content of User Interface

Number of tests (times)	Test content	Test result
2	Is the page link normal	1 failure, 1 pass
1	Is the display of text and images normal	Pass
2	Can audio files be tried, downloaded, and collected	2 pass
1	Can video files be previewed, downloaded, and collected	Pass

4.2 Performance Testing

The experimental results show that after being fully put into use, the system can meet the needs of various users and meet the expectations of developers. Performance tests were conducted on the system from the following aspects.

(1) Network connection speed test

Nowadays, users can choose to make phone calls, broadband networks, or wireless networks. The running speed of the system depends on the internet method, and most people have good patience to wait for a long time when downloading a program. When browsing a webpage, people become very impatient and constantly refresh the interface,

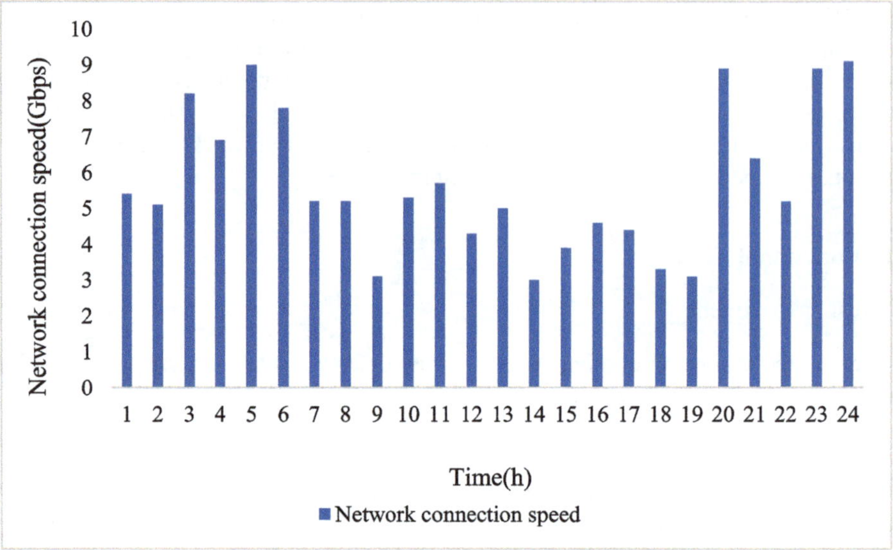

Fig. 1. Network Connection Speed Test Results

making the webpage load quickly. If the reaction time is too long, they would lose patience and leave the system interface.

The network connection speed test results are shown in Fig. 1. The 24-h network connection speed monitoring found that the highest value was 9.1 Gbps and the lowest value was 3.0 Gbps, fully meeting the internet speed requirements.

(2) System load testing

Load testing is the performance testing of a networked shared system under a specific load to ensure that it can run under a certain load. The size of the load can be determined based on how many users simultaneously access the system at a certain time. This article divides users into 150250350 for testing. When the number of users is 350, the system takes 9.66 s, which is less time. The system load test results are shown in Fig. 2.

(3) System data recovery test

System data recovery testing is a test of the data recovery ability of a system under human or non human failure conditions, which involves detecting the conditions of system failure, assessing the degree of system data recovery, and performing disaster recovery backups on the system. After testing, the system has the ability to recover disaster recovery data, and its recovery time varies depending on the resource capacity in the database. The larger the capacity, the less time it takes to recover. While building and improving the sharing system, it is required that system management personnel regularly do a good job in system security backup to improve the efficiency of disaster recovery backup.

Here, the test cases for this article include restoring files, restoring images, restoring videos, restoring documents, restoring audio, and restoring databases. The recovery time

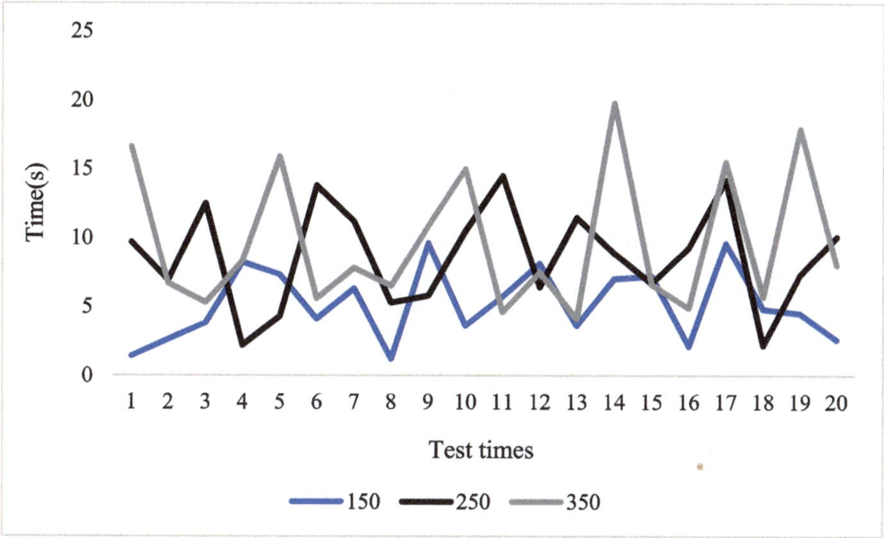

Fig. 2. System Load Test Results

for the database is 5.2 s, and the recovery time for other test cases is very short. The system data recovery test results are shown in Fig. 3.

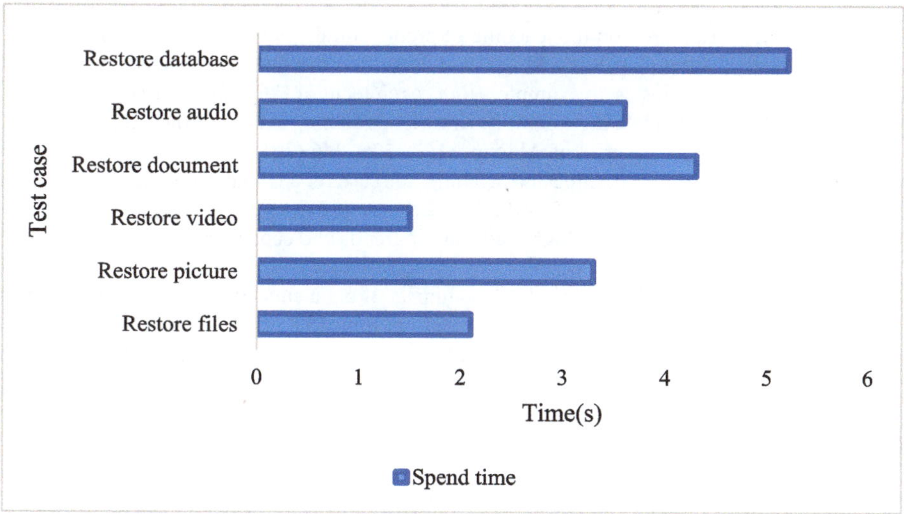

Fig. 3. System Data Recovery Test Results

5 Conclusions

The recovery and testing results of the vocational education resource sharing system based on wireless networks in this article are good, and the system load testing takes less time. In the future, with the development of technology, faster and more reliable wireless network products would emerge. They are in line with the development of educational informatization and provide strong support for building an advanced, efficient, convenient, and free educational network platform.

References

1. Zhang, Z.: Research on the sharing system of urban and rural educational resources under the collectivized school-running model based on "Internet+." Western Qual. Educ. **4**(5), 86–87 (2018)
2. Guo, Y., Zhang, X., Lei, C.: Research and design of a software system for jointly building and sharing educational resources in the region. Softw. **41**(8), 21–23 (2020)
3. Sun, S.: Research on the construction of a library information resource sharing system based on cloud computing–taking the library of Tianjin Vocational College as an example. Libr. Res. **43**(3), 26–29 (2019)
4. Hong, Z., Fei, J.: Analysis on the new model of educational resource construction in the era of "Internet+." Res. Electro-Chem. Educ. **41**(7), 48–54 (2020)
5. Liu, H., Yue, Y., Wansheng, J.: Blockchain technology, differences in educational resources and high-quality economic development–an empirical analysis based on the allocation of higher education resources in our country. J. Shaanxi Normal Univ. Philosophy Soc. Sci. Edn. **49**(1), 145–158 (2020)
6. Liu, X., Wu, F.: Research on the teaching of architectural design courses based on the network resource sharing platform–taking the architectural design of provincial-level boutique resource sharing courses as an example. High. Architectural Educ. **28**(2), 131–137 (2019)
7. Al-Kurdi, O., El-Haddadeh, R., Eldabi, T.: Knowledge sharing in higher education institutions: a systematic review. J. Enterp. Inf. Manag. **31**(2), 226–246 (2018)
8. Bovill, C.: Co-creation in learning and teaching: the case for a whole-class approach in higher education. High. Educ. **79**(6), 1023–1037 (2020)
9. Castro, R.: Blended learning in higher education: trends and capabilities. Educ. Inf. Technol. **24**(4), 2523–2546 (2019)
10. Colvard, N.B., Watson, C.E., Park, H.: The impact of open educational resources on various student success metrics. Int. J. Teach. Learn. High. Educ. **30**(2), 262–276 (2018)
11. Hamidi, H., Chavoshi, A.: Analysis of the essential factors for the adoption of mobile learning in higher education: a case study of students of the University of Technology. Telematics Inform. **35**(4), 1053–1070 (2018)
12. Musselin, C.: New forms of competition in higher education. Soc. Econ. Rev. **16**(3), 657–683 (2018)
13. Paseka, A., Schwab, S.: Parents' attitudes towards inclusive education and their perceptions of inclusive teaching practices and resources. Eur. J. Spec. Needs Educ. **35**(2), 254–272 (2020)
14. Penprase, B.E.: The fourth industrial revolution and higher education. High. Educ. Era Fourth Ind. Revolution **10**(1), 978–981 (2018)
15. Redmond, P., Abawi, L., Brown, A., et al.: An online engagement framework for higher education. Online Learn. J. **22**(1), 183–204 (2018)
16. Shaturaev, J.: A comparative analysis of public education system of Indonesia and Uzbekistan. Biosci. Biotechnol. Res. Commun. **14**(5), 89–92 (2021)

17. Zulaikha, Z.F., Hariri, H., Rini, R., et al.: Analysis of vocational education curriculum in ASEAN economic community: a literature review. J. Soc. Hum. Educ. **1**(3), 157–170 (2021)
18. Alshurideh, M.T., Al Masaeid, T., Alzoubi, H.M., et al.: Components determining the behavior and psychological impact of entrepreneurship among higher vocational students. J. ReAttach Therapy Dev. Diversities **5**(2s), 189–200 (2022)
19. Triyono, B., Mohib, N., Kassymova, G., et al.: The profile improvement of vocational school teachers' competencies. High. Educ. Russia **29**(2), 151–158 (2020)
20. Apriana, D., Kristiawan, M., Wardiah, D.: Headmaster's competency in preparing vocational school students for entrepreneurship. Int. J. Sci. Technol. Res. **8**(8), 1316–1330 (2019)

Prediction and Optimization of Mechanical Manufacturing Quality Based on Deep Learning

Dong Han[1(✉)], Tao Dong[2], and Songyu Zhu[3]

[1] Industrial Center, Harbin Vocational and Technical College, Harbin 150000, China
hd67776069@126.com

[2] Research Office, Harbin Vocational and Technical College, Harbin 150000, China

[3] School of Electronics and Information Engineering, Harbin Vocational and Technical College, Harbin 150000, China

Abstract. The application of deep learning technology in mechanical design parameter optimization has opened up new research directions. The limitations of traditional design methods, especially in the search for global optimal solutions, have prompted researchers to explore more efficient solutions. Deep learning, with its outstanding performance in pattern recognition and predictive analysis, is considered a powerful tool for mechanical design optimization. Through comparative analysis and case studies, the accuracy of deep learning models in predicting mechanical manufacturing quality and their potential application in design parameter optimization were demonstrated. Through experiments, it was found that the deep learning technology quality prediction system predicted 187 perfect products, 145 medium products, and 147 defective products for mechanical quality, with a prediction error of 21, and a high overall judgment accuracy rate. It can be seen that deep learning methods can significantly improve design efficiency, reduce costs, and improve the overall performance of mechanical systems. This study aimed to provide an efficient optimization strategy for the field of mechanical design to promote the improvement of performance and reliability.

Keywords: Deep Learning · Mechanical Manufacturing · Quality Prediction · System Optimization · Data Collection

1 Introduction

In the era of big data, the rapid development of information technology has profoundly changed various fields, including mechanical manufacturing. With the vigorous development of the global economy and continuous technological progress, traditional manufacturing technology can no longer meet the needs of modern society for efficient and high-quality manufacturing. Therefore, mechanical intelligent manufacturing technology has emerged with the goal of improving the level of intelligence in the manufacturing process and achieving higher production efficiency and product quality.

In the era of big data, the rise of deep learning technology has brought revolutionary changes to multiple fields. Wu Yanxia pointed out that deep learning plays a crucial role

Y. Pei et al. (Eds.): IC 2024, LNEE 1214, pp. 194–203, 2024.
https://doi.org/10.1007/978-981-97-4193-9_21

in extracting valuable information from massive data, which is particularly prominent in the field of mechanical manufacturing [1]. In terms of machinery, Yang Sheng emphasized that the development of mechanical design, manufacturing, and automation is gradually shifting towards high-precision and high-efficiency [2]. This ability is crucial for understanding and predicting complex data patterns in manufacturing processes.

With the development of the economy and technological progress, traditional manufacturing technology can no longer meet the needs of modern society. GUAN Peng mentioned that this has promoted the development of mechanical intelligent manufacturing technology. The core of intelligent manufacturing technology is to improve the level of intelligence in the manufacturing process, in order to achieve the goal of optimizing production efficiency and product quality. In this process, mechanical inspection analysis plays a key role [3]. Chen Nanping pointed out that mechanical inspection and analysis is a highly specialized and wide-ranging engineering technology, which is crucial for ensuring and improving product quality [4].

In the field of quality prediction, XIA Mingze believed that finding a suitable prediction model is the core task. This model needs to be able to accurately learn and predict future quality trends from historical data [5]. CHENG Wen added that error log information in storage systems is very helpful for optimizing system availability and stability, which also applies to the manufacturing field, where log information can be used to predict equipment failures and product defects [6]. Finally, Zhai Li emphasized the value of data analysis in assisting decision-making or management, which is crucial for extracting valuable information from data to optimize the mechanical manufacturing process [7].

2 Current Situation of Mechanical Manufacturing Quality Prediction

2.1 Significance of Mechanical Manufacturing Quality

With the continuous improvement of technological level in various industries, the development of mechanical design, manufacturing, and automation is gradually shifting towards high-precision and high-efficiency. However, product manufacturing quality generally refers to the degree to which the product meets the functionality understood by the designer and meets the product design standards. The quality of accessory manufacturing and product assembly are often different factors. The quality of the accessory manufacturing industry directly affects the quality of products, but it also affects the quality of product assembly. Therefore, the quality of the accessory manufacturing industry is not only the core of product quality, but also the foundation. The composition of product manufacturing quality is shown in Fig. 1.

Quality prediction is crucial for design [8]. Based on the analysis in Fig. 1, it can be seen that processing quality indicators are key parameters for measuring the processing quality of parts. These indicators collectively determine the performance and service life of parts. Below is a brief analysis of some indicators.

Dimensional accuracy: It refers to the difference between the actual size and design size of a part, including geometric accuracy indicators such as linearity, roundness, and

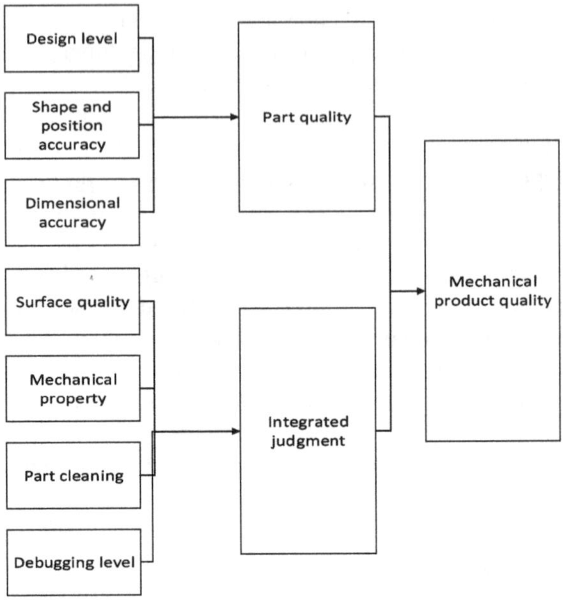

Fig. 1. Composition of mechanical manufacturing quality

cylindricity. The higher the dimensional accuracy, the closer the dimensions of the parts are to the design requirements, and the correct fit can be achieved during assembly.

Accuracy of form and position: It refers to errors in the shape and relative position of parts, such as flatness, coaxiality, verticality, symmetry, etc. The higher the accuracy of shape and position, the more accurate the direction and position of the parts after assembly.

Surface quality: It reflects the roughness, flatness, and machining marks of the part surface. High quality surface processing can improve the fatigue life of parts, reduce wear, and improve sealing performance.

Mechanical properties: It includes the strength, hardness, toughness, etc., of the parts, usually improved through post-processing processes such as heat treatment and surface treatment. Good mechanical properties are crucial for designing mechanical parts, as they can improve their wear and corrosion resistance.

2.2 Manufacturing Quality Prediction

Quality prediction technology is an effective strategy for reducing software testing costs and improving quality [9]. Moreover, the quality prediction technology of modern mechanical manufacturing is a modern technology that combines industrial innovation and modern science and technology. It is an important symbol of national scientific and technological progress, which is also a higher requirement put forward by social development and the growing needs of the people.

There are currently many machine learning algorithm models that can predict the future trends and situations of various problems, and different prediction models have application scenarios and advantages and disadvantages [10]. Until the 1970s, the manufacturing technology for products was very single, and the successful launch of a new product could lead to the launch of many homogeneous products. The trend of world integration and economic globalization is unstoppable, and market competition would become more intense. However, as the role of highly integrated systems becomes increasingly prominent, quality prediction has become an effective means to improve system reliability [11].

The recent international manufacturing quality prediction data is shown in Table 1.

Table 1. Technical forecast data for international manufacturing quality

Year	Technology type	Forecast accuracy (%)	Prediction effect evaluation	Average scrap rate	Automation level
2020	Machine learning model	74	Good	2%	Partial automation
2020	Sensor technology	75	Good	3%	Partial automation
2020	Real-time monitoring system	78	Good	4%	Partial automation
2021	Machine learning model	72	Good	1.50%	semi-automatic
2021	Sensor technology	77	Good	2.50%	Partial automation
2021	Real-time monitoring system	81	outstanding	3%	Partial automation
2022	Machine learning model	74	Good	1%	semi-automatic
2022	Sensor technology	78	Good	2%	Partial automation
2022	Real-time monitoring system	70	Good	2.50%	Partial automation

Table 1 summarizes the manufacturing quality prediction data for three different years. According to Table 1, it is not difficult to find that the accuracy of various technologies is not particularly high, and it may be found that quality prediction is very important but there are some errors [12].

3 Deep Learning Quality Prediction System

3.1 Deep Learning Technology

Deep learning has been widely applied in various fields [13], including prediction. The quality prediction system can correctly judge the quality of products and provide real-time prediction and feedback during the production process, supporting enterprises to optimize production engineering and improve product quality. This paper introduces the principle, application examples, and future development directions of a quality prediction system based on deep learning. The deep learning technology model is shown in Fig. 2.

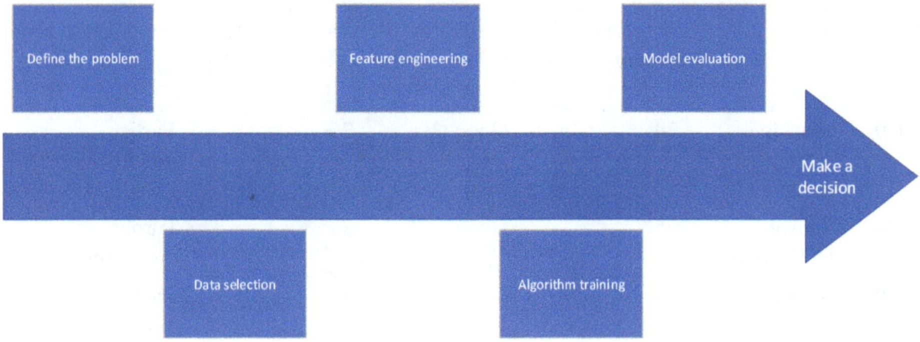

Fig. 2. Deep learning operation model

The deep learning operation model in Fig. 2, combined with quality prediction, is mainly as follows:

Deep learning has powerful feature representation and function fitting capabilities, and can extract hierarchical features from massive data. Deep learning also has great potential and application prospects [14]. Moreover, in the field of manufacturing quality management, the use of sensor data and deep learning algorithms enables real-time prediction of product quality, immediately capturing any potential abnormal situations, which not only improves production efficiency but also ensures the firm stability of product quality.

The revolutionary progress in the field of medical imaging has enabled deep learning models to automatically analyze medical images, providing doctors with a powerful and convenient tool to evaluate image quality. Doctors can rely on the feedback and suggestions of these models to improve the accuracy and efficiency of imaging diagnosis, thereby improving the medical experience of patients.

Product defect detection adopts deep learning technology, which quickly detects any potential product defects through advanced analysis of product images and sensor data. Once a problem is identified, the production team can take immediate measures to quickly rework or replace it to ensure product quality is not compromised, maintaining brand reputation and customer satisfaction.

In the field of natural language processing, the application of deep learning models has brought new possibilities for text quality assessment. These models can automatically detect issues such as consistency and grammar errors in text, providing strong support for editing and proofreading work. This technology not only improves work efficiency, but also ensures consistency and high-level standardization of text, thereby improving the quality of text processing. From this, it can be seen that the success of deep learning also requires correct calculation methods, and increasing the degree has a positive effect on target recognition [15].

3.2 Quality Prediction System Model

Deep learning has strong self-learning ability [16], and the deep learning mechanical quality system is shown in Fig. 3.

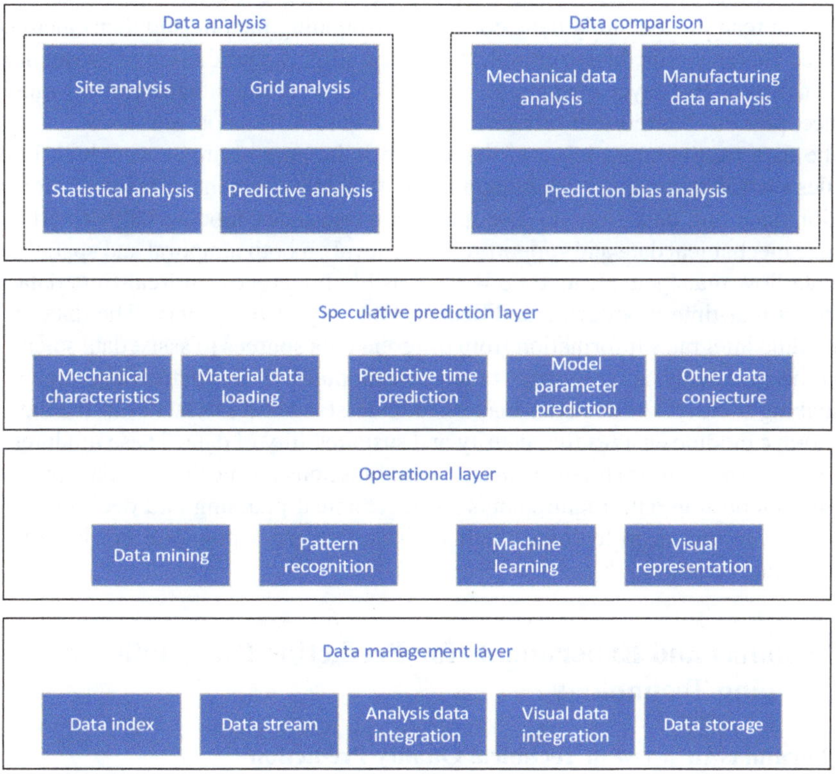

Fig. 3. Framework of deep learning technology quality prediction system

The introduction of the quality prediction system framework for deep learning technology in Fig. 3 is as follows:

The system is an important component of a data processing system. Its main task is to detect and obtain the data required by the platform through monitoring equipment and sensors, including but not limited to weather features, traffic flow related features, human motion features, road network structure features, etc. The rapid development of artificial intelligence technology and machine learning algorithms has provided new solutions for product quality prediction problems [17].

The data management layer is responsible for effectively managing the collected mechanical data. This includes utilizing methods such as spatiotemporal indexing, stream data, orbit data management, graph data management, and heterogeneous data indexing to ensure efficient storage and retrieval of data.

In terms of calculating the support population, the system provides various auxiliary tools for data analysis and calculation. These tools utilize mathematical methods to analyze the acquired and managed data, including data mining, pattern recognition, machine learning, and visualization methods.

The system achieves the prediction of mechanical quality through steps such as loading and road network data, mechanical data, collecting and calculating time points, and setting model parameters. Based on the mechanical quality prediction results, the system can read the assembly data, consider surrounding factors, and then calculate the predicted characteristic data to complete the prediction and prediction of air quality. The functionality of this layer enables the system to predict the situation of air quality in advance, providing important reference information for decision-makers.

The data management layer of the system is its core component, which includes modules such as spatiotemporal indexing, data flow management, data integration, visualization data, and data storage. The spatiotemporal index module helps to efficiently organize mechanical data and supports data retrieval on both temporal and spatial scales. The data flow management module is responsible for processing real-time data flows to support real-time monitoring and decision-making of machinery. The data integration module integrates information from different data sources to assist data scientists in comprehensive analysis. The visualization data module presents data in a graphical manner, making it easier for decision-makers to understand and interact with the data. The data storage module ensures the security and sustainability of data. These modules work together to provide powerful data management functions for mechanical data processing systems, supporting data requirements for mechanical planning and decision-making. It can be seen that deep learning methods have become a powerful driving force for technological progress [18].

4 Formulas and Experiments for Predicting the Quality of Deep Learning Technology

4.1 Formulas Involved in Technical Quality Prediction

Formula (1) is obtained by dividing the cost of waste materials by the total production cost or by dividing the total quantity of waste by the total production quantity of the product. It is mainly used as a result measure to determine whether the process is producing and assembling parts according to standardization.

F is the scrap rate.

$$F = \frac{D}{Z} \times 100\% \tag{1}$$

Formula (2) is used for the inspection pass rate or process pass rate, which is obtained by dividing the one-time qualified quantity of parts in the production process by the total production quantity of parts or weighted average. The defect rate of part processing is obtained by dividing the total number of unqualified parts by the number of production inputs. This formula is used as a process metric to determine the quality level of the production process.

$$X = \left(\frac{G}{L} + \frac{B}{T}\right) \times 100\% \tag{2}$$

In Formula (2), X represents the inspection pass rate or process pass rate; G represents the one-time pass quantity of the process; L represents the total production quantity of parts; B represents the unqualified quantity of parts; the latest T represents the production input quantity of parts.

4.2 Comparative Experiment on Quality Prediction of Deep Learning Technology

The experiment randomly set up 500 manufacturing machinery with three types of quality, namely perfect product (no mechanical problems), medium product (the machinery can be used normally but has slight defects in appearance), and defective product (the equipment cannot be used normally). The production materials selected for the experiment can only produce 200 manually set perfect products, 150 medium products, and 150 defective products. The experimental personnel conducted advance inspections of all instruments and made relevant records. Finally, mechanical quality prediction was performed using a deep learning technology quality prediction system (Group A), and

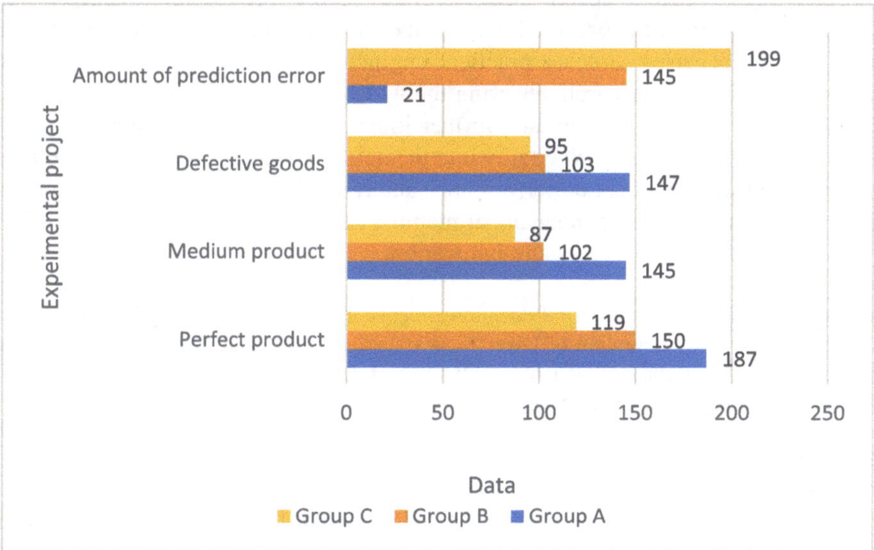

Fig. 4. Three types of mechanical quality prediction data

sensor technology was used to collaborate on prediction (Group B) and traditional manual prediction (Group C). Finally, by comparing two sets of data, the system effectiveness analysis was conducted. The experimental data is shown in Fig. 4.

According to the data analysis in Fig. 4, the deep learning technology quality prediction system predicted 187 perfect products, 145 medium products, and 147 defective products for mechanical quality, with a prediction error of 21. Sensor technology collaborated to predict 150 perfect products, 102 medium products, 103 defective products, and 145 prediction errors, with a relatively high overall error level. The traditional manual prediction of 119 perfect products, 87 medium products, and 95 defective products resulted in 199 prediction errors, with the highest prediction error rate.

5 Conclusions

It is not difficult to discover the enormous potential of deep learning based mechanical manufacturing quality prediction and optimization through research. This innovative method, like a breeze, brings new dawn to the field of mechanical design. It not only exhibits astonishing accuracy in optimizing mechanical design parameters, but also provides a more efficient way to find the global optimal solution, abandoning various limitations of traditional design methods. Through experimental verification, deep learning technology has demonstrated high accuracy in predicting mechanical quality, providing a feasible quality prediction system for engineering practice. This not only means that people may predict the quality of mechanical manufacturing in a more reliable way, but also means that they can provide higher performance and lower costs for engineering projects. Future research can further investigate the applicability of deep learning in different types of mechanical manufacturing, optimizing models to adapt to the diversity of various design parameters and improving the intelligence level of the system. In addition, the organic integration of deep learning technology with actual manufacturing scenarios to solve practical engineering problems would become the core direction of future research, ensuring the widespread application of deep learning technology in mechanical design and further improving reliability. The results of this research undoubtedly inject new vitality into the field of mechanical design, opening the door for further improvement of performance and reliability. Deep learning has become a clear stream in the field of mechanical manufacturing, bringing infinite possibilities for future technological progress and innovation.

Acknowledgment. Key Project of Education Science Planning of Heilongjiang Province in 2023: Research on the Integration of Diversified Vocational Education Resources with Chinese Characteristics and Enterprise Technological Innovation (Project No.: GJB1423535).

First-level Natural Science Project of Harbin Vocational and Technical College in 2020: Design and Implementation of Fire Sprinkler and Alarm System Based on Triz Theory (Project No.: HZY2020ZY004);

Key Project of the 14th Five-Year Plan of Education Science in Heilongjiang Province in 2022:Research on Teaching Mode Based on Integration of TRIZ Theory and Thinking Orientation (Project No.:ZJB1422141).

References

1. Yanxia, W., et al.: Progress and trend of deep learning FPGA accelerators. Chin. J. Comput. **42**(11), 2461–2480 (2019)
2. Sheng, Y.: Application analysis of CNC machining technology in mechanical design. Pap. Making Equipment Mater. **52**(4), 121–123 (2023)
3. Peng, G.: Application of industrial robot in intelligent manufacturing. Heilongjiang Metall. **039**(003), 163–164 (2019)
4. Nanping, C.: Failure analysis of mechanical parts. Mater. Rail Transit. **2**(2), 54–59 (2022)
5. Mingze, X.I.A., Chunpeng, S.H.I., Zhengyu, L.I.U., et al.: Research on quality prediction method in product production process. Manuf. Autom. **42**(4), 147–149 (2020)
6. Yan, L.I., et al.: Error log analysis and system optimization for Lustre cluster storage. Comput. Sci. **49**(10), 1–9 (2022)
7. Li, Z.: Design and implementation of visual analysis system based on civil service job list and profession. Comput. Knowl. Technol. **49**(22), 45–48 (2022)
8. Linlan, L., Rong, S., Jian, S.: Link quality prediction based on random forest. J. Commun.Commun. **40**(4), 202–211 (2019)
9. Chuan, T., Chuang, W., Can, L., et al.: Software quality prediction method based on improved pigeon flock RBF network. Aeronaut. Comput. Technol. **51**(5), 24–28 (2021)
10. Xia, L., Chunwang, H.: Prediction of air quality in Beijing based on machine learning. Inf. Record. Mater. **22**(10), 150–153 (2021)
11. Can, L., Chuan, T., Chuang, W., et al.: Software quality prediction method based on improved stacked self-coding network. Yangtze River Inf. Commun. **34**(12), 4–7 (2021)
12. Xiaozong, J., et al.: Analysis on the influencing factors of Coke quality prediction. Baotou Steel Technol. **46**(2), 5–24 (2020)
13. Yongsheng, W., Siyu, J.: Review of object detection algorithms based on deep learning. Comput. Digital Eng. **51**(6), 1231–1237 (2020)
14. Janiesch, C., Zschech, P., Heinrich, K.: Machine learning and deep learning. Electron. Mark. **31**(3), 685–695 (2021)
15. Zheyu, L., Hong, W., Xiaoxin, H.: Deep learning under complex tasks: mechanism and optimization strategy. Mod. Educ. Technol. **028**(008), 12–18 (2018)
16. Wang, B., Zhang, N., Lu, W., Geng, J., Huang, X.: Intelligent missing shots' reconstruction using the spatial reciprocity of green's function based on deep learning. IEEE Trans. Geosci. Remote Sens.Geosci. Remote Sens. **58**(3), 1587–1597 (2020)
17. Feng Xiaobin, X., Mingjiang, T.Y., et al.: Forecasting Industrial product quality based on the integration of multiple models in stacking. Math. Pract. Underst. **51**(22), 267–276 (2021)
18. Liu, B., Ding, M., Shaham, S., Rahayu, W., Farokhi, F., Lin, Z.: When machine learning meets privacy: a survey and outlook. ACM Comput. Surv.Comput. Surv. **54**(2), 1–36 (2021)

Intelligent Design of Interior Space Based on Topology Information Region Matching

Xiaoxing Zhang[1,2](✉) and Thirayu Jumsai Na Ayudhya[1](✉)

[1] School of Architecture, Art, and Design, King Mongkut's Institute of Technology Ladkrabang, No.1 Chalongkrung alley, Ladkrabang District, Bangkok 10520, Thailand
stuartpp@163.com, thirayu.ju@kmitl.ac.th

[2] School of Communication and Design Art, Yunnan University of Finance and Economics, Long Quan Road, Kunming, China

Abstract. Interior design, as a concern for homebuyers, has long plagued them because it requires consideration of practicality, aesthetics, and cost-effectiveness. Under the influence of various factors, intelligent technology-based interior space design has emerged. However, this article believes that this is not enough to solve some of the difficulties faced by interior design. Therefore, this article proposes to use topology information region matching to improve interior design. The main principle is to analyze the characteristics of interior space in order to develop more reasonable design schemes. This article conducts comparative experiments, comparing intelligent design based on topology information region matching with conventional design. In the light coverage rate performance of fixed lighting fixtures, the average coverage of the former is 91.03%, and the average coverage of the latter is only 79.33%. The significant gap proves that the method proposed in this article can indeed significantly improve the resource utilization rate of interior design.

Keywords: Topology Information Region Matching · Intelligent Interior Space Design · Intelligent Lighting System · Lighting Coverage Rate

1 Introduction

Regardless of the type of building, interior decoration must be carried out after completion to meet their respective functional requirements, and corresponding interior design has become a necessary task. The research objective of this article is to optimize the intelligent design of interior spaces based on topology information region matching, and improve the design ideas of interior design from the perspectives of practicality and aesthetics.

This article first studies the principles of interior design, analyzing from multiple perspectives such as planning, layout, materials, and lighting. Taking lighting systems as an example, it can be found that intelligent technology has matured and been applied in interior design today. Then, a method of topology information region matching is proposed to improve the intelligent design of interior spaces, and the feasibility of this method is further verified through comparative experiments.

Y. Pei et al. (Eds.): IC 2024, LNEE 1214, pp. 204–211, 2024.
https://doi.org/10.1007/978-981-97-4193-9_22

2 Related Work

Many researchers have put forward their own opinions on the improvement of interior design. Sydor M believed that current interior design needs the help of materials science to enhance its artistic value and durability, and proposes a composite material based on mycelium for design [1]. Kim J pointed out that current interior design has gradually begun to adopt sustainable design concepts, and he believed that using green design and energy-saving materials would be a good choice [2]. Murali P K, taking the rapidly developing intelligent autonomous vehicle as an example, proposed that the design of the interior of the vehicle has begun to get people's attention, and user experience, acceptance and trust can be improved through intelligent interior decorations and functions [3]. Charissis V proposed the design of an intelligent in car system to further enhance the driver's driving experience in the car [4]. Bao Z considered that interior design is a complex and lengthy process, and intelligent technology can effectively improve this phenomenon [5]. Intelligent technology is ultimately a broad concept that is not sufficient for the specific work of interior design.

This article believes that interior design can be improved through topology information region matching. Zhang C believed that topology information technology can be used in intelligent transportation systems to predict the speed of vehicles on the road for control purposes [6]. Wazirali R considered that obtaining accurate network topology information from the data plane is crucial, and he designed a software defined network to separate the control plane from the data plane [7]. Luppi A I found that studying network topology can provide great assistance for the study of brain networks [8]. It can be seen that this method is indeed applicable to interior design.

3 Method

3.1 Interior Design Principles

In traditional interior space design concepts, the fundamental purpose is to plan, layout, and decorate the interior space of a building to meet the functional needs of users, enhance the beauty and practicality of the space. Mei Z S believed that interior space design involves comprehensive consideration of various aspects of the space, including the selection and combination of elements such as layout, furniture, color, lighting, materials, etc. [9]. A comfortable, fully functional, and user-friendly interior environment can be created through these methods.

Table 1 lists many considerations for interior design to ensure that the design meets the actual functional needs of users, including the purpose and functional zoning of each area. Considering functional requirements cannot be separated from interior space planning. Designers need to develop a reasonable layout to ensure the full utilization of space, and to set up channels and circulation spaces reasonably. In order to meet different functional requirements, appropriate color schemes are needed, utilizing natural light and artificial lighting to create an ideal atmosphere and visual effect. Secondly, designers also need to choose appropriate materials, consider their texture, durability, ease of cleaning, and ensure coordination with the overall design style. There are many uses for materials, such as furniture decoration and style themes. Generally speaking, it is necessary to

Table 1. Factors to consider in interior design

Total factor	Dividing factors	Factor details
Requirements	Spatial planning	Ensure full utilization of space
	Color and lighting	Create an ideal atmosphere and visual effect
Material	Furniture decoration	Make it consistent with the overall design and meet the needs of users
	Style theme	Ensure visual consistency of the design
Special design	Personalization	Make the design reflect the user's personality, taste, and interests
	Technology Integration	Integrate modern technology to enhance the convenience of life
	Environmental protection	Consider environmental factors, improve energy efficiency, and reduce waste
Communication		Effective communication with other stakeholders to ensure the interests and expectations of others

choose suitable furniture and decorations, determine the overall style and theme of the design, make many design elements consistent with the overall design concept, and meet the needs of users. In addition to these basic designs, special factors need to be added. This article believes that personalized design, modern technology integration, and environmental factors can be considered. Different customers have different needs, and this design approach can reflect the user's personality, taste, and interests. Integrating modern technology, such as smart home systems and audio devices, can improve the convenience of life. As for considering environmental factors, the specific way can be to choose sustainable and environmentally friendly materials so as to improve energy efficiency and reduce waste. As a designer, necessary communication and interaction also need to be carried out, effectively communicating with clients, architects, and other stakeholders to ensure that the design meets the expectations of all stakeholders.

3.2 Intelligent Technology in Modern Interior Design

There are many aspects of modern interior design that can be improved using intelligent technology. In the eyes of many, the use of intelligent devices in interior design aims to enhance the functionality, comfort, and efficiency of interior spaces. Most people believe that the use of intelligent devices in interior design can improve the functionality, comfort, and efficiency of interior spaces. One of the more common ones is intelligent lighting systems. Gao M Z proposed the design of an interior intelligent lighting system using a microcontroller [10].

Figure 1 shows the framework of the intelligent lighting system. In its overall framework, the core undoubtedly relies on intelligent control strategies such as self-adaptive control and distributed control. This framework has three core components, namely the

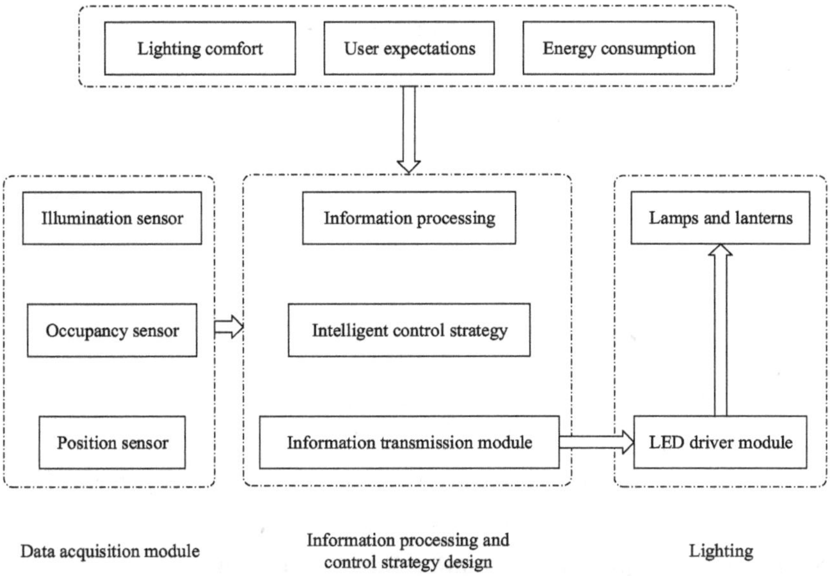

Fig. 1. Framework of intelligent lighting system

data acquisition module, information processing and control strategy design, and lighting system. The data acquisition module mainly includes a series of sensors such as illumination sensors, occupancy sensors, and position sensors, which are responsible for data acquisition. Of course, if there are special requirements, additional sensors can also be arranged. The information processing and control strategy design part needs to rely on requirements such as lighting comfort, user expectations, and energy consumption. The information processing is carried out, and control information is output through intelligent control strategies, which are then transmitted to the next part through information transmission modules. The lighting system sends information to the lamps through LED driver modules to complete the output of the lights, and subsequent intelligent adjustments are delivered through this channel.

Jin Q Z elucidated the relationship between visual art and psychological cognition in his research, and this article demonstrates it through Fig. 2 [11]. In this relationship, visual art conveys information to cognition, and human psychology gains experience through cognition. The human psyche is the basis for cognitive judgment of likes and dislikes, thus generating interest in visual art through cognition. Intelligent lighting systems play a crucial role in interior design. Intelligent technology improves spatial comfort and experience by adjusting light brightness, color temperature, and color. It can enhance spatial sense and visual effects through flexible control, highlighting key areas. It can be said that intelligent lighting systems can play a great role in interior design, meeting the psychological needs of users and the artistic atmosphere of the interior through flexible light color adjustment. Xu X Y agreed with this point, pointing out that the good lighting environment brought by intelligent interior design can improve people's visual comfort [12].

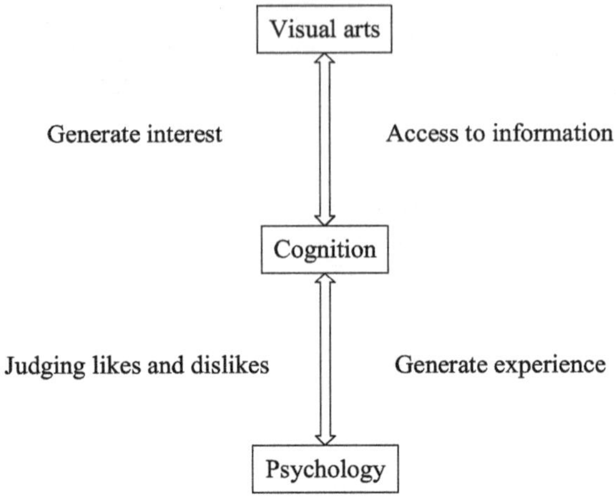

Fig. 2. Relationship between visual art and psychological cognition

3.3 Application of Topology Information Region Matching

Jiang H L considered that in the study of spatial structures, topology can analyze the spatial properties between objects and their relationships while considering shape invariance [13]. The interior scene areas have the characteristics of large color differences and relatively rich colors. This article believes that using topology information region matching method can effectively perform area segmentation and color matching.

Table 2. Role of topology information region matching in interior design

Role	Specific details
Space connection and layout optimization	Make the space smoothly connected and improve overall utilization efficiency
Homomorphic relationships and design consistency	Ensure structural similarity between different regions to maintain design consistency
Construction and Analysis of Topology Maps	Identify key nodes, channels, connectivity, and other information to provide strong support for design decisions
Intelligent planning and adaptive design	Make interior design more intelligent

Table 2 shows the role of topology information region matching in interior design. This article believes that the greatest role of topology information region matching is to fully utilize spatial connectivity and layout optimization. By utilizing the concepts of topology, designers can analyze the connectivity of interior spaces and understand the topological relationships between different areas. By matching regions with similar topological structures, designers can optimize the spatial layout, making the space smoothly connected and improving overall utilization efficiency. Considering the homomorphic relationship, designers can ensure structural similarity between different areas, thereby maintaining design consistency. This is crucial for large building projects or interior spaces with multiple functional areas, helping to create an overall harmonious environment. Furthermore, using topology diagrams, designers can more intuitively represent the structure and relationships of interior spaces. Similarly, matching based on topology information can make interior design more intelligent. By considering the topological relationships between different areas, the design can better adapt to user needs and changes in spatial functionality, providing more flexible design solutions. It can be said that from a certain perspective, the application of topology information region matching can promote the intelligence of interior design.

4 Results and Discussion

4.1 Comparative Experiments on Light Coverage

The method used in this article mainly improves the effectiveness of interior intelligent design from the perspectives of color and lighting. Therefore, this article aims to examine the method based on the utilization rate of lighting. The specific method is to select two sets of design schemes. One is a conventional design scheme, and the other is an intelligent design scheme based on the method proposed in this article. Both schemes can only use a fixed number of identical lighting equipment, and the final statistics of the lighting coverage under both schemes are conducted.

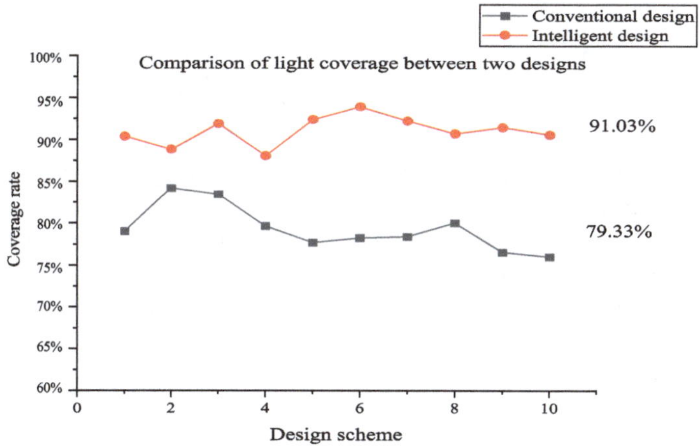

Fig. 3. Comparison experiment of light coverage between two designs

The experimental results of comparing the light coverage of two designs are shown in Fig. 3. In this experiment, based on the light coverage of 10 sets of each of the two designs, the average light coverage of the conventional design is only 79.33%, and the average light coverage of the intelligent design is 91.03%, which is a huge difference. It can be seen that intelligent design solutions based on topology information region matching can significantly improve the utilization of lighting.

4.2 Comparative Experiments on Popularity

Wu Y's research was focused on interior intelligent design, but his proposed approach was to use the Internet of Things for design [14]. In order to further verify the innovation of the method proposed in this article, a comparison is made between the former as the experimental group and the latter as the control group. A house is designed using two options and the willingness of respondents is collected to verify the popularity of each option.

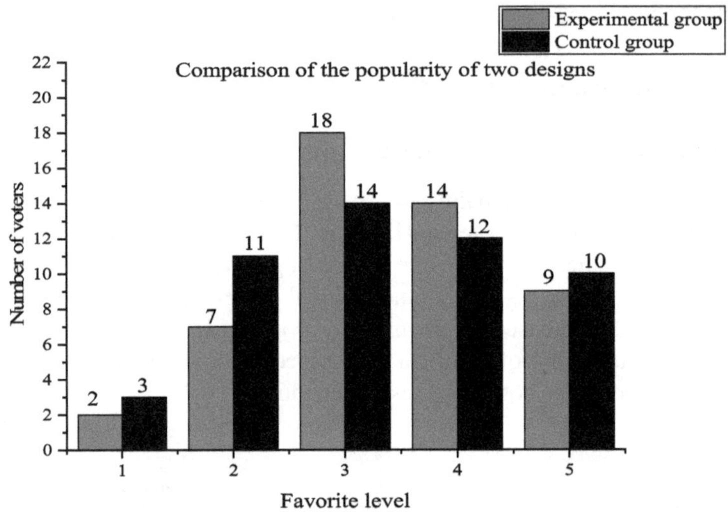

Fig. 4. Comparative survey on popularity level

Figure 4 summarizes the comparison survey results of the popularity of two design schemes. It can be seen from the results that the difference in popularity between the two design schemes is very small, and most respondents gave a score of 3 or 4 to the two schemes. However, the method proposed in this article receives more votes with a score of over 3, indicating that it has a certain degree of innovation compared with other mainstream intelligent design solutions.

5 Conclusion

The overall idea of this article is to analyze the principles of interior design and some points that need attention, in order to analyze the application of many intelligent technologies represented by intelligent lighting systems, laying a foundation for the application of topology information region matching. Finally, this article demonstrates the feasibility and innovation of the proposed method through comparative experiments and surveys. However, there are also shortcomings in this article. For example, a comparative survey of popularity is essentially subjective, and this subjective survey requires a larger scope of investigation to be universal, otherwise it is difficult to be convincing.

Overall, with the increasing development of intelligent technology, interior design is bound to develop rapidly, meeting the needs of users from multiple perspectives such as aesthetics, practicality, economy, and environmental protection.

References

1. Sydor, M., Bonenberg, A., Doczekalska, B., Cofta, G., et al.: Mycelium-based composites in art, architecture, and interior design: a review. Polymers **14**(1), 1–21 (2021)
2. Kim, J., Heo, W.: Interior design with consumers' perception about art, brand image, and sustainability. Sustainability **13**(8), 1–20 (2021)
3. Murali, P.K., Kaboli, M., Dahiya, R.: Intelligent In-vehicle interaction technologies. Adv. Intell. Syst. **4**(2), 1–27 (2022)
4. Charissis, V., et al.: Employing emerging technologies to develop and evaluate in-vehicle intelligent systems for driver support: infotainment AR HUD case study. Appl. Sci. **11**(4), 1–29 (2021)
5. Bao, Z., Laovisutthichai, V., Tan, T., Wang, Q., Lu, W., et al.: Design for manufacture and assembly (DFMA) enablers for offsite interior design and construction. Build. Res. Inf. **50**(3), 325–338 (2022)
6. Zhang, C., Zhang, S., James, J.Q., Yu, S., et al.: FASTGNN: a topological information protected federated learning approach for traffic speed forecasting. IEEE Trans. Ind. Inf. **17**(12), 8464–8474 (2021)
7. Wazirali, R., Ahmad, R., Alhiyari, S.: SDN-openflow topology discovery: an overview of performance issues. Appl. Sci. **11**(15), 1–31 (2021)
8. Luppi, A.I., Stamatakis, E.A.: Combining network topology and information theory to construct representative brain networks. Network Neurosci. **5**(1), 96–124 (2021)
9. Mei, Z.S.: Discussion on homestay architecture and interior space design. Cult. Des. **40**(13), 66–67 (2018)
10. Gao, M.Z., Hong, J.P.: Design of indoor intelligent lighting system based on single chip microcomputer. J. Hubei Normal Univer. (Nat. Sci.) **39**(3), 96–100 (2019)
11. Jin, Q.Z., Zhang, Q.L.: Intelligent lighting application of interior design based on visual art psychology. Light Lighting **46**(3), 33–37 (2022)
12. Xu, X.Y., Liu, M.J., Fu, B.C.: A review on indoor intelligent lighting control system. J. Suzhou Univ. Sci. Technol. (Nat. Sci. Ed.) **40**(2), 11–19 (2023)
13. Jiang, H.L., Quan, L.J., Wu, T.F., Qie, Q.: Multi task graph neural network algorithm for enhancing molecular topology information. Comput. Eng. Appl.. Eng. Appl. **59**(10), 86–93 (2023)
14. Wu, Y.: Application and development of internet of things technology in indoor barrier-free design. Shoes Technol. Des. **3**(2), 174–176 (2023)

Construction of Ancient Literature Shared Resource Database Based on Image Scanning and Recognition (OCR) Technology

Changqian Chen(✉)

The College of Literature and Journalism of Sichuan University, Chengdu, Sichuan Province, China
Changqian_Chen@21cn.com

Abstract. With the rapid advancement of information technology and the Internet, the digitization and sharing of ancient literary works have become an important research field. This study is dedicated to using image scanning and recognition (OCR) technology to build a comprehensive, efficient, and easily accessible ancient literature shared resource library. Initially, a broad literature review was conducted to understand the latest advancements and applications of OCR technology in document digitization. Subsequently, a series of representative and academically valuable ancient literary works were selected for scanning and recognition. During the scanning process, the system utilized high-resolution scanners and multi-angle lighting technology to ensure image quality. Experimental data reveals that the satisfaction levels of five parameters of the resource library, including recognition accuracy satisfaction, recognition speed satisfaction, resource database data satisfaction, resource detection speed satisfaction, and overall system satisfaction, ranged between 92.15% and 98.23%. This indicates that OCR algorithms and technologies like artificial intelligence can enhance the accuracy and efficiency of text recognition.

Keywords: OCR Technology · Ancient Literature · Resource Sharing · Data Collection · System Security

1 Introduction

Based on image scanning and recognition (OCR) technology, modern technology provides a powerful tool for digitizing classic works of ancient literature, thereby providing a more convenient way for readers to access them. OCR technology has been successfully applied in many fields [1] and has played an indispensable role in many fields [2]. Ancient literature, as a precious cultural heritage, contains the emotional experience and aesthetic spirit of the nation [3]. In order to better inherit and popularize these works, resource sharing is particularly important. It is not only an important lever for supply side structural reform, but also a key factor in innovation driven development in the modernization process [4]. With the vigorous development of sharing economy, various forms of sharing education have sprung up based on the Internet plus model [5]. However, in this process, research on data collection is not limited to machine learning, but also

Y. Pei et al. (Eds.): IC 2024, LNEE 1214, pp. 212–223, 2024.
https://doi.org/10.1007/978-981-97-4193-9_23

involves multiple considerations of data management systems [6]. More importantly, in terms of information security and privacy, continuous efforts must be made to ensure a secure and effective environment for users [7].

Through scholars' research, it can be concluded that the construction of an ancient literature shared resource library based on OCR technology is not only a fusion of technology and culture, but also a concrete manifestation of modern governance and innovation driven development. It covers multiple levels from technology application, cultural inheritance, resource sharing and data management, and even information security and privacy protection. This comprehensive project aims to better protect and disseminate the cultural heritage of the nation, while also providing a safe, efficient, and convenient resource sharing platform for modern society.

2 Construction of Shared Resource Library for Ancient Literature

2.1 Construction of Shared Resource Library

Nowadays, resource sharing has become a new mode of social and economic development [8], and in the information age of the 21st century, society is surrounded by various data, knowledge, and information. The value of this information lies not only in its content, but also in how it is effectively organized, stored, and shared. For enterprises, academic institutions, government departments, and even ordinary individuals, a structured and easily accessible shared resource library has become a key tool for improving efficiency, promoting innovation, and strengthening cooperation. However, with the explosive growth of information volume, building and maintaining a resource library that meets both current needs and can adapt to future changes has become increasingly challenging.

Resource sharing is a brand new carrier in the information age [9]. However, a successful shared resource library is not only a place to store information, but also an ecosystem that covers multiple stages of information collection, classification, storage, retrieval, and distribution. It needs to consider the needs of different users, provide flexible access permissions, and ensure the security and integrity of information. In addition, with the advancement of technology and changes in user needs, the resource library itself also needs to be constantly updated and optimized. The shared resource library framework is shown in Fig. 1.

In Fig. 1, the shared resource library construction framework is divided into client side, management side, and other functional side. Here is a brief introduction to these terminals:

The resource library primarily offers an intuitive user interface (UI), serving as the initial communication bridge between the system and users. It allows direct user access, enabling them to browse, search, upload, download, and share various data resources through its interface. Utilizing the most advanced algorithms, this interface can intelligently recommend relevant resources based on users' historical behavior and preferences. Additionally, the resource library includes a management interface specifically designed for administrators, forming the control core of the resource library. In this interface, administrators can set user permissions, ensure real-time security of resource access, and have the authority to categorize, label, and review resources, ensuring the

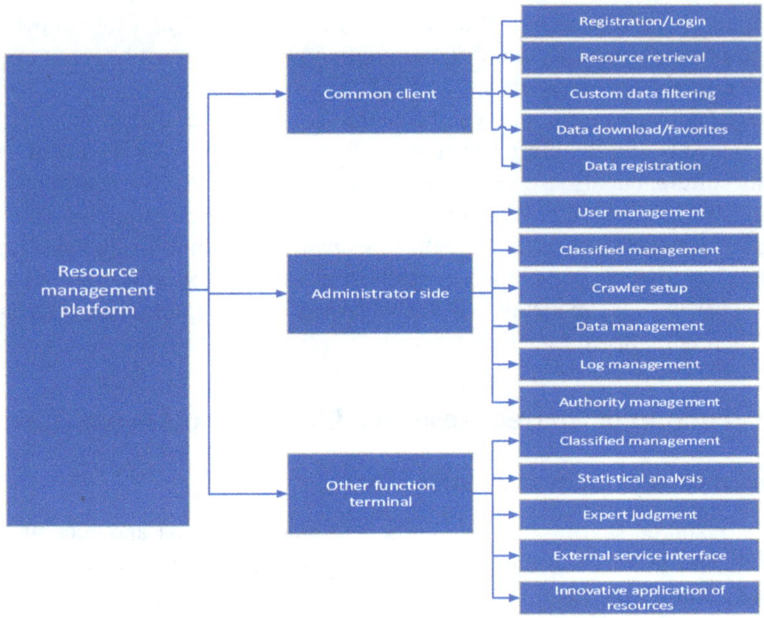

Fig. 1. Framework for building a shared resource library

stable operation of the resource library. To enhance external interaction and real-time monitoring, the resource library has further expanded mobile app support and opened API interfaces, allowing users to easily access it in various situations and providing interoperability opportunities with other applications. This not only improves the accessibility of the resource library but also strengthens its integration capabilities with external systems. Administrators can also use built-in data analysis tools to conduct in-depth analysis of the actual operational status and usage of the resource library. The resource library functions like an efficient machine, essentially meeting various user needs and providing convenient services. With its expandable features, it can also be upgraded at any time to adapt to future changes.

2.2 Key Points of Sharing Ancient Literature

The development theory of ancient literature includes the development theory of literary language, the development theory of literary form, the theory of era style, etc. [10], and ancient literature is an important heritage of human civilization, recording the thoughts, culture, and history of ancient society. In order to protect, inherit, and study these precious texts, sharing ancient literature has become an important task. The purpose of sharing ancient literature is to enable more people to have access to, learn and understand these texts, and promote the inheritance and development of ancient culture. In the process of sharing ancient literature, there are some main focuses that need to be paid attention to and valued.

Ancient literary works contain rich elements of moral education. The classic chapters and moral essence [11] are selected, and the sharing of ancient literature mainly focuses on protecting, inheriting and studying the cultural heritage of ancient literature. The following are the main focuses of ancient literature sharing:

At present, data security governance is in its infancy [12], but the protection and digitization of textual data: the protection of ancient literature is the foundation of sharing. Through digital technology, ancient literature, inscriptions, and epigraph are transformed into electronic documents or images to ensure their preservation and inheritance. This can prevent ancient literature from being damaged or lost due to time, environment, or human factors.

Construction of a text sharing platform: A dedicated sharing platform has been established for researchers, scholars, and enthusiasts to upload, download, and exchange resources related to ancient literature. These platforms can provide various forms of sharing, such as electronic documents, images, databases, etc., to meet the needs of different users.

Text translation and interpretation: The translation and interpretation of ancient literature are important contents for sharing. By sharing research results and academic papers, the research and understanding of ancient literature are promoted. Researchers can share their translation and interpretation results so that others can better understand and utilize ancient literature.

Academic exchange and cooperation: The sharing of ancient literature also encourages academic exchange and cooperation. Through academic conferences, seminars, and journals, researchers can share their research results, engage in academic discussions and collaborative research, and promote the development of ancient literature research.

Education and popularization: The sharing of ancient literature is also committed to promoting the education and popularization of ancient literature. By sharing teaching resources, textbooks, and course designs, students and the public are provided with more opportunities to understand and learn about ancient literature, thereby promoting the inheritance and dissemination of ancient literature.

Through the sharing of ancient literature, it can promote the protection, research, and inheritance of ancient literature, promote the tradition and innovation of ancient culture, and enrich people's understanding of history and culture.

2.3 Data Collection Ancient Literature Sharing System

By innovating the methods of big data analysis, contemporary organizational problems have been better studied. From this, it can be seen that the shared resource library of ancient literature is a treasure trove in today's digital era, which gathers rich ancient literature, literary works, and historical documents, providing convenient access for researchers, students, and literary enthusiasts. With the continuous development of technology, more and more ancient literature and literary works are digitized and opened to the world through the Internet. These resource libraries not only provide digital versions of ancient literature, but also provide various search, screening, and reading tools, making it easier for people to explore and study the precious heritage of ancient literature.

The learning of data collection is a greater trend and opens up many opportunities for new research. In the resource library for data collection, various types of ancient literature and literary works, including ancient poetry, prose, historical records, philosophical works, etc., can be found. Whether interested in ancient Greek and Roman literature, ancient Chinese literature, or literature from other regions and periods, these resource libraries can meet the demand. Interested content can be found through keyword searches, author names, work titles, or specific time periods. The directory of ancient literature shared resource library is shown in Table 1.

Explanation of the fields in Table 1:

Under the detailed and complex setup by engineering designers, users can conveniently access information about various ancient literary shared resource libraries. In addition to common digital libraries and well-known academic databases, there are also some resource libraries specifically dedicated to ancient literature that are worth mentioning. The Perseus Digital Library, for instance, focuses on providing digital versions of ancient Greek texts. Similarly, different countries and even university libraries have their own digitization projects, further enriching the online accessibility of ancient literary works. There are also specialized ancient literature websites and open educational resource platforms, providing rich platforms for exchange for researchers and learners. Such a diversified array of resources not only enriches the study of ancient literature but also offers more possibilities for academic exchange in related fields.

3 Construction of OCR Technology Ancient Literature Shared Resource Library

3.1 OCR Technology Ancient Literature Identification

In recent years, with the development of modernization, OCR technology has emerged. Ancient literature recognition is a task that utilizes OCR technology to convert ancient literature into editable text. OCR technology is implemented through steps such as image preprocessing, feature extraction, character recognition, text correction, and result output. However, the recognition of ancient literature may be more challenging compared to modern writing, as ancient literature may have issues with font variations and diverse writing styles. Therefore, in order to improve accuracy, ancient literary recognition may require more training data, specialized algorithm models, and manual intervention and correction.

OCR technology intelligent recognition: With the rapid development of big data artificial intelligence, machine learning, deep learning, and reinforcement learning technologies are gradually being widely applied. OCR technology can be used to identify ancient literature through the following steps, as shown in Fig. 2.

OCR is a recognition technology used to recognize images and characters in graphics. In Fig. 2, the main process of OCR ancient literature recognition is divided into:

By collecting images and documents containing ancient characters, the next step is to enhance these images, remove noise, and binarize them to improve the recognition rate of the text. A character database for ancient text is established, collecting and labeling samples of ancient characters, which may come from existing ancient books, dictionaries, or specific ancient text databases. Utilizing image processing and computer vision technology, the shape and stroke features of the text are extracted from the images. Through trained model algorithms, the extracted features are matched with the character database to identify each character. Considering the possible character shape variations

Table 1. Catalog of ancient literature shared resource library

Resource name	Document type	Language	Year of publication	Storage location	Access authority	Contact information	Main research field	Extra function
Library of Alexandria	Synthesis	Multilingual	Ancient times	Digital/physical	Publicity	Website/Email	Synthesis	Online forum
The Perseus library	Ancient Greece/Rome	Greek/Latin	Ancient times	Figure	Publicity	WEBSITE	Ancient Greece/Rome	Annotation tool
Ancient books network	Ancient Chinese	Chinese	Ancient to modern times	Figure	Partial disclosure	Website/Phone	Ancient Chinese	Search engine
JSTOR	Academic paper	English	Modern times	Figure	Subscribe to	Website/Email	Academic research	Citation management
National library	Synthesis	Multilingual	Ancient to modern times	Digital/physical	Partial disclosure	Website/Email	Synthesis	Reading space
Project MUSE	Ancient Chinese	English	Modern times	Figure	Subscribe to	Website/Email	Literature and Social sciences	Citation generator

and unknown typographical errors in ancient documents, it is necessary to correct the OCR results, which can be done by comparing with a known ancient text database to ensure accuracy. Finally, the OCR system converts the recognized text into an editable text format for further use by the user. This process provides an effective method for the digitization of ancient literature, making it more accessible for research.

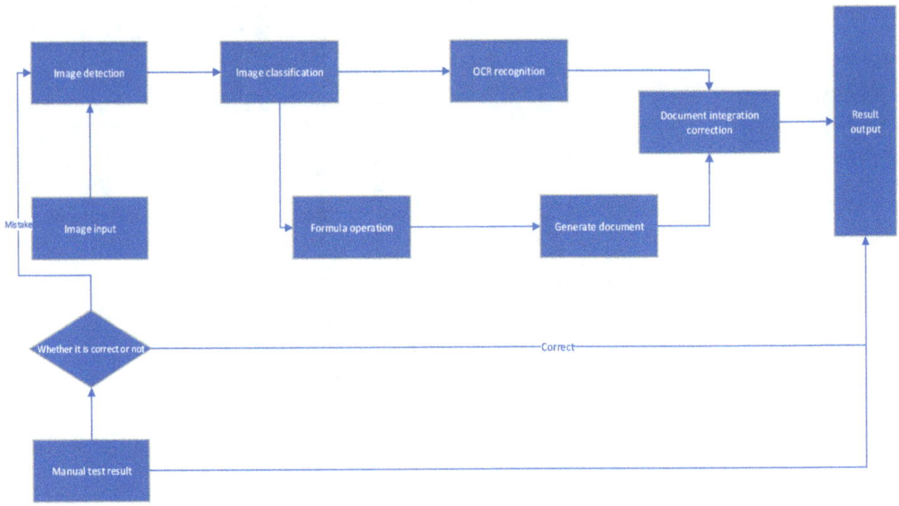

Fig. 2. OCR technology ancient literature recognition process

3.2 OCR Technology Ancient Literature Recognition Formula

The formula in the system shown in Fig. 2 is calculated as follows:

Image preprocessing formula (based on deep learning):

Inputting: original image (I).

Outputting: preprocessed image (I′)

$$I' = \text{DeepPreprocessing}(I, P) \tag{1}$$

Function: Through deep learning models, the original image is preprocessed, including image enhancement, denoising, smoothing, and other operations, to improve the accuracy and robustness of subsequent processing steps.

Feature extraction formula (based on convolutional neural network):

Inputting: preprocessed image (I′).

Outputting: feature vector (F)

$$F = \text{CNNFeatureExtraction}(I', M) \tag{2}$$

Function: A convolutional neural network model is used to extract semantic features from images. The image is transformed into high-dimensional feature vectors through multiple convolutional layers, pooling layers, and fully connected layers for subsequent character segmentation and recognition.

Character segmentation formula (based on conditional random field):

Inputting: preprocessed image (I′), feature vector (F).

Outputting: character positions and bounding boxes

$$\mathrm{B} = \mathrm{CRFSegmentation}\left(\mathrm{I}', \ \mathrm{F}, \ \mathrm{S}\right) \tag{3}$$

Function: Based on the conditional random field model, utilizing the pixel relationships and contextual information of the image, character segmentation is performed on the preprocessed image and feature vectors to obtain the position and bounding box of characters, providing accurate input for subsequent character recognition.

Character recognition formula (based on cyclic convolutional neural network):

Inputting: character image (C), feature vector (F).

Outputting: character recognition results

$$\mathrm{R} = \mathrm{CRNNRecognition}(\mathrm{C}, \ \mathrm{F}, \ \mathrm{T}) \tag{4}$$

Function: Based on a cyclic convolutional neural network model, taking into account the contextual information and local features of characters, character images and feature vectors are sequentially modeled and recognized to obtain the final character recognition results.

The steps in these formula examples are interrelated, and ancient literature is gradually extracted and recognized through steps such as preprocessing, feature extraction, character segmentation, and character recognition. The specific design needs to be adjusted and optimized according to the actual situation, taking into account various technologies such as image processing, feature extraction, character segmentation, and character recognition.

3.3 System Construction

In today's era, where OCR (Optical Character Recognition) technology is increasingly maturing, its use in the research and preservation of ancient literary digitization is becoming more important. Due to the complexity of ancient literature, traditional manual research methods can no longer meet the growing demands. To better utilize modern technology in interpreting and processing ancient literature, OCR technology is widely applied for the automatic recognition and digital processing of ancient literary works. Building an efficient and reliable ancient literature recognition system is key to achieving this goal. Through system construction, we can better protect and inherit the cultural heritage of ancient literature, promoting the development of ancient literature research. The system is as illustrated in Fig. 3.

The OCR technology used to construct an ancient literary recognition system in Fig. 3 mainly includes the following steps:

Data collection and preprocessing:

Image datasets containing ancient literature are collected, which can be obtained through channels such as the internet and libraries.

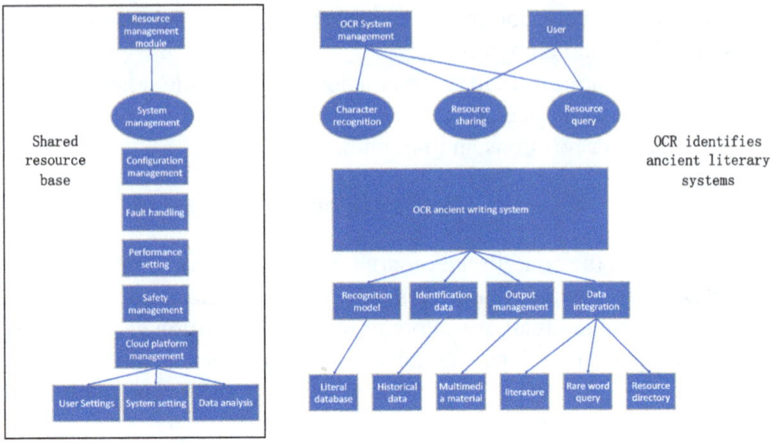

Fig. 3. OCR ancient literature shared resource database architecture

The collected images are preprocessed, including image enhancement, denoising, smoothing, and other operations to improve the accuracy and robustness of subsequent processing steps.

Feature extraction and selection:

Deep learning models such as convolutional neural networks (CNN) are utilized to extract features from preprocessed images and convert them into high-dimensional feature vectors.

Based on the importance and discrimination of features, appropriate feature subsets are selected to reduce computational complexity and improve recognition accuracy.

Character segmentation and positioning:

Based on models such as Conditional Random Fields (CRF), the pixel relationships and contextual information of the image are utilized to segment and locate characters in the preprocessed image, obtaining the positions and bounding boxes of the characters.

Character segmentation and localization can be achieved using methods such as pixel values, edge detection, and connected regions.

Character recognition and correction:

Based on models such as convolutional recurrent neural network (CRNN), character images are sequentially modeled and recognized to obtain the final character recognition results.

Language models, dictionary correction, and other methods can be used to correct and optimize the recognition results, improving recognition accuracy and semantic consistency.

The issue of data security not only involves the vast databases owned by enterprises, but also involves data at the national level. In Fig. 3, modules such as system security are added, and system security is crucial for the construction of OCR technology's ancient literary recognition system. Ensuring the security of the system can protect the data and information of ancient literature from unauthorized access and tampering. This helps prevent potential data leakage and information damage, and protects the confidentiality and integrity of ancient literature. By adopting secure data transmission and storage methods, as well as strengthening system authentication and access control, malicious attacks and illegal access can be effectively prevented, ensuring the reliability and stability of ancient literary recognition systems.

The research and application of OCR have made significant progress. Through Fig. 3, the system can achieve automated processing and analysis of ancient literature, providing a powerful tool for the research and protection of ancient literature. This system can help researchers quickly and accurately identify and interpret ancient literature, inscriptions, cultural relics, etc., thereby promoting the in-depth development of ancient literature research.

4 System Usage Experiment

The experiment randomly selects 500 students majoring in ancient literature, 200 teachers majoring in ancient literature, and 100 professors majoring in ancient literature from a certain university. By first using the traditional ancient literature recognition system (the first experiment), a comparison is made with the ancient literature shared resource library system based on image scanning and recognition technology (the second experiment). Five parameters, namely satisfaction with the recognition accuracy, recognition speed, resource database data, resource detection speed, and overall system satisfaction,

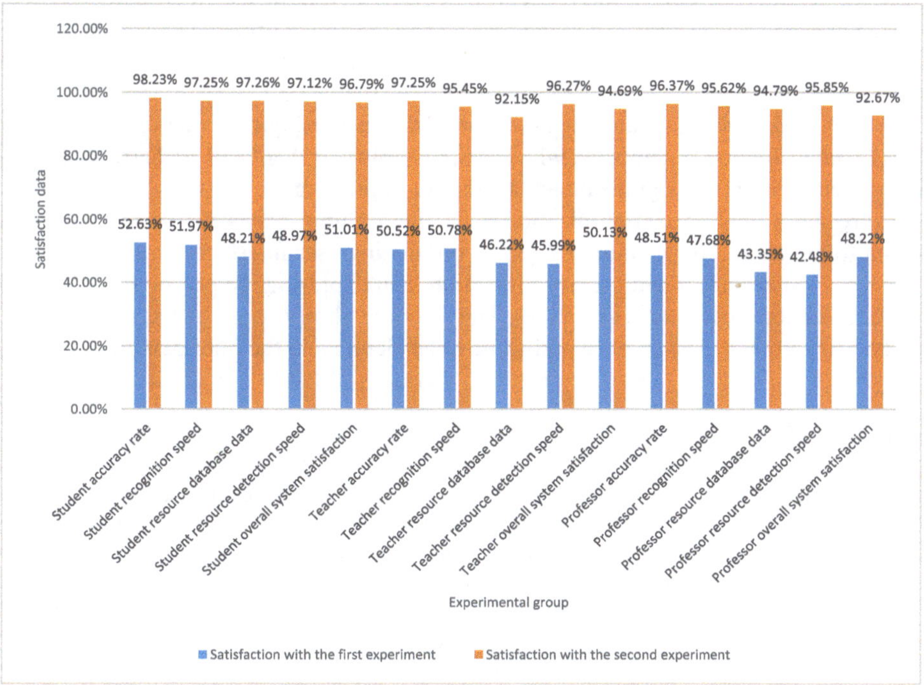

Fig. 4. System usage experimental data

are tested by the experimental subjects. The practicality of the system is judged by collecting experimental parameters from the experimental subjects. The experimental data is shown in Fig. 4.

The analysis of the experimental data in Fig. 4 indicates that among students, teachers, and professors specializing in ancient literature, the ancient literature shared resource library system based on image scanning and recognition technology (i.e., the second experiment) showed significant improvement in various metrics compared to the traditional ancient literature recognition system (i.e., the first experiment).

By analyzing the experimental data in Fig. 4, we can see that the satisfaction levels in the first experiment ranged between 42.48% and 52.65%, while in the second experiment, the satisfaction levels were significantly higher, ranging between 92.15% and 98.23%. This represents a substantial improvement.

In summary, the ancient literature shared resource library system based on image scanning and recognition technology is significantly better in practicality than traditional ancient literature recognition systems. This new system provides more efficient and accurate tools for students and educators majoring in ancient literature, significantly improving their work and learning efficiency.

5 Conclusions

By utilizing image scanning and recognition (OCR) technology, a comprehensive, efficient, and easily accessible ancient literature shared resource library can be constructed. After extensive literature review and experimental verification, the enormous potential of OCR technology in the digitization of ancient literature has been confirmed. The experimental data revealed that the new system achieved a high satisfaction rate of over 92% in all evaluation parameters, which clearly indicated that OCR and artificial intelligence technology had significant advantages in improving the accuracy and efficiency of text recognition. Further optimization of OCR algorithms can be achieved by expanding multilingual and multicultural support, optimizing user interfaces and experiences, enhancing data security and privacy protection, and exploring the potential application of OCR technology in other disciplines. The resource library based on OCR technology would bring greater convenience and value to academia and the public in the future.

References

1. Li, S., Li, F., Lei, P., et al.: Analysis on improving financial audit quality and efficiency based on OCR technology. China Manag. Inf. Technol. **25**(7), 103–106 (2022)
2. Wang, H., Dong, T., Zhang, X., Li, C.: Application of OCR technology in accounting automation. Sci. Technol. Econ. Guide **732**(34), 41–42 (2020)
3. Cao, M.: A new exploration on the reading teaching of classical works of ancient Chinese literature. J. Inner Mongolia Normal Univ. Educ. Sci. Edn. **32**(5), 103–108 (2019)
4. Liu, Q.: Construction of "Mass Innovation" service resource sharing mechanism in the process of national governance modernization. Adm. Forum **26**(02), 42–49 (2019)
5. Xu, Q., Li, X., Bulebrihan, I.: Research on course resource sharing mechanism based on NERC Nantong Branch Center platform. J. Xinjiang Radio Telev. Univ. **23**(4), 12–15 (2019)

6. Wei, L., Kuan-Ching, L., Long, J., et al.: An industrial network intrusion detection algorithm based on multifeature data clustering optimization model. IEEE Trans. Industr. Inf.Industr. Inf. **16**(3), 2063–2071 (2019)
7. Jalali, M.S., Landman, A., Gordon, W.J.. Telemedicine, privacy, and information security in the age of COVID-19. J. Am. Med. Inf. Assoc. **28**(3), 671–672 (2021)
8. Zhao, X.: Research on the development status quo and development dilemma of sharing economy – taking shared bicycles as an example. J. Hebei Youth Manag. Inst. **34**(6), 88–93 (2022)
9. Zou, M.Y.: Digital archive resource sharing strategy based on cloud platform. Vis. View **000**(005), 1 (2021)
10. Xie, W.: The development of ancient Chinese literature and the construction of marxist literature theory with Chinese characteristics. J. Zhejiang Ocean University Hum. Edn. **37**(1), 38–44 (2020)
11. Jin, Z.: The significance of infiltrating literary works in moral and rule of law teaching: a case study of ancient Chinese literature. Educ. Rev. **000**(035), 132–134 (2020)
12. Sheng, J., Amankwah-Amoah, J., Khan, Z., et al.: COVID-19 pandemic in the new era of big data analytics: methodological innovations and future research directions. Br. J. Manag.Manag. **32**(4), 1164–1183 (2021)

A Research on the Academic System in Universities Based on the One-Hot Encoding PAC Fuzzy Comprehensive Evaluation Algorithm

Zhiqian Lu[1], Yihan Liu[2](✉), and Qiang Li[3]

[1] College of Computer and Cyber Security, Hebei Normal University, Shijiazhuang, Hebei, China

[2] College of China Combustion Engineering, Hebei Normal University, Shijiazhuang, Hebei, China

LiuYihaninfo@21cn.com

[3] College of Career Technology, Hebei Normal University, Shijiazhuang, Hebei, China

Abstract. The influence of contemporary learning exhibits complex and multifaceted characteristics. This study, building upon a detailed explanation of the model, constructs a comprehensive set of evaluation indicators. By selecting survey questionnaire data from the 15th Electrical Engineering Cup on "Evaluation of the Impact of Artificial Intelligence on College Students' Learning" as a sample, this research proposes specific steps for a study on the impact of artificial intelligence on college students' learning based on the fuzzy comprehensive evaluation method. Initially, the one-hot encoding method 'get dummies' in pandas is used to process non-continuous numerical features, thereby fuzzifying the data. Subsequently, Principal Component Analysis (PCA) is employed to centralize the data, reducing its dimensionality by calculating covariance, eigenvalues, and eigenvectors. The primary components are then selected, and data is projected onto these components to obtain a dimensionally reduced dataset. The main method used in this article is PCA algorithm implementation through SVD decomposition of the covariance matrix for the purpose of data classification. The research results reveal that the fuzzy comprehensive evaluation model not only integrates subjective and objective factors but also allows for qualitative adjustments and batch processing of data, making the evaluation results more comprehensive and accurate. This effectively aids in a deeper understanding of the impact of artificial intelligence on college students' learning and provides robust support for related decision-making outcomes.

Keyword: Fuzzy Comprehensive Evaluation · One-Hot Encoding · Principal Component Analysis · Artificial Intelligence

1 Introduction

With the rapid development of information technology, artificial intelligence, as a cutting-edge technology, has shown enormous potential and prospects for application in various fields. The field of education is facing new challenges and opportunities. "Artificial Intelligence + Education" offers more possibilities at the teaching level, primarily

Y. Pei et al. (Eds.): IC 2024, LNEE 1214, pp. 224–235, 2024.
https://doi.org/10.1007/978-981-97-4193-9_24

focusing on intelligent analysis technology for assisting student learning, enhancement of teachers' intelligent technology application capabilities, development and sharing mechanisms for teaching resources, construction of intelligent platforms for teaching management and evaluation, technological support and upgrades for the learning environment, and many other areas [1]. For college students, it provides new ways of learning, resources, and personalized support, to some extent promoting the improvement of learning outcomes and the transformation of education. Accurately evaluating the impact of artificial intelligence technology on college students' learning is significant for fully leveraging the advantages of artificial intelligence, improving the learning efficiency of college students, and promoting the coordinated development of artificial intelligence and education.

Currently, numerous studies at home and abroad on the impact of artificial intelligence technology on learning achievements have been conducted, often using methods like classroom observation and survey questionnaires to identify influencing factors and propose corresponding strategies. However, due to the randomness in different technological application methods, application environments, and students, the research conclusions drawn from different research methods vary significantly [1], making it difficult to reach accurate conclusions. This study aims to design an evaluation indicator system for the impact of artificial intelligence on college students' learning based on fuzzy comprehensive evaluation, using survey questionnaire data from the 15th Electrical Engineering Cup topic "Evaluation of the Impact of Artificial Intelligence on College Students' Learning" as a sample to explore the effects of artificial intelligence on student learning outcomes.

2 Preliminary Work

Governments and researchers both domestically and internationally have conducted numerous studies on the impact of artificial intelligence applications on student learning outcomes. Domestic researcher Zhang Huihua collected data on the offering of artificial intelligence courses in primary and secondary schools nationwide, aiming to identify numerous influencing factors in course offerings and using the interpretative structural modeling method for structured analysis of these factors.

Boyi Xiao proposed a comprehensive status evaluation model for hydroelectric unit sets based on hierarchical analysis and improved fuzzy evaluation with Gaussian thresholds, effectively addressing issues of indicator weight and limit determination in existing unit status evaluation models [1].

Patel, A., and Jana, S. believe that intuitionistic fuzzy sets, as an extension of fuzzy sets, better represent and address the uncertainties in real-world problems. The proposed IFSM has been implemented in several pattern classification issues to prove its effectiveness and proposed a new facial recognition method. Using the proposed IFSM and entropy measure, the MADM technique IF-E-TOPSIS was constructed. This method was applied to software quality assessment, verifying its effectiveness [2].

Shengnan Ban's proposed method has significance for the integrity of underground gas storage in depleted natural gas reservoirs, where a good sealing cap rock is crucial. The weights of these factors were analyzed using hierarchical analysis. The Fuzzy Comprehensive Evaluation Method (FCEM) was applied to the sealability evaluation of cap rocks in Zhujiadun, Xu2, and Xing9, three typical underground gas reservoirs. Among the three gas fields, Zhujiadun gas storage had the best sealing cap rock. The evaluation procedure and method proposed in this paper have reference value for evaluating the tightness of cap rocks in depleted gas reservoirs [3].

Huang Bowen's team proposed a well-organized and comprehensive system for ecological fragility assessment and analysis methods. Taking Fujian Province as an example, by establishing a fuzzy evaluation model, index data was fuzzily mapped to evaluation scores for the calculation of the Ecological Vulnerability Index (EVI). Based on the results, spatial clustering analysis based on K-means, driving factor analysis based on GeoDetector, and spatial autocorrelation analysis were conducted. Finally, the ecological vulnerability under various decision-making attitudes was analyzed using the ordered weighted averaging operator. The research results have important guiding significance for the ecological protection and sustainable development of Fujian Province [4].

3 Research Design

3.1 Research Design Process

Initially, before analyzing the survey questionnaire data, systematic data preprocessing and cleaning were conducted to ensure data quality and accuracy. For vague responses in the questionnaire, we adopted a numeric strategy, converting vague descriptions into specific numerical representations, laying a solid foundation for subsequent analysis (Fig. 1).

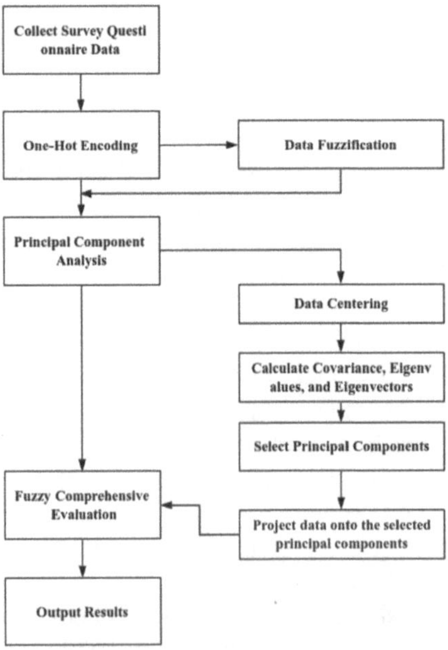

Fig. 1. Model Building Process

Next, to extract principal information from multiple related variables and simplify complexity, we applied Principal Component Analysis (PCA) to the numerically processed data for dimensionality reduction. The aim of this step was to reduce data dimensions while retaining as much useful information as possible, providing a clearer data basis for subsequent evaluations.

Finally, using the fuzzy comprehensive evaluation method, we comprehensively and accurately assessed the impact of artificial intelligence on college students' learning by considering the dimensionality-reduced data and related factors. By establishing clear computational processes and decision rules, we could better understand the role of artificial intelligence technology in college students' learning, providing a scientific basis for educational practice and policy-making.

3.2 One-Hot Encoding

One-Hot Encoding, simply put, also known as one-bit effective coding, is used in various fields and is a fail-safe method for encoding categorical variables. The process of one-hot encoding can be described through the following formulaic steps:

Suppose there is a feature element set containing n samples, comprising m distinct elements, i.e., the feature contains m different categories.

(1) Identify categories in the element set

Sequentially identify all different categories in the data, divided based on category, denoted as C1, C2, C3, C4…, Cm.

(2) Assign integer coding within categories

Assign a unique integer code to each category, with each category's code being distinct. Generally, coding starts from 0 and increments, denoted as 0, 1, 2, 3,…, m − 1.

(3) Create one-hot encoding matrices for each feature in the sample

Typically, each feature in the sample is processed, and a binary array of length m is created. Each unique category's index position is set to 1, with the remaining positions set to 0. For the jth feature of the ith sample, the corresponding one-hot encoding OHE(i, j) can be represented.

Usually, OHE(i, j)[k] = 1, though cases of overfitting may occur. If the feature's integer code is k, in remaining cases, OHE(i, j)[k] = 0, where k = 0, 1, 2, 3…, m − 1.

(4) Integrate encoding for multiple features

If feature values require encoding, traditional methods may be insufficient to solve the feature encoding issue, necessitating the merging of one-hot encodings into a matrix. This not only facilitates input for machine learning models but also allows for rapid encoding of the model. In this matrix, each row represents a basic sample, and each column represents a feature encoding.

(5) Apply an appropriate machine learning model

First, input the generated one-hot encoding matrix into a machine learning model. Choosing a suitable machine learning model becomes especially important. Use these encodings to understand the relationships between features, clarify the model, and assist in classification, regression, or other tasks.

The core idea is to use an N-bit state register to encode N possible states, each state having its independent register bit. At any time, only one of these bits is effective. The advantage of this encoding method is that it can effectively handle non-continuous numerical features, and its design and implementation are relatively simple. A notable advantage is that it does not require additional decoders or complex operations while extending the expressive capability of features to some extent, providing a powerful tool for data analysis and modeling [6]. Below is an example of data after one-hot encoding, as shown in Table 1:

Table 1. Data after One-Hot Encoding

Question	Original Data	After One-Hot Encoding
1. Gender	Female	[0, 1]
2. Major	Liberal Arts	[0, 1, 0, 0]
3. Common Means of Internet Access	Accessing the Internet via Mobile Phone	[1, 0, 0, 0, 0]
4. Weekly Internet Usage Duration	Less than 7 h	[0, 0, 1, 0, 0]
5. Do You Use Learning Software Tools	Yes	[1, 0]
6. When Do You Use Learning Software Tools	Whenever I Have Time	[1, 0, 0, 0]
7. Your Personality	Other	[1, 0, 0, 0, 0, 0]
8. Do You Access Learning Resources from Universities Nationwide	Yes	[1, 0]
9. Has Your Teacher Ever Recommended Learning Software	Recommended	[1, 0]
10. Would You Like to Use AI Learning Tools	Yes	[1, 0]

In summary: For categorical characteristic variables such as gender, major, grade, and personality, we can use the one-hot encoding method to convert them into numerical features. In this process, we normalized each feature dimension, mapping it to the [0, 1] range.

3.3 Principal Component Analysis (PCA)

Principal Component Analysis (PCA) is a statistical method used for data dimensionality reduction and feature extraction. Its main process involves transforming high-dimensional data into low-dimensional data through linear transformation, while retaining as much of the original data's information as possible. The main steps of PCA are outlined as follows:

(1) Data Centering

Generally, there are significant differences between different data features. To eliminate these differences, data centering is the first chosen method. Data centering allows each feature value to be subtracted from its mean, ideally aiming for a zero mean value. This not only simplifies subsequent computations and helps eliminate the impact of data translation, but also makes large-scale data easier to understand. The advantages of data centering are prominent, and the process can be described as follows:

$$x' = x - \mu \tag{1}$$

Here, x represents the original data, x′ the centered data, and μ the mean of the initial data.

(2) Covariance Matrix After Centering

In most cases, after centering, the data usually becomes more complex and rarely simple. It's necessary not only to calculate the covariance matrix C but also to compute the covariance matrix itself. Essentially, a covariance matrix is a symmetric matrix and not very complex. Its elements not only cover eigenvalues, but elements c_ab also represent the covariance between the ath and bth features, reflecting the variations and correlations of data across different feature values. The process of covariance matrix computation can typically be calculated using the following formula:

$$C = \frac{1}{n} X^T X \tag{2}$$

(3) Eigenvalue Decomposition of the Covariance Matrix

Usually, the covariance matrix undergoes eigenvalue decomposition, resulting in a set of eigenvalues that represent the data, along with their corresponding eigenvectors. Different eigenvectors constitute the transformation matrix Q, and the eigenvalues of Q represent the variance of data in the direction of the principal components. Therefore, the advantage of decomposing the covariance matrix lies in using eigenvector decomposition, which helps identify the main directions of data variation, eventually leading to the calculation of p eigenvalues and corresponding eigenvectors.

$$C = Q \Lambda Q^{-1} \tag{3}$$

where is the eigenvector matrix of the covariance matrix, Λ is the diagonal matrix of eigenvalues of the covariance matrix, and Q^{-1} is the inverse matrix of Q.

(4) The original data is multiplied by the principal component matrix W, resulting in an n × k matrix Y, which is the dimensionally reduced data. These data are the projections in the direction of the principal components, i.e., the new coordinate values. This paper completes the process of dimensionality reduction from 30 dimensions to 2 dimensions.

$$Y = XW \tag{4}$$

The data after Principal Component Analysis is shown in Table 2:

Table 2. Data after Principal Component Analysis

Question	Original Data	Principal Component Analysis
1. Gender	Female	[1.01051283]
2. Major	Liberal Arts	[1.010512]
3. Common Means of Internet Access	Accessing the Internet via Mobile Phone	[0.579412]
4. Weekly Internet Usage Duration	Less than 7 h	[0.57941231]
5. Do You Use Learning Software Tools	Yes	[−0.371491]
6. When Do You Use Learning Software Tools	Whenever I Have Time	[−0.37149114]
7. Your Personality	Other	[−0.37149114]
8. Do You Access Learning Resources from Universities Nationwide	Yes	[−0.37149114]
9. Has Your Teacher Ever Recommended Learning Software	Recommended	[−0.37149114]
10. Would You Like to Use AI Learning Tools	Yes	[−0.37149114]

In summary, PCA can map high-dimensional data to a low-dimensional space, preserving the main patterns of variation in the data, which is conducive to data analysis, visualization, and model building. The dimensionally reduced dataset can be used more effectively for various tasks, such as clustering, classification, and regression analysis, while also helping to understand the hidden associations and structures in the data.

3.4 Algorithm Design in Fuzzy Comprehensive Evaluation

In practical problem-solving, decision-making often involves a comprehensive consideration of multiple influencing factors. However, some attributes are fuzzy and difficult to quantify precisely, and cannot be simply judged as "good" or "bad." To address this challenge, fuzzy comprehensive evaluation decision-making has become an effective solution. It quantifies data using the fuzziness of membership functions and then selects appropriate quantitative evaluation methods for comprehensive assessment. Its uniqueness lies in its ability to uniformly handle quantitative factors in the problem while flexibly addressing differences between fuzzy attributes, making it highly reliable in practical applications [5]. The main parts of its process are as follows:

(1) Establishing the Factor Set and Evaluation Set

The factor set is a collection of various influencing factors that affect the evaluation object. Based on the results of the principal component analysis, we constructed the factor set as follows:

$$U = \{u_1, u_2\} \tag{5}$$

The evaluation set includes the collection of various possible outcomes for the evaluated object. Therefore, this paper constructs the evaluation set based on the above results, where v_1, v_2, v_3, v_4 represent the levels of excellent, good, medium, and poor:

$$U = \{v_1, v_2, v_3, v_4\} \tag{6}$$

(2) Establishing the Fuzzy Comprehensive Evaluation Matrix

We chose to use a fuzzy matrix to describe the membership degrees of different evaluation levels, reflecting the importance of different levels. Therefore, based on the aforementioned factor set and evaluation set, we reconstruct the fuzzy comprehensive evaluation matrix as shown below (Table 3):

Table 3. Fuzzy Matrix

Evaluation Indicator	Evaluation Level			
	v1	v2	v3	v4
u1	0.87	0.04	0.03	0.05
u2	0.05	0.91	0.02	0.007

In the constructed fuzzy matrix above, the evaluation levels and evaluation indicators form a table, where uij represents the membership degree of factor ui at some evaluation level vj. The fuzzy comprehensive evaluation matrix helps decision-makers understand the degree of impact of each influencing factor on different evaluation levels, facilitating optimal decision-making.

(3) Fuzzy Rule Operation

Often, calculating membership functions can result in information loss. To reduce information loss during calculation, we decided to use the balanced operator method, where secondary indicators and their sub-items can be well handled. The following formula represents the balanced operator operation process well:

$$\mu_c(x) = [\mu_A(x) \cdot \mu_B(x)]^{1-\gamma} \cdot [1 - (1 - \mu_A(x))(1 - \mu_B(x))]^{\gamma} \tag{7}$$

Here, γ represents the balanced operator.

(4) Defuzzification of Operation Results

To obtain accurate control quantities, we use the centroid method to solve the output of the membership function, namely:

$$v_0 = \frac{\sum_{k=1}^{m} v_k \mu_k(v_k)}{\sum_{k=1}^{m} \mu_k(v_k)} \tag{8}$$

x represents the final defuzzification result, and xi is the value point of the membership function. Table 4 is the final output of the fuzzy evaluation after fuzzification and defuzzification:

Table 4. Fuzzy Comprehensive Evaluation Output

Question	Normalization
1. Gender	[1.]
2. Major	[0.]
3. Common Means of Internet Access	[1.]
4. Weekly Internet Usage Duration	[0.]
5. Do You Use Learning Software Tools	[0.6880613018]
6. When Do You Use Learning Software Tools?	[1.]
7. Your Personality	[0.]
8. Do You Want to Access Learning Resources from Universities Nationwide	[0.2228133559]
9. Has Your Teacher Ever Recommended Learning Software	[0]
10. Would You Like to Use AI Learning Tools	[0.2228133559]

Through fuzzy comprehensive evaluation decision-making, it helps us better handle complex decision problems, especially those with fuzziness and uncertainty. It transforms fuzzy and subjective information into a mathematically operable format, making the decision process more scientific. By flexibly using membership functions and quantitative evaluation methods during the evaluation process, we can more comprehensively consider various influencing factors, providing more accurate bases for future decision-making and enhancing the reliability of decisions.

4 Conclusion

4.1 Research Analysis

The study applied fuzzy comprehensive evaluation methods to deeply analyze the impact of artificial intelligence on university students' learning in a certain province. The research results indicate that the impact of artificial intelligence technology on university students' learning is positive. According to the comprehensive evaluation results in Table 5:

Table 5. Comprehensive Evaluation Results

Evaluation Level	v1	v2	v3	v4
Result	0.42	0.52	0.03	0.03

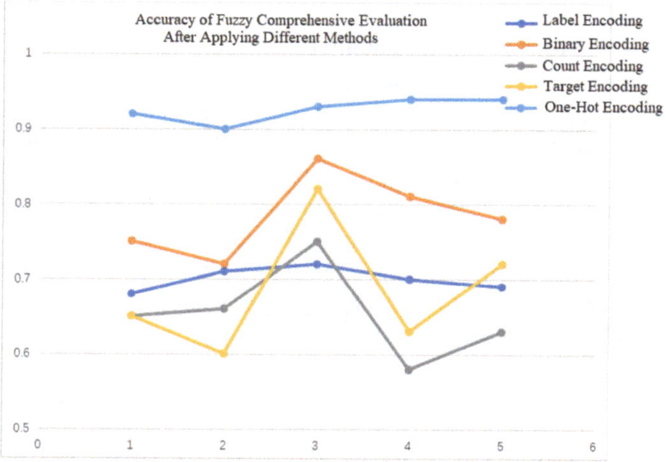

Fig. 2. Accuracy of Fuzzy Comprehensive Evaluation with Different Methods Applied

We compared the results of fuzzy comprehensive evaluation after using one-hot encoding with those of various other types of encodings. To minimize potential errors in machine learning, we conducted a full process five times. The results showed that the comprehensive evaluation value of one-hot encoding could reach up to 0.94, while the highest accuracy of other types of encodings was only about 0.85, demonstrating a clear advantage in accuracy (Fig. 2).

In summary, this study delved deeply into the role of artificial intelligence technology in university students' learning, highlighting its positive effects. The comprehensive evaluation results affirm the significant role of artificial intelligence technology in enhancing university students' learning outcomes, providing useful guidance and basis for further optimizing learning experiences and creative outcomes.

4.2 Development and Prospects

Fuzzy comprehensive evaluation has many potential applications in the field of education, offering numerous actionable aid plans for educators and boosting students' learning efficiency. The following case illustrates some potential applications of fuzzy comprehensive evaluation in education:

Dynamic academic early warning is an artificial intelligence-based academic alert system. In a student's academic life, each semester's grade for a subject is usually composed of many parts, such as homework, class attendance, classroom performance, mid-term exams, final exams, etc., each with different weights in the total grade. The goal of dynamic academic early warning is to timely collect the status of various grade indicators for students, calculate the current total score by multiplying each factor by its weight, and alert students accordingly. For example, if a student has a low homework completion rate, infrequent class attendance, and average classroom performance, the system detects a significant risk of failing and issues a timely warning, advising the student of relevant solutions.

Challenges and Considerations in the Development of Artificial Intelligence Technology:

The rapid development of artificial intelligence has gradually exposed both its advantages and disadvantages. For instance, data privacy and security is a critical area. National laws need to be strictly formulated to regulate the behavior and development of artificial intelligence. Additionally, the fairness of artificial intelligence warrants national consideration. Different algorithms have their advantages and disadvantages, and algorithmic bias is unavoidable. Therefore, minimizing the impact of algorithmic bias on student groups is a problem that still needs to be considered. While promoting the coordinated development of artificial intelligence and education, we must also properly handle issues of privacy and fairness, ensuring that technology can maximize its effectiveness without harming people's interests and privacy.

Acknowledgements. This thesis was jointly authored by Liu Yihan and Lu Zhiqian, both of whom contributed equally and are the co-first authors. Additionally, the successful completion of this thesis could not have been achieved without the care and assistance of various teachers. Special thanks are due to Professor Li Qiang and Professor Qiao Liyong for their guidance and help. They have contributed significantly to this topic, helping me overcome numerous challenges. I am deeply grateful to them!

References

1. Xiao, B., Zeng, Y., Zou, Y., Hu, W.: Hydropower unit state evaluation model based on AHP and Gaussian threshold improved fuzzy comprehensive evaluation. Energies **16**(15), 116–119 (2023)
2. Patel, A., Jana, S., Mahanta, J.: Construction of similarity measure for intuitionistic fuzzy sets and its application in face recognition and software quality evaluation. Expert Syst. Appl. **237**(Part B), 121491 (2024)
3. Ban, S., et al.: The application of the fuzzy comprehensive evaluation method in the sealing evaluation of caprocks in underground gas storage. Appl. Sci. **13**(17), 206–210 (2023)

4. Huang, B., Ruibo, Z., Shifa, C., Xuan, Z., Xingxue, J.: Fuzzy evaluation of ecological vulnerability based on the SRP-SES method and analysis of multiple decision-making attitudes based on OWA operators: a case of Fujian province. China. Ecol. Ind. **153**, 312–318 (2023)
5. Liang, J., Jiahao, C., Xueqin, Z., Yue, Z., Jiajun, L.: Anomaly detection based on one-hot encoding and convolutional neural networks. J. Tsinghua Univ. (Sci. Technol.) **59**(07), 523–529 (2019)
6. Gu, W., Wang, J.: A study on the impact of artificial intelligence technology on learning achievements—based on a meta-analysis of 40 experimental and quasi-experimental studies. China Adult Educ. **01**, 50–55 (2021)

Student Behavior Based on Machine Learning Algorithms and Big Data Technology

Jie Zhao and Xiaowei Dong(✉)

Sichuan University Jinjiang College, Meishan, Sichuan, China
DongXiaoweisc@21cn.com

Abstract. The traditional method of analyzing student behavior usually requires manual collection of student behavior data, or extraction of relevant data from school management systems and student information systems, and storage in databases for processing. However, this method is relatively time-consuming and cumbersome. With the development of machine learning algorithms and big data technology, today's student behavior analysis can be processed and analyzed using corresponding methods and tools. The Hadoop Distributed File System (HDFS) was used to collect and store student behavior data, while Decision Tree and Random Forest were selected to train and test student behavior data to improve the accuracy and effectiveness of student behavior analysis. Finally, student behavior was predicted. This article analyzed student behavior based on machine learning algorithms and big data technology, with a prediction accuracy of up to 95% for student grades. It can also detect and solve student behavior problems in a timely manner, providing targeted guidance and suggestions for students. Student behavior analysis based on machine learning algorithms and big data technology can help students and educators understand their individual differences and needs, provide personalized education and guidance, and promote the optimization of education and the comprehensive development of students.

Keywords: Decision Tree · Random Forest · Hadoop Distributed File System · Big Data Analysis · Student Behavior Analysis

1 Introduction

Learning management systems play a fundamental role in educational and teaching models, and analyzing and studying student behavior within the learning management system can improve the learning process model [1]. Student behavior analysis collects, processes, and analyzes data on students' learning and behavior, extracting information about their behavior and learning patterns. This can help schools and educators understand students' learning needs and behavioral habits, thus developing personalized educational strategies and support measures.

This article used machine learning algorithms and big data technology for student behavior analysis. Firstly, the source of student behavior data was clarified, and student grades were obtained from the academic administration system. The online learning

Y. Pei et al. (Eds.): IC 2024, LNEE 1214, pp. 236–244, 2024.
https://doi.org/10.1007/978-981-97-4193-9_25

trajectory of students was obtained from the learning management system, and a large amount of behavioral data obtained was stored in the Hadoop Distributed File System (HDFS). This system allows data to be distributed across multiple nodes for high availability. Subsequently, decision trees and random forest algorithm models were selected to train the data, and the model parameters were continuously adjusted during the training process to adapt to the characteristics of student behavior data and improve the accuracy of student behavior analysis. Then, the trained model was used to predict students' performance and behavior, such as grades, potential needs, etc., providing personalized guidance for students and more accurate decision support for educators.

2 Related Work

Through student behavior analysis, it is possible to predict students' academic achievements in advance, identify their needs and psychological states, and provide personalized education. Xiaoyong Li and others used a questionnaire survey method to collect student behavior data and analyze their behavior status. They proposed an unsupervised ensemble clustering framework that used student behavior data to discover behavior patterns and verified the relationship between different behavior patterns and student GPA [2]. T. S. Ashwin and others proposed a framework for analyzing inconspicuous student participation using nonverbal cues, and trained and tested the facial, gesture, and body postures of multiple students in a classroom environment. It was found that there is a positive correlation between students' level of engagement and exam scores [3]. Fuzheng Zhao and others proposed a learning behavior analysis framework based on result confirmation to explore the hidden reasons behind learning patterns and strategies, and conducted case studies on e-book reading data, proposing corresponding learning behavior strategies, and proving their effectiveness [4]. However, the structure of student behavior data is complex and diverse, with a large amount of data. The use of traditional data analysis methods is inefficient, and personalized guidance and suggestions cannot be provided.

This article analyzed student behavior through machine learning algorithms and big data technology. Machine learning algorithms make decisions and predictions by learning patterns and patterns from data. Big data technology is a method used to process and analyze large-scale datasets. Suresh Neethirajan studied the significant progress made by sensors, big data technology, and machine learning algorithms in animal husbandry and agriculture, helping animal farmers reduce production costs, improve efficiency and animal health, and increase the value of reducing environmental footprint [5]. Wei Li and others reviewed the application of machine learning technology in big data analysis in the healthcare field, emphasizing the various challenges faced by existing technologies. The research provided insights for healthcare practitioners and government agencies, enabling them to grasp the latest trends in intelligent medical big data analysis based on machine learning [6]. Anirbid Sircar and others reviewed machine learning solutions to problems in the oil and gas industry and introduced various types of machine learning algorithms that can process data from different departments in the oil and gas industry. They also demonstrated the benefits of big data storage capabilities and improved the performance of the oil and gas industry [7]. This article used machine learning algorithms and big data technology to process and analyze a large amount of student behavior data, providing feedback and improvement suggestions to students and educators.

3 Methods

3.1 Collection of Student Behavior Data

The recent grades of students are obtained from the academic administration system, and the online learning trajectory of students is obtained from the learning management system, such as learning duration, learning preferences, subject orientation, learning progress, and completion status. Table 1 shows the recording of student behavior data.

Table 1. Record table of student behavior data

Student number	Learning time	Learning preference	Learning progress	Number of questions	Number of responses
1001203	2 h and 35 min	Chemistry	48%	8	12
1003547	4 h and 56 min	Physics	67%	15	16
2006539	6 h and 6 min	Chinese	89%	23	18
2003412	3 h and 23 min	English	58%	18	20

3.2 Data Preprocessing

Data preprocessing is an indispensable part of machine learning, as the quality of data directly affects the learning ability of the model [8]. Firstly, data cleaning is required. Data cleaning uses various methods to identify and handle errors, inconsistencies, omissions, and outliers in the dataset, such as estimating erroneous values, standardizing inconsistent values, deleting missing values, and replacing outliers with mean values. Then, data integration is required, which involves merging data from different data sources and formats into a consistent dataset. Data integration can improve the efficiency of managing data access. Finally, for data conversion, it is necessary to transform the original data, change the distribution of the data, and make the data more in line with the requirements of the machine learning algorithm model. Through data preprocessing, raw data can be made more suitable for machine learning and other data analysis tasks, improving the quality and accuracy of analysis results.

3.3 Big Data Storage

After preprocessing student behavior data, it is necessary to store the data, as the structure of student behavior data is complex and the amount of data is large. This article uses HDFS to store the data on multiple nodes, achieving high availability of the data.

HDFS is a Master/Slave structure [9], mainly composed of DataNode, NameNode, SecondaryNameNode, and Client. When using HDFS to store data, the general process is as follows: first, the client needs to communicate with NameNode; the client sends a message to NameNode, and NameNode checks whether the client has write permission. If there is permission, NameNode checks whether a file with the same name already exists. If there is no file with the same name, NameNode generates a new file record, and then NameNode sends the corresponding DataNode list and a token to the client. The DataNode list is the address where the data is stored. Token is the key that the client needs to send before writing data to the DataNode. Next, after receiving the DataNode list and token, the client finds the first DataNode in the list and write data. After completion, the root replication parameter copies the data block to the second DataNode through a pipeline, and then gives it to the third DataNode. After all is completed, each DataNode sends block information to the NameNode, which records it and sends confirmation messages to the client one at a time. Finally, the client shuts down the process. Figure 1 shows the process of storing data in HDFS.

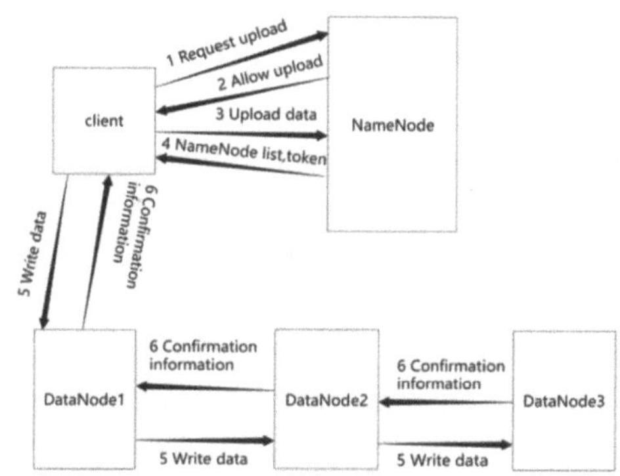

Fig. 1. HDFS storage data flow

HDFS can replicate multiple replica storage on different machines, so even if a node fails, the system can still provide reliable data storage and recovery mechanisms through backup replicas, reflecting its high availability characteristics.

3.4 Machine Learning Model Training

After storing data through the HDFS cluster, a total of 104572 valid student behavior data are obtained. These data are divided into 8 training sets and 2 testing sets for machine learning model training and testing. All machine learning technologies fit models with data, and specific algorithms are also diverse [10]. The machine algorithms used in this article to analyze student behavior are decision trees and random forests.

3.4.1 Decision Tree Training

Decision tree is a tree structure used to divide a dataset into multiple small decision units. Common decision tree algorithms include ID3, C4.5, and CART (classification and compression tree), and this article uses the CART algorithm. CART uses Gini exponent as the impure variable, reducing a significant amount of logarithmic operations.

In the classification problem, if there are K classes and the probability of the sample points belonging to the k-th class is P_k, then the Gini index of the probability distribution is defined as:

$$\text{Gini}(P) = \sum_{k=1}^{K} P_k(1 - P_k) = 1 - \sum_{k=1}^{K} P_k^2 \tag{1}$$

For a given sample set D, its Gini index is:

$$\text{Gini}(D) = 1 - \sum_{k=1}^{K} (\frac{|C_k|}{|D|})^2 \tag{2}$$

Among them, C_k is a subset of samples belonging to the k-th class in D, and K is the number of classes.

The Gini index is an indicator that measures the effectiveness of decision tree partitioning. By calculating the Gini coefficients of the current node and its child nodes, the purity and degree of separation of feature partitioning can be evaluated. The smaller the Gini coefficient, the higher the purity of the partition.

By constructing a decision tree model, it is possible to analyze and understand which factors interfere with students' learning and behavior, as well as predict and intervene in students' future behavior and learning situation.

3.4.2 Random Forest

Random forest is an ensemble learning model composed of multiple decision trees, which combines the advantages of randomness and decision trees. The random forest adopts the Bagging idea, which randomly selects a subset of n1 samples from the training set with size n, where ($n1 < n$). As the sample at the root node of the decision tree, m features are randomly selected from the features of sample M, satisfying $m < M$. Then, the optimal feature is selected from these m features as the splitting feature until it cannot be split. Finally, a large number of decision trees are established and a random forest is formed. Compared with decision trees, the random forest method can improve the prediction performance of imbalanced data due to its randomness [11]. Random forests are suitable for parallel computing, big data processing, and to some extent avoid overfitting, resulting in stable performance.

The use of random forest models can identify abnormal situations in students' learning and behavior, predict their academic performance, and help them achieve better learning experiences and grades.

4 Experimental Results

4.1 Decision Tree Gini Index

During the training process of the decision tree, as the number of iterations increases, changes in the Gini index value can be observed, as shown in Fig. 2.

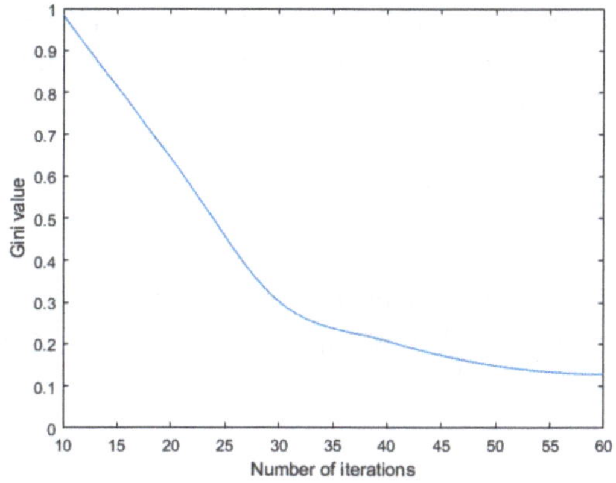

Fig. 2. Gini index change curve

As shown in Fig. 2, as the number of iterations increases, the Gini index value of the decision tree gradually decreases, indicating that with the learning of the model and data segmentation, the newly added nodes and branches can better partition the samples. The learning effect of decision trees is getting better and the prediction results are becoming more accurate.

4.2 Big Data Technology Throughput

HDFS supports continuous write and read operations of data, providing high throughput and high efficiency in processing and analyzing large-scale student behavior data. Figure 3 shows the throughput of HDFS when storing data.

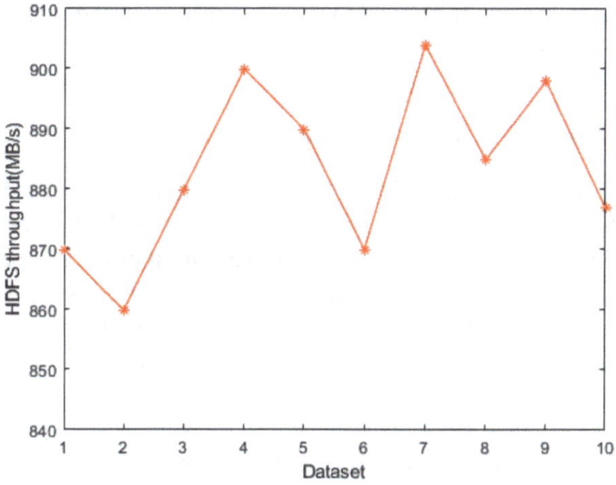

Fig. 3. HDFS storage data throughput

Figure 3 shows the throughput of 10 datasets stored in the HDFS cluster. The throughput is the read and write speed of the HDFS cluster over a period of time, which can be used as a measure of HDFS performance.

4.3 Performance Prediction Effect

This article uses trained decision trees and random forest models to test 9573 students from 10 schools and predict their final exam scores. Figure 4 shows the results of randomly selecting 10 students to compare their predicted grades with their actual grades.

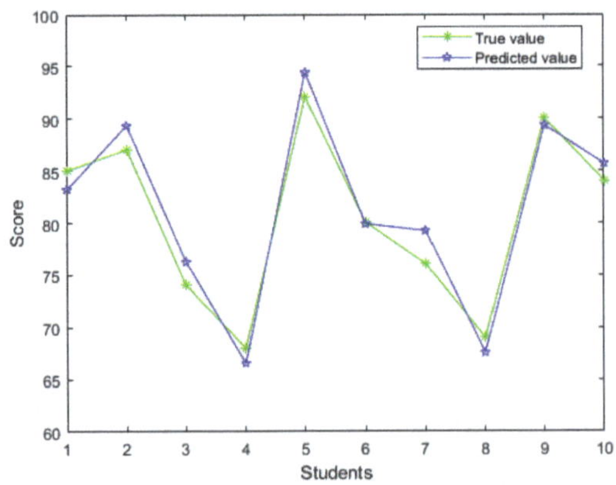

Fig. 4. Comparison between predicted and actual grades

From Fig. 4, it can be observed that there is not much difference between the predicted score and the actual score, indicating that the model has a good prediction effect, with an average prediction accuracy of 95%.

4.4 Effectiveness of Guidance and Suggestions

The trained model is used to analyze the behavioral data of 9573 students and provide each of them with 5 guidance and suggestions. Students evaluate these 5 guidance and suggestions based on their own actual situation. Among them, 6879 people believe that all 5 suggestions are valid; 1574 people believe that 4 suggestions are valid; 562 people believe that 3 suggestions are valid; 287 people believe that 2 suggestions are valid; 160 people believe that 1 suggestion is valid; 111 people believe that all suggestions are invalid. The results are shown in Fig. 5.

As shown in Fig. 5, most students believe that the suggestions provided by decision trees and random forests are both accurate and effective.

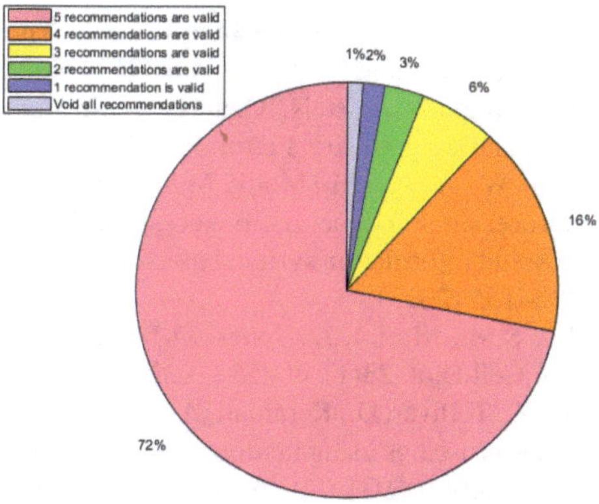

Fig. 5. Model guidance effect

5 Conclusions

Machine learning algorithms and big data technology have achieved certain results in analyzing student behavior. The use of big data technology can store massive amounts of student behavior data and has the advantage of high availability. Machine learning algorithms can be used to predict student behavior, learning outcomes, and learning needs, providing targeted support for students, and allowing educators to adjust teaching strategies in a timely manner to improve students' learning outcomes. However, when using machine learning algorithms and big data technology to analyze student behavior, it is necessary to pay attention to protecting student privacy and security, and avoid infringing on individual rights and interests of students. In the future, corresponding measures should be developed to address areas that need improvement, and machine learning algorithms and big data technologies should be applied in a reasonable and compliant manner. Machine learning algorithms and big data technology can continuously improve the education process and learning experience, making education more flexible and adaptable to students' needs.

References

1. Cantabella, M., Martínez-España, R., Ayuso, B., Yáñez, J.A., Muñoz, A.: Analysis of student behavior in learning management systems through a big data framework. Future Gener. Comput. Syst. **90**, 262–272 (2019)
2. Li, X., Zhang, Y., Cheng, H., Zhou, F., Yin, B.: An unsupervised ensemble clustering approach for the analysis of student behavioral patterns. IEEE Access **9**, 7076–7091 (2021)
3. Ashwin, T.S., Guddeti, R.M.R.: Unobtrusive behavioral analysis of students in classroom environment using non-verbal cues. IEEE Access **7**, 150693–150709 (2019)
4. Zhao, F., Hwang, G.-J., Yin, C.: A result confirmation-based learning behavior analysis framework for exploring the hidden reasons behind patterns and strategies. Educ. Technol. Soc. **24**(1), 138–151 (2021)
5. Neethirajan, S.: The role of sensors, big data and machine learning in modern animal farming. Sens. Bio-Sens. Res. **29**, 100367 (2020)

6. Li, W., et al.: A comprehensive survey on machine learning-based big data analytics for IoT-enabled smart healthcare system. Mob. Netw. Appl. **26**, 234–252 (2021)
7. Sircar, A., Yadav, K., Rayavarapu, K., Bist, N., Oza, H.: Application of machine learning and artificial intelligence in oil and gas industry. Petrol. Res. **6**(4), 379–391 (2021)
8. Bilal, M., Ali, G., Iqbal, M.W., Anwar, M., Malik, M.S.A., Kadir, R.A.: Auto-prep: efficient and automated data preprocessing pipeline. IEEE Access **10**, 107764–107784 (2022)
9. Qianru, W.: Design of hospital information system based on HDFS architecture. Inf. Comput. (Theor. Edn.) **35**(04), 133–135 (2023)
10. Greener, J.G., Kandathil, S.M., Moffat, L., Jones, D.T.: A guide to machine learning for biologists. Nat. Rev. Mol. Cell Biol. **23**(1), 40–55 (2022)
11. Zhou, X., Pan, L., Zheng, Z., Tolliver, D., Keramati, A.: Accident prediction accuracy assessment for highway-rail grade crossings using random forest algorithm compared with decision tree. Reliab. Eng. Syst. Saf. **200**, 106931 (2020)

Construction of a Rural Digital Governance Platform Based on Artificial Intelligence Technology

Guowei Li[1,2] and Zhe Luo[1](✉)

[1] School of Public Administration, Sichuan University, Chengdu, Sichuan, China
LuoZhescu@21cn.com
[2] Civil Aviation Flight University of China, Guanghan, Sichuan Province, China

Abstract. At present, China's rural society is in a period of profound changes and adjustments, and some new problems and contradictions have emerged. The original traditional means of rural governance are no longer able to effectively solve the newly generated problems and contradictions, and rural governance urgently needs to be transformed. The emergence of artificial intelligence not only effectively solves the problem of insufficient effectiveness of rural governance, but also promotes the transformation and development of rural governance subject, structure, scope and means. The emergence of artificial intelligence not only effectively solves the problem of lack of effectiveness of rural governance, but also promotes the transformation and development of rural governance subject, structure, scope and means. However, artificial intelligence is not a panacea, and there are still technical, ethical, and practical dilemmas to be cracked. Based on the above analysis, this paper put forward countermeasure suggestions such as establishing the concept of technical governance, building a data governance platform, innovating the mode of intelligent governance, and improving the application of governance technology, which provide experience and reference for the transformation of modern rural governance empowered by artificial intelligence and the construction of a new pattern of rural governance.

Keywords: Digital Life · Digital Governance · Management and Maintenance · Rural governance · Artificial intelligence · Creative application

1 Introduction

The new technological revolution represented by AI technology innovation marks our entry into the era of artificial intelligence. The development of AI is not only a simple technological innovation, but also changes the way of human production and life, gives birth to new economic and social relationships, and reshapes the social structure and order. The application of artificial intelligence in rural governance activities is a natural trend and is also the right thing to achieve the modernization of national governance system and governance capacity [1]. In order to accelerate the implementation of rural revitalization strategy, adapt to the new requirements of modern governance, respond to

Y. Pei et al. (Eds.): IC 2024, LNEE 1214, pp. 245–254, 2024.
https://doi.org/10.1007/978-981-97-4193-9_26

the aspirations of farmers for a better life, and respond to the new trend of integrating artificial intelligence into the operation of rural society, it is necessary to break through the traditional governance paradigm by means of artificial intelligence, and continuously improve the level of wisdom and security and resilience of rural governance. However, at the same time, we should also pay attention to the risks brought by AI, such as "algorithmic black box", data information security, technical dependence dilemma, etc., to avoid the deformation of AI into artificial "failure" [2]. This requires that in the process of AI-enabled rural governance, we should have the concept of people first, adhere to the problem-oriented, key breakthroughs, deep integration of big data, artificial intelligence, Internet of Things and other cutting-edge technologies, organic integration of multi-channel data resources, and give full play to the role of digital technology empowerment. It is necessary to focus on education, medical care, social security services and other livelihood protection areas, focus on the key aspects of rural governance and key ROC, thematic analysis, dynamic monitoring and active governance, and promote agile governance, scientific governance, and intelligent rural governance. The construction of a digital governance system in the countryside has become an unstoppable trend, allowing artificial intelligence to better empower rural governance, and leading the modernization of rural governance with technological innovation applications. Artificial intelligence provides great imagination and development opportunities for the transformation of rural governance. Integration of Artificial Intelligence in Rural Governance (See Fig. 1).

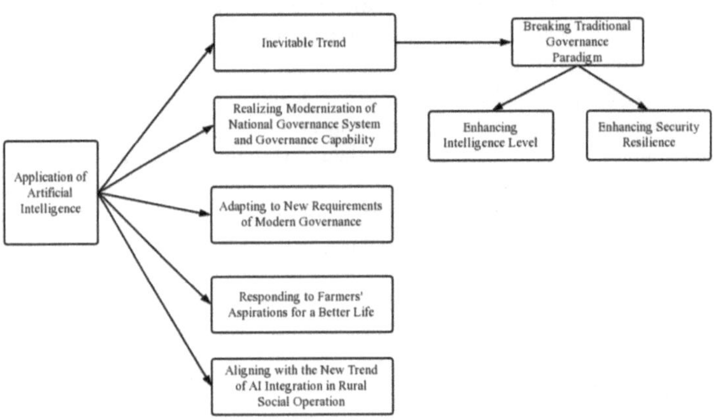

Fig. 1. Integration of Artificial Intelligence in Rural Governance

2 Governance Revolution: Artificial Intelligence for Rural Governance Transformation

2.1 The Main Body of Governance from "People-Oriented" to "Human-Machine Co-management" Change

The main body of rural governance has undergone the transformation from "official government" to "township governance" and from "mono-government" to "multi-government". The essence of the transformation is the change from "sole policy" to "group policy". The political and academic discourse on the main body of rural governance is mostly based on the plurality of subjects, reflecting the scientific and democratic nature of governance, and the main body of governance is essentially composed of people and human organizations [3]. In the era of artificial intelligence, governance objects are becoming increasingly complex, which puts forward higher requirements for governance decision-making and implementation. Human-machine co-governance becomes an inevitable choice. The "human-machine co-governance" of human-machine cooperation presents several characteristics: First, the governance decision-making is more scientific and democratic, artificial intelligence unblocks the information transmission channels and makes up for the information asymmetry and knowledge gap of the governance subjects, which can better improve the quality of the participation of multiple subjects in governance; second, the governance means are more scientific and the governance technology is more sophisticated. The application of artificial intelligence can promote more convenient and efficient governance and better realize public values. As shown in Fig. 2, the smart greenhouse control system, aided by artificial intelligence, helps to address the relatively backward production methods in rural areas.

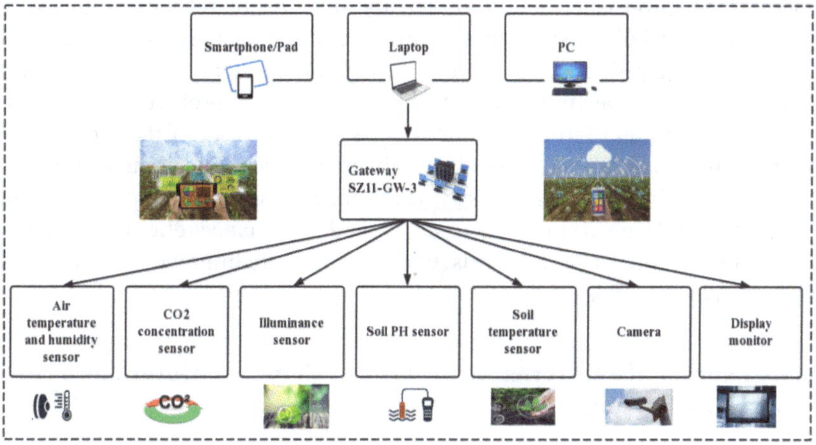

Fig. 2. Smart Greenhouse Control System

2.2 The Change of Governance Structure from "Hierarchical Control" to "Affirmative Decentralization"

For a long time, under the influence of the hierarchical organizational structure, the Party committee government of the township, which has a higher level of authority, has taken the central position in village management activities, guiding the village Party branches and villagers' autonomous organizations to carry out rural management activities. This hierarchical structure, while ensuring the orderly operation of rural governance activities, also poses a number of problems: On the one hand, social organizations, enterprises and institutions are easily excluded from the management system; on the other hand, village cadres are too tired to cope with various administrative affairs arranged by their superiors and cannot spare their energy to truly serve the villagers [4]. In the era of artificial intelligence, by building an open sharing platform for big data and integrating data resources held by multiple subjects and industries, the information gap between different subjects such as grassroots government, villagers, social organizations and enterprises and institutions is effectively reduced, and their awareness and ability to participate in rural governance is significantly enhanced, effectively easing the work pressure of village cadres and forming a more egalitarian and decentralized governance structure.

2.3 The Change of Governance Scope from "Fragmented Governance" to "Holistic Governance"

Vigorously promoting the modernization of rural governance system and governance capacity is an inevitable requirement to achieve comprehensive rural revitalization, consolidate the Party's ruling base in rural areas, and meet the needs of rural people for a better life [5]. Traditional rural governance decision-making mainly relies on the experience-based governance tradition, and the division of responsibilities is mostly based on the historical division of labor experience. The decision-making inertia of rural governance under empirical governance is strong, and farmers are more adaptable and face serious information asymmetry problems, and may face problems such as unscientific decision-making and farmers' numbness. With the advent of the information age, farmers' information channels are gradually diversified, and the amount of information received and output has greatly increased, so relying on traditional empirical governance can no longer meet the needs of governance. Digital governance relies on big data, information technology, and decision models, which will greatly improve the scientificity and feasibility of decision-making.

2.4 The Change of the Governance Tool from "Empirical Governance" to "Data Governance" Change

In order to meet the trend of whole government reform since the 21st century, Western countries are vigorously promoting the digital transformation of government. At the same time, the digitalization of government is necessary for the transformation of the scope of governance from "fragmented governance" to "holistic governance" [6]. At present, the problem of fragmentation of rural digital services is highlighted, with a wide gap in information services between urban and rural areas and between regions, and uneven digital construction conditions. The theory of holistic governance requires rural governance to achieve cross-regional integration, cross-sectoral integration and cross-functional integration to build a new system of rural governance. The construction of digital government under artificial intelligence and big data inevitably requires the transformation of the scope of government governance from "fragmented governance" to "holistic governance".

3 The Governance Dilemma: Barriers to Artificial Intelligence Embedded in Rural Governance

With the development of economy and society and the progress of technology, artificial intelligence has played a great role in rural governance. The popularization of the application of technical devices has greatly improved the efficiency of rural governance. However, there are still some obstacles to the embedding of AI in rural governance, mainly in the areas of technical generalization and subject controversy, data information security, and insufficient governance foundation.

3.1 Ethical Dilemma: Technology Generalization and Subjectivity Controversy

Since ancient times, China has formed a deep-rooted bloodline governance model based on clan blood relations, which has a strong color of human governance. Along with the widespread use of artificial intelligence, the main body of governance has shifted from "human-oriented" to "human-machine governance", which poses a challenge to the traditional main body of rural governance. First, the technology generalization and subject function degradation. In the era of artificial intelligence, rural governance relies on the results of big data processing, and the scientific nature of governance is manifested through big data operation, weakening the basis of human-centered governance. This is essentially the process of governance subjects gradually accepting the rigid results of big data processing. In this process the subject function of human as the subject of governance is gradually degraded. Second, the challenge of fairness of participation of governance subjects. The application of artificial intelligence in rural governance relies on various intelligent devices, data software and other technical carriers, while there are differences in the wealth of rural residents, and differences in individual understanding and learning of intelligent devices, which may lead to natural disparities in the ability and level of participation of governance subjects in rural governance, and then affect the willingness of governance subjects to participate in rural governance, resulting in the participation and effectiveness of rural governance. This will affect the willingness of governance subjects to participate in rural governance and lead to a lower participation and effectiveness of rural governance, which is not conducive to the role of grassroots self-government.

3.2 Technology Dilemma: Data Security Risks

Protection of privacy is a respect for human freedom and dignity, and is an important feature of the progress of human social civilization. Collecting massive personal information and behavioral records, and then running calculations through big data for decision making is a distinctive feature of rural governance in the era of artificial intelligence, but it also brings a series of data information security risks [7]. First, the technical risk of data privacy leakage. Although artificial intelligence systems have made great progress with the advancement of technology and technological development, there are still objective risks such as computer viruses, hacker attacks and damage to storage media, which pose security risks to the personal information stored in the system. Second, human intervention in data privacy. The collection of data information on the one hand is conducive to promoting the efficient operation of rural governance, but on the other hand, it also puts forward higher requirements on the quality of data managers. Currently, China is vigorously promoting rural revitalization, and the rapid development of rural areas has multiplied the opportunities and challenges of governance, but data management still cannot avoid human intervention, thus laying the hidden danger of malicious human intervention in data privacy. Rural governance is essentially the provision of facilitated services to rural residents through systematic and scientific management, with the aim of better promoting rural development and enhancing the well-being of residents. However, the security risks posed by the large number of personal behavior records collected by big data pose a challenge to the core concept of people-centered rural governance. As shown in Fig. 3, Data Information Security Risks and their Challenges to Rural Governance.

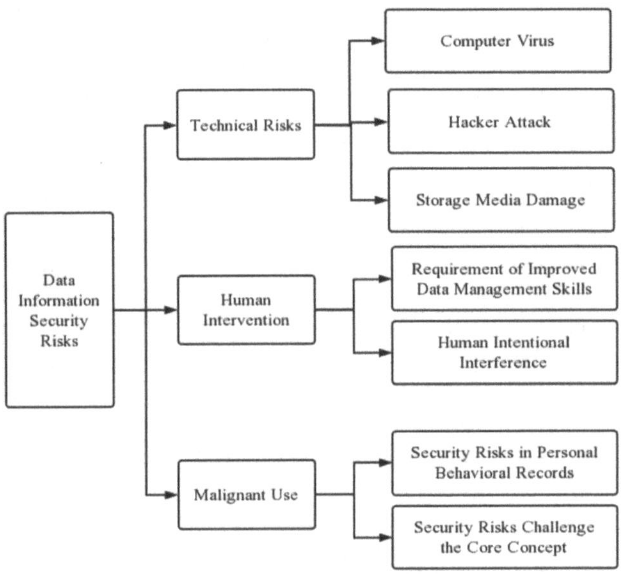

Fig. 3. Data Information Security Risks and their Challenges to Rural Governance

3.3 Realistic Dilemma: The Lack of Governance Concepts, Infrastructure, and Talent Pool

In the era of artificial intelligence, rural governance faces many changes. However, due to historical and practical reasons, rural governance in the era of artificial intelligence is very scarce in terms of governance concept, infrastructure, and talent team [8]. First, the governance concept is backward. In recent years, both the theory and practice of governance have evolved rapidly. Relying on information technology and intelligent devices, urban governance has rapidly completed iterative innovation. Compared with urban areas, rural areas are lagging behind in the popularization of governance theories and practices, and the governance concepts of governance subjects are relatively backward, making it difficult to adapt to the rapidly changing needs of modern governance.

Second, the lack of infrastructure. Due to the historical reasons such as the dual structure of urban and rural areas and the "price scissors" between workers and farmers, as well as the practical reasons such as backward economic and social development and backward public services, rural areas are more backward in economic and social aspects compared with cities, and the infrastructure construction is lagging behind. In addition, despite the development of technology and the increasing civilianization of technical equipment, rural economies are more backward and under greater pressure to procure equipment and purchase technical services, making it difficult to effectively support the construction needs of rural governance in the era of artificial intelligence. Third, there is a shortage of talent. For a long time, a large number of young and prime labor force outflow, resulting in the increasingly "hollowing out" and "aging" of the countryside. There is already a shortage of talent. Limited by a lack of job opportunities and poor public services, the number of new and outgoing talent is close to zero. In recent years, although China has been vigorously promoting rural revitalization, focusing on revitalizing rural stock talents and further promoting incremental and flow talents to build the countryside, it has achieved very little. Overall, the rural governance talent pool is still difficult to meet the demand for governance.

4 Governance Innovation: Artificial Intelligence Empowers Modern Rural Governance

4.1 Strengthen AI Awareness and Education and Establish the Concept of Data Governance

Strengthening the publicity and education of emerging technologies such as artificial intelligence in rural areas of China is the key to promote the transformation of rural governance from "human governance" to "human-machine governance". It is also the way to promote the transformation of rural governance in the new era [9]. For a long time, due to the dual structure of urban and rural areas, the concept, education and technology in rural areas in China are lagging behind in different degrees compared with urban areas, resulting in the application of new technologies in rural areas. To promote the transformation of rural governance in the era of artificial intelligence, first of all, the government should establish the concept of data governance, and establish a management system of "talking with data, managing with data, making decisions with data, and innovating with data". Multiple governance bodies should open and share data, break the barriers of data governance, and fully realize system integration, data sharing and function linkage. Secondly, the degree of understanding and ability to use emerging technologies such as artificial intelligence among rural cadres is enhanced through special lectures, course training and other forms of education. There is a need to transform their governance concepts and enhance their ability to perform their duties to facilitate the transformation of rural governance and promote rural economic and social development.

4.2 Increase the Construction of New Infrastructure and Build a Data Governance Platform

At present, the digitalization of China's rural areas has achieved certain results despite the accelerated construction of 5G and other information technology infrastructure. However, compared with urban digitalization and smart city construction, the gap between rural informatization is large, the urban-rural digital divide persists, and the digitalization process in rural areas still lags much behind [10]. Therefore, we should strengthen the digital infrastructure of the countryside, take building information model data as the skeleton, integrate and converge data resources such as basic geographic information and IOT perception, overlay thematic data such as population data and public services, integrate government data, social data and Internet data, and complete the construction of a data governance platform for the rural areas. We need to build a rural governance perception system with "all-time perception and multi-dimensional monitoring", and realize the digitization of resources through three-dimensional modeling, data identification, real-time state perception, real-time data analysis, and model scientific decision-making, so as to form a digital twin space with all-round "mirror mapping and real-time interaction" between the real countryside and the information model. The basic framework for digital rural governance is shown in Fig. 4.

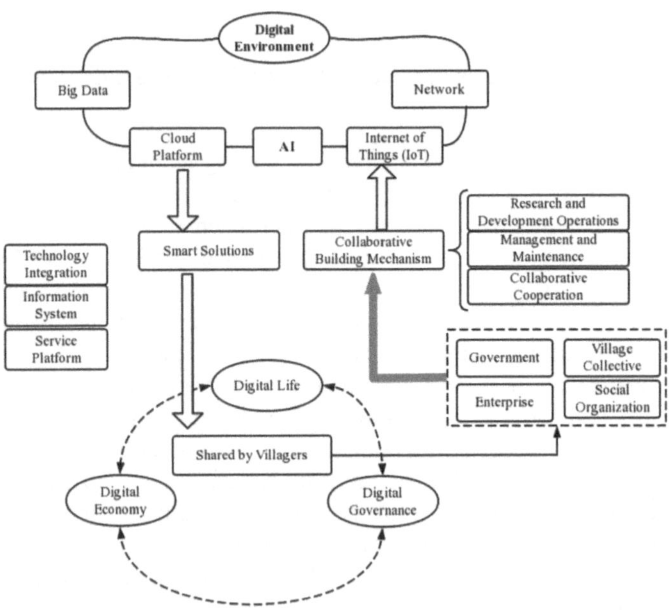

Fig. 4. Basic Framework for Digital Rural Governance

4.3 Innovative Rural Intelligent Governance Model to Achieve Accurate Problem Identification

The core of rural governance in the new era lies in the precise identification of problems. Under the premise of "human-machine governance", artificial intelligence brings technical empowerment to the realistic scenario of accurate identification of rural governance problems. We should rely on data platforms, intelligent models and perception systems to achieve accurate prediction and identification of problems, with emphasis on strengthening accurate monitoring technology for the habitat environment and intelligent analysis technology for agricultural-related public opinion, and establishing a closed-loop disposal mechanism for the whole process of "prediction-identification-disposal" to ensure accurate and efficient disposal of problems. We should deeply integrate big data, artificial intelligence, Internet of Things and other cutting-edge technologies, organically integrate multi-channel data resources, focus on education, medical care, social security services and other areas of livelihood protection, conduct thematic analysis, dynamic monitoring and active governance of the "hot events" that have been complained about many times, and innovate a rural intelligent governance model.

4.4 Establish Intelligent Governance Decision-Making Mechanisms and Improve the Application of Governance Technologies

The continuous iterative development of artificial intelligence technology will promote the continuous improvement of intelligent decision-making. One of the keys is to establish an intelligent governance decision-making mechanism. First, to strengthen data collection, to achieve intelligent decision-making needs to have a comprehensive, massive data as support. We should innovate and explore the application of "one sheet", build a pool of data resources, data resource library, integrate the collection of all kinds of data reports filled out manually to form "one sheet" for all the data collection, so as to achieve a clear base of all kinds of data, and promote the collection of data work orders, the intelligent and convenient, to realize the existing data without secondary collection, incremental and updated data with regular reminders, automatic generation and intelligent pushing of collected data, and "one collection, multiple sharing". Second, the application of technology should be strengthened. In the process of China's digital countryside construction, the scope of application of big data technology and the quality of application have increased significantly. In order to ensure the effective implementation of intelligent decision-making functions, it is necessary to pay attention to the application of various advanced information technology, especially to strengthen the application of artificial intelligence. We should use artificial intelligence to analyze the data resources of rural society in order to grasp the characteristics and development demands of rural governance itself, establish a scientific decision-making platform relying on big data, promote the precision of rural governance, and improve the effectiveness of governance.

5 Conclusion

The transformation of rural governance is a long-term task, in which artificial intelligence will play an important role. In recent years, with the promotion of digital village construction and the implementation of rural revitalization strategy, many regions in

China have made use of digital empowerment to promote the organic coupling of artificial intelligence and rural governance multi-systems, and the transformation of rural governance has achieved phased results. However, we also need to see the obstacles to the transformation of rural governance in the era of artificial intelligence, and need to promote the deep integration of artificial intelligence and rural governance through the innovative application of governance concepts, models and mechanisms.

General Secretary Xi Jinping pointed out in the report of the 20th Party Congress that "from now on, the central task of the CPC is to unite and lead the people of all nationalities to build a strong socialist modern state, achieve the second hundred-year goal, and promote the great rejuvenation of the Chinese nation with Chinese modernization". China has entered a new journey of building a socialist modernized country and marching towards the second century goal. In the process of promoting the modernization of the country, rural areas are still the difficult areas to crack the problem of unbalanced and insufficient development, and are the short board in the construction of common wealth. To a certain extent, the modernization of rural governance is a necessary condition for national modernization and an important indicator for the completion of a strong socialist modernization. To realize the modernization of rural governance, the key is to embed new technologies such as artificial intelligence in all aspects of governance and promote the transformation of "intelligent" rural governance.

Acknowledgement. This article is a general research project of the Ministry of Education's Humanities and Social Sciences, titled "Research on the Path of Improving the Happiness of Low-Income Rural Groups from the Perspective of Equalization of Basic Public Services" (20YJAZH074); It is a phased research achievement of the China Civil Aviation Flight College's general project (J2021-117).

References

1. Wei, B., Ang, Q., Li, A., et al.: Study on the path of digital village construction. Hubei Agricult. Sci. **61**(14), 183–187 (2022)
2. Lin, P.: Risks of educational artificial intelligence and its governance in the context of risk society. China Adult Educ. **04**, 19–24 (2022)
3. Li, Z., Han, Q.: Holistic governance: the way of rural governance. J. Tianjin Univ. Commer. **42**(04), 61–66 (2022)
4. Wang, X.: Rural revitalization and rural governance modernization. J. Shanxi Normal Univ. (Soc. Sci. Edn.) **49**(01), 53–60 (2022)
5. Huang, B.: The dilemma of rural governance and the way out. Agricult. Econ. **01**, 83–85 (2022)
6. Qiu, Z., Li, Y., Xu, W.: Digitalization and the change of rural governance structure. J. Xi'an Jiaotong Univ. (Soc. Sci.) **42**(02), 74–84 (2022)
7. Wang, C., Liu, N.: Privacy crisis and trust reconstruction in the era of artificial intelligence. Editorial Friend **08**, 101–106 (2021)
8. Zimin, L.: Artificial intelligence + rural social governance coupling and practice - taking "Zheli Xing village common wealth" as an example. Reg. Gov. **01**, 21–24 (2022)
9. Huang, S.: The dilemma and governance of digital identity anxiety - in the context of digital village strategy. Soc. Sci. **04**, 82–89 (2022)
10. Liya, W., Ye, J.: On artificial intelligence embedded in rural construction. J. Soochow Univ. Philosophy Soc. Sci. Edn. **42**(05), 34–41 (2021)

Design of a Tennis Training Assistance System Under Sensory Recognition Technology

Chengmei Wang(✉)

Chongqing Vocational Institute of Engineering, Chongqing, China
Wang_Chengmei@21cn.com

Abstract. With the gradual scientific development of current tennis training methods, tennis training assistance systems play an important role in improving training effectiveness and enhancing training efficiency. The traditional tennis training assistance system still has problems such as low accuracy of motion recognition and long response time. This article aims to use somatosensory recognition technology to design a tennis training assistance system to better meet the training needs of students and trainers. The article first introduces the overall structure of the tennis training assistance system. Then, this article collected data using KinectV2 cameras. Afterwards, it utilizes multi-target tracking to obtain joint position information, and then uses the DeepSort multi-objective algorithm to collect joint position information. Finally, this article uses a human pose map and a stacked model to achieve real-time recognition of continuous actions in the system. To test the effectiveness of somatosensory recognition technology in tennis training assistance systems, this article compared it with traditional systems. The research results show that for dataset 10, the action recognition accuracy of the paper's system is 95.4%, and the response time is only 0.013 ms. The accuracy of action recognition in traditional systems is only 80.8%, and the response time reaches 0.052 ms. The results indicate that the tennis training assistance system under the proposed method has the best accuracy in motion recognition and the shortest system response time. This study highlights the important impact of somatosensory recognition technology on the accuracy and response time of action recognition in tennis training assistance systems, providing more basis for promoting the widespread application of somatosensory recognition technology in sports training assistance systems.

Keywords: Tennis Training Assistance System · Somatosensory Recognition Technology · System Design · Deep Sort Multi-objective Algorithm · Action Recognition

1 Introduction

With the rapid development of technologies such as wireless communication and image processing, a new generation of sports equipment is changing people's exercise habits through intelligent interaction [1]. Traditional sports equipment simply records physical parameters like the athlete's movement time and distance, and then provides exercise

Y. Pei et al. (Eds.): IC 2024, LNEE 1214, pp. 255–264, 2024.
https://doi.org/10.1007/978-981-97-4193-9_27

recommendations through simple statistical analysis. However, the new generation of sports equipment, represented by Kinect and virtual reality, can capture human motion postures, enrich sports scenarios, and make exercise more enjoyable. Motion sensing technology, as an important component of these technologies, is widely used in medical, entertainment, sports, and other fields. In tennis training, motion sensing technology has also gradually gained attention and application. In existing research, tennis training assistance systems based on motion sensing technology are still relatively few and require further development and study. Motion sensing technology plays an important role in tennis training assistance systems. Through such systems, training efficiency can be improved, training quality can be enhanced, and most importantly, personalized training, reduced sports injuries, and increased participation in training can be achieved. With the continuous technological advancement in recent years, the application of motion sensing technology in tennis training will become more widespread. By continuously adjusting and perfecting motion sensing technology, the system's accuracy and reliability can be improved, allowing it to play a greater role in tennis training. Research has found that the system is also very effective when applied to the training of other sports, promoting the overall progress of sports technology. The study of tennis training assistance systems based on motion sensing technology has significant practical and theoretical value, promoting the continuous development of tennis technology.

This study provides an overall system framework and combines motion sensing technology to optimize motion detection and target tracking in tennis training, thereby improving the performance of the tennis training support system. Ultimately, comparative analysis shows that the tennis training assistance system developed in this study has the fastest response time and the best motion detection accuracy. The innovation of this system lies in its focus on the importance of motion sensing technology compared to traditional systems and its application in the optimized design of the tennis training support system, which helps to improve system performance and assist in tennis training.

2 Related Work

With the continuous progress and development of technology, auxiliary training systems would play an increasingly important role in sports training, providing athletes and coaches with more efficient, accurate, and personalized training plans and guidance [2]. Therefore, many scholars have conducted research on auxiliary training systems and proposed different methods for optimizing system design. Chong-gao Chen discussed the design of a sports teaching and training action correction auxiliary system based on the NET platform, and stated that the system can collect training actions and mark key points, making corresponding frame matching accurate and improving training action satisfaction [3]. Talha Muhammad simplifies motion training using 3D modeling and motion capture techniques, and states that it can effectively improve the effectiveness of motion training and the standardization of player movements, reducing the difficulty of motion training [4]. Min Y conducted research on a multi Kinect based motion assisted training system, which improved the accuracy of Tai Chi training and is of great significance for both physical education teaching and scientific training [5]. Chen Qiuju discussed the posture and action capture in aerobics training based on mathematical similarity matching statistical analysis. The proposed auxiliary training method for posture and action capture in aerobics can provide higher accuracy for dance training and theoretical support for dance training [6]. Although there are obvious drawbacks, these researchers' work on supplementary training systems can strengthen their theoretical base and offer other design concepts. The research focus of scholars has not been combined with somatosensory recognition technology, which leads to significant problems in the response time and accuracy of action recognition of the designed system. Its research lacks significant reference value and is limited.

Body sensation recognition technology is a technology that utilizes machine equipment and sensors to perceive and recognize information such as human body movements, postures, and positions. This technology can be combined with devices such as computers and smartphones to achieve human-computer interaction, allowing users to control and operate devices through their own actions and postures. Sensory recognition technology is widely used in games, entertainment, sports, medical and other fields, bringing many conveniences and innovations to people's lives and work [7]. Li Wenbo proposed a pull-up training system, which combines IoT technology and somatosensory recognition technology in machine vision, and uses cloud services to build a complete solution architecture. It realizes the automation and dataization of daily pull-up training for students, while making management more scientific and simple [8]. The Kinect camera designed by Li Jiayi achieves tactile control of a two wheeled balanced vehicle. When the human body makes corresponding actions, the system can automatically obtain bone information and convert it into control commands to send to the balance car, achieving sensory control of the car's forward, backward, left and right turns, etc. [9]. Scholars' research on somatosensory recognition technology can provide a certain theoretical basis for this study. Because the research focus is not on tennis training assistance systems, this makes it difficult to apply the research results effectively.

3 Design Process of Tennis Training Assistance System

3.1 Overall System Framework

In response to the current problems in tennis training assistance systems, this article designs a tennis training assistance system based on somatosensory recognition technology. Among them, the user terminal is mainly responsible for real-time action recognition, behavior analysis, user information collection, data storage and transmission. The server is centralized on the cloud server, responsible for system updates and tennis training history on the user side. This system adopts the USB3.0 (Universal Serial Bus) interface to achieve communication between software and hardware; Through TCP/IP (Transmission Control Protocol/Internet Protocol) communication protocol, communication between users and the business side has been achieved. The overall framework structure of the system in this article is shown in Fig. 1:

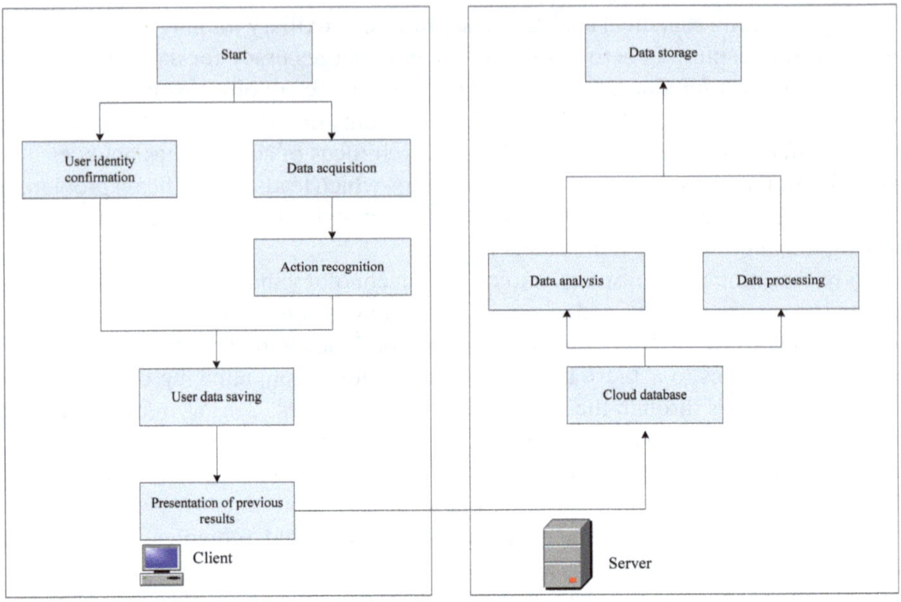

Fig. 1. System Block Diagram Structure

3.2 Implementation of Physical Perception Recognition in Tennis Training

Sensory recognition is the use of specific devices to recognize the human body for specific purposes, and it has a wide range of applications in many cases [10]. At present, the industry divides body sensing recognition technology into various types based on implementation principles, including inertial sensing, optical sensing, inertial and optical sensing.

3.2.1 Data Collection and Target Tracking

During tennis training on the human body, video information obtained from KinectV2 cameras often contains a large amount of information, making it difficult to use for behavioral analysis. In addition, due to the complex environment in which the athlete operates, filtering out the background can improve the accuracy of human behavior recognition. KinectV2 uses an infrared light source as the light source and projects the modulated near-infrared light source onto the KinectV2 camera. It determines the distance between the target and the KinectV2 camera by analyzing the reflection time of the measured object on the target surface. In addition, in order to reduce the computational complexity of human action recognition, a separation strategy can be used to segment characters from the background during human action recognition, and the subsequent recognition only needs to retain the character image.

This article uses cameras to collect 6 basic training movements and 18 joint positioning information of tennis trainers, and uses them as basic motion data. The position information of each joint point is a two-dimensional coordinate, and by combining the position information of each joint point within a certain period of time, data templates for basic movements such as jumping and standing are obtained. It is necessary to use multi-target tracking to obtain joint position information in complex tennis training for tennis trainers. The specific process of multi-target tracking is shown in Fig. 2.

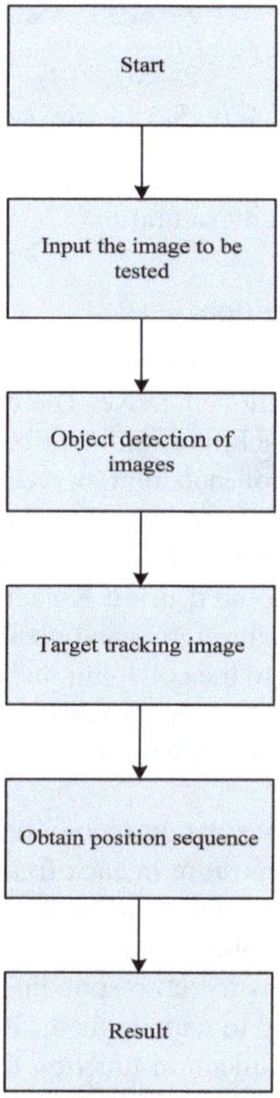

Fig. 2. Multi target tracking process

3.2.2 DeepSort Multi-objective Algorithm

This article would use the DeepSort multi-objective algorithm to collect joint position information. The DeepSort multi-objective algorithm is a new image feature-based adaptive tracking method, which detects objects in each pixel image in the image, and then processes the obtained detection results through a weighted matching algorithm and compares them with existing motion trajectories to obtain joint motion trajectories [11]. Among them, the weight can be calculated by the Mahalanobis distance between the point and the motion trajectory, and the Mahalanobis distance needs to be obtained through the Kalman filtering function. By setting a threshold, the time interval required for matching the motion trajectories of two images can be compared. If the time difference exceeds the critical value, it indicates that the trajectory matching in adjacent time frames has failed for a long time, indicating that the motion has ended. When matching, weights can be assigned to each characteristic based on its maximum response value. The specific calculation method is as follows:

$$x = \frac{\max(g(S_n))}{\max(g(S_n)) + \max(g(S_z))} \tag{1}$$

Among them: $g(S_n)$, $g(S_z)$ are the relationship between the gradient angle of image edge characteristics and chromaticity saturation.

3.2.3 Real Time Action Recognition

This article establishes a multi-objective human tennis training motion model using the multi-objective bone data mentioned above. The main joints of the human body are extracted using a bottom-up approach, and automatic recognition of human movements is achieved based on the distribution of each joint. In order to achieve more accurate human motion recognition results, this paper combines visual geometry and convolutional neural networks to detect key joints, in order to obtain more accurate human motion recognition results. Firstly, images can be captured using KinectV2 cameras, and feature vectors can be extracted from the images through visual geometry. Then, the extracted feature vectors can be regarded as inputs to the convolutional neural network, and key joints in the image can be extracted through multi-stage regression; Then, using greedy reasoning, the relative positions of each key joint are determined to construct a human tennis training motion model.

To achieve real-time continuous action recognition, it is necessary to comprehensively analyze the human motion posture in each frame. This article relies on a human pose map and utilizes a stacked model to achieve real-time continuous action recognition. The specific steps are as follows: Firstly, this article preprocesses the human movements of each frame, extracts the corresponding digital image features, and uses single frame encoding technology to encode them. It then uses Relu as the activation function and Softmax as the classification function to jointly learn the linear stacking model and convolutional neural network. This article uses scaling to set system parameters and adjusts them during the training process of sample data. Finally, the accuracy of system action recognition is tested using sample data.

Introducing Batch Normalization into the stack model can normalize the data, resulting in a mean distribution of 0 and a variance of 1 for each digital image feature. In order to enhance the system's real-time performance, this paper decouples several levels of visual geometry groups from convolutional neural networks using techniques like data normalization and linear transformation. When standardizing data in a stacked model, the specific expression is:

$$p_i = \frac{1}{j}\sum\nolimits_{m=1}^{j} S_i^m \tag{2}$$

$$q_i^2 = \frac{1}{j}\sum\nolimits_{m=1}^{j} \left(S_i^m - p_i\right)^2 \tag{3}$$

Overall, using somatosensory recognition technology, it is possible to capture the full body movements of tennis players in real-time, including hitting movements, footwork, body posture, etc., and accurately recognize their movements. By comparing with standard movements, athletes can identify their technical shortcomings and identify areas for improvement.

4 Actual Testing of Tennis Training Assistance System

4.1 Experimental Design and Data Sources

This research compares the somatosensory recognition-based tennis training aid system with conventional systems to validate its overall performance. This article takes 10 students majoring in tennis from a local sports university as the research objects, collects tennis training data from 10 students, preprocesses the training data, and finally divides the data into 10 datasets. This article compares the accuracy of tennis training motion recognition and the real-time response ability of system feedback and correction between two systems, in order to provide more basis for the application of this system.

4.2 Experimental Results

4.2.1 Accuracy Results of Action Recognition

Accuracy is an important performance measure for tennis training assistance systems. A system with high accuracy in motion recognition has better performance, which can better correct the problems that students encounter during tennis training. This is of

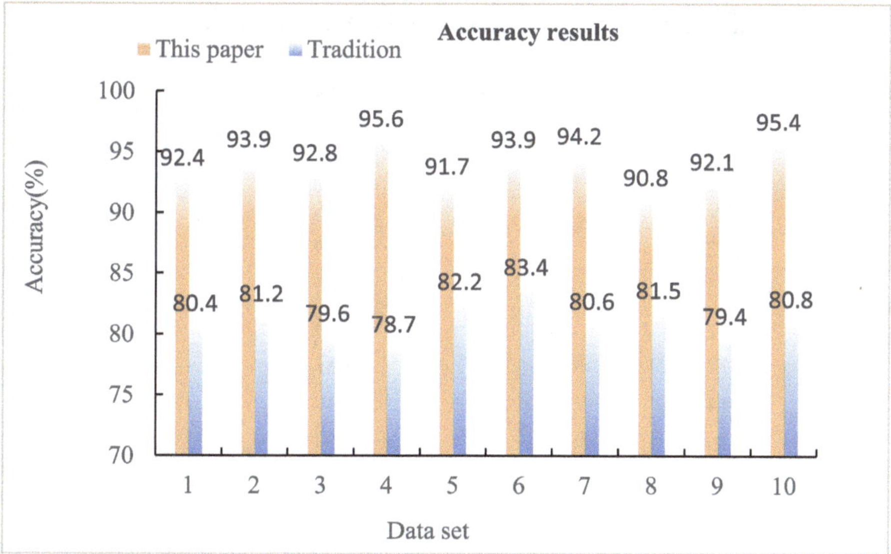

Fig. 3. Comparison of action recognition accuracy between two systems

great significance for improving training efficiency and effectiveness. In this study, the accuracy of action recognition between two systems is compared, and Fig. 3 displays the findings:

As demonstrated in Fig. 3, the system used in the paper for dataset 1 has a recognition accuracy of 92.4%, whereas traditional systems have an accuracy of 80.4%; the system used in this article for dataset 5 has an accuracy of 91.7%, whereas traditional systems have an accuracy of 82.2%; and the system used in this article for dataset 10 has an accuracy of 95.4%, versus 80.8% for traditional systems. It can be seen from this that the action recognition accuracy of the system in this article is significantly higher than that of traditional systems. Its recognition accuracy for 10 datasets is also above 90%, with the highest recognition accuracy reaching 95.6% and the lowest reaching 90.8%; The recognition accuracy of traditional systems for 10 datasets ranges from 78% to 84%, with the highest recognition accuracy being only 83.4% and the lowest being only 78.7%. From this, it can be seen that the tennis training assistance system based on somatosensory recognition technology in this article has high accuracy, which can accurately recognize the tennis movements of students and correct incorrect tennis movements. This is conducive to improving the training efficiency of students, improving their training results, and enhancing their competitive ability.

4.2.2 Real Time Response Capability for System Feedback and Correction

Response capability is also an important indicator for measuring system performance and a key part of tennis training assistance systems. A responsive tennis training assistance system can provide faster feedback on problems during tennis training and promptly correct them. Therefore, this article tested the response time of two systems, and the results are shown in Table 1:

Table 1. Comparison of Response Times between Two Systems

Data set	This paper system (milliseconds)	Traditional system (milliseconds)
1	0.011	0.059
2	0.013	0.046
3	0.009	0.058
4	0.012	0.049
5	0.015	0.062
6	0.007	0.053
7	0.012	0.057
8	0.014	0.049
9	0.010	0.054
10	0.013	0.052

From the data in Table 1, it can be seen that the system studied in this paper has a response time of 0.011 ms for Dataset 1, while the traditional system's response time is 0.059 ms. For Dataset 5, our system's response time is 0.015 ms, compared to 0.062 ms for the traditional system. For Dataset 10 discussed in this paper, the response time of our system is 0.013 ms, whereas it is 0.052 ms when using the traditional system. Comparative analysis reveals that the response time of our system is significantly faster than that of traditional systems. This clearly demonstrates the superior real-time performance of our system, which can provide quicker feedback and timely solutions to problems encountered in tennis coaching. This is beneficial for students to conduct correct tennis training more efficiently and with improved quality, making it of significant practical value.

5 Conclusions

With the development of science and technology, somatosensory recognition technology has also been widely applied in various fields and has brought different impacts to people's daily lives. The main focus of this article is the design of a tennis training assistance system based on somatosensory recognition technology. Initially, this piece offers a concise overview of the pertinent research background and its importance. It also reviews and evaluates the benefits and drawbacks of earlier studies conducted by academics. Then, the overall design of the tennis training assistance system was carried out, which achieved tactile recognition of tennis sports training by introducing target tracking, Deep Sort multi-objective algorithm, and real-time action recognition. This ultimately formed the tennis training assistance system in this article. To verify the system of this article, it would also be compared with traditional systems. The research results show that the performance of the paper's system is better than traditional systems. The paper's system has higher accuracy in recognizing tennis training movements and shorter response time. This helps students better understand the shortcomings in training, provide timely feedback and correction, improve the efficiency and quality of tennis training for students, better meet the current demand for tennis training, and have good promotion prospects. There are also some shortcomings in this study. Due to practical limitations, the research on system functions is relatively single, and further research on the system is needed in the future.

References

1. Ngugi, L.C., Abelwahab, M., Abo-Zahhad, M.: Recent advances in image processing techniques for automated leaf pest and disease recognition–a review. Inf. Process. Agric. **8**(1), 27–51 (2021)
2. Leng, X., Jiang, H., Dai, G.: Martial arts auxiliary training method based on ecological science. Ekoloji **28**(108), 523–527 (2019)
3. Chong-gao, C., Połap, D.: Design of action correction assistant system in physical education teaching and training based on. NET platform. Mobile Netw. Appl. **27**(3), 1228–1239 (2022)
4. Talha, M.: Research on the use of 3D modeling and motion capture technologies for making sports training easier. Revista de Psicología del Deporte (J. Sport Psychol.) **31**(3), 1–10 (2022)
5. Min, Y., Mou, L.: Research on sports auxiliary training system based on multiple Kinect. Electron. Des. Eng. **29**(18), 115–120 (2021)

6. Chen, Q., Alsemmeari, R.A.: Research on aerobics training posture motion capture based on mathematical similarity matching statistical analysis. Appl. Math. Nonlinear Sci. **7**(2), 203–216 (2021)
7. Wang, M.: Gesture recognition using a bioinspired learning architecture that integrates visual data with somatosensory data from stretchable sensors. Nat. Electron. **3**(9), 563–570 (2020)
8. Li, W., Lei, Z.: A pull-up training system for students based on somatosensory recognition. Internet Things Technol. **8**(2), 18–20 (2018)
9. Li, J.: Design and implementation of balance car control system for somatosensory recognition. Digit. Technol. Appl. **37**(5), 7–8 (2019)
10. Xu, H., Wang, Y., Geng, H., Feng, Y., Zhang, S.: A virtual sand table display technology based on somatosensory interactive equipment. Aircr. Des. **40**(4), 45–48 (2020)
11. Liu, H.: Improved DeepSORT algorithm based on multi-feature fusion. Appl. Syst. Innov. **5**(3), 55 (2022)

Advanced Voice Analysis Platform Based on MATLAB Software

Lei Guo(✉), Zhiwen Zhang, and Yingxin Gao

Key Laboratory of Linguistic and Cultural Computing, Ministry of Education, Northwest Minzu University, Lanzhou 730000, Gansu, China
guolei_xbmu@163.com

Abstract. This study focuses on the phonetic development of an English-Chinese bilingual infant under the age of two years. A speech analysis platform is developed using MATLAB, which can automatically segment speech samples and analyze related acoustic parameters. By tracking and recording children's speech and using the method of speech experiment, this paper studies the vocal expression of infants before the age of two, expands the vision of infant speech research, and summarizes the phased characteristics and rules of the development of bilingual infants' speech on the basis of large corpus analysis.

Keywords: Bilingualism · Infants and Young Children · Speech · Acoustic Analysis

1 Introduction

In 1982, Grosjean's global survey revealed that over half of the world's population is engaged in learning two languages, highlighting the widespread nature of bilingualism. This study delves into the technical aspects of bilingual language acquisition, particularly in the context of English-Chinese bilingualism. Bilingual language acquisition, a significant cognitive achievement, involves the native language and a secondary language. This process is especially intriguing in young children, who demonstrate the ability to master a language rapidly, often before reaching three years of age. The mechanisms and developmental patterns behind such accelerated language learning in bilingual environments are a focal point of linguistic research [1].

This paper presents a technical analysis of the phonological development in a Chinese child raised in the UK, in an English-Chinese bilingual setting. This child, despite being exposed to both languages from birth and having English as the first language, did not differentiate between the two languages cognitively at an early age. The study explores the stages of phonetic differentiation, imitation, and assimilation in such a bilingual context. The learning process in this environment is predominantly instinctual, occurring without deliberate teaching or conscious effort. The research also investigates the phonetic development disparities between monolingual and bilingual children, given the variations in linguistic environments. This analysis is crucial in understanding the

Y. Pei et al. (Eds.): IC 2024, LNEE 1214, pp. 265–274, 2024.
https://doi.org/10.1007/978-981-97-4193-9_28

unique phonological adaptations and developmental trajectories in bilingual children, particularly in the early years of their language acquisition journey [2].

To further delve into the phonetic development of bilingual children, this study employed an advanced voice analysis platform based on MATLAB software. This platform is capable of automatically segmenting voice samples and analyzing related acoustic parameters. With this approach, we can accurately track and record the vocal changes in children, thus gaining a deeper understanding of the phonetic characteristics in a bilingual environment. The platform utilizes complex signal processing and acoustic analysis algorithms, enabling the identification and quantification of children's speech patterns at different ages, including parameters like fundamental frequency, formants, and sound intensity. The application of these technologies provides us with a richer and more systematic perspective to comprehend the uniqueness and patterns in the phonetic development of bilingual children. Through in-depth analysis of this data, we are better positioned to reveal the impacts of a bilingual environment on children's speech acquisition and the vocal changes that occur during this process.

2 Experimental Methods

At present, there is no unified model for the research methods of infant phonetics. Linguists and psychologists have their own research methods according to different research topics, among which the two typical methods are natural sampling method and experimental method.

2.1 Source of Corpus

The research data in this paper comes from the corpus collected by the author through tracking recordings, of which the corpus collected by the recordings is selected by the subjects. An infant under 2 years old is selected as the research object. The infant was born in London, UK, female, and the parents are both Chinese with the higher education level, their Mandarin is very standard. In daily communication with children, Mandarin is also used, and English is the main language communication when going out. After the subjects entered the preschool in the UK after one year old, teachers and other children are all communicate in English. Subject's parents pay attention to early education, often reading Chinese and English stories, and teaching English songs. The subject is lively and active, sensitive to music, she also has a strong language expression ability, who likes to talk [3].

2.2 Design of Corpus

The establishment of the corpus took two years, during which the infant's vocal expressions were systematically recorded. The original data in the corpus are as follows:

First, establish a voice expression corpus for infants under the age of two. The content of the corpus is divided into emotional expression and vocal expression, in which emotional expression includes crying signal and laughter signal, while verbal vocal expression is mainly speech signal. Crying is mainly physiological crying, laughter is divided into spontaneous vocal laughter and induced laughter.

Second, the number of recordings of the child's voice in the corpus is 461, and the duration of each voice fragment is different. Each corpus is not less than 5 min, and the longest time is not more than 30 min.

2.3 Determination and Extraction of Data

In this study, the determination and extraction of parameters should also be combined with the content and progress of the research, and the basic acoustic parameters are mainly here, as shown in Fig. 1.

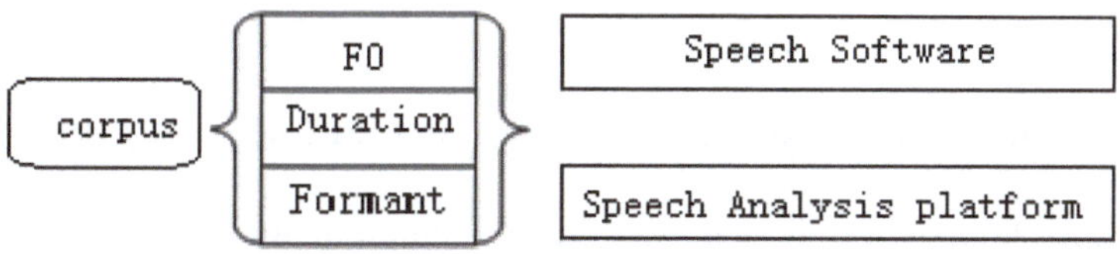

Fig. 1. Basic acoustic parameters

Infants' emotional voice expression is different in pitch and intensity. The difference in pitch is reflected by comparison of fundamental frequency, and the difference in breathing is reflected by the intensity and fluctuation of energy. The study of speech mainly focuses on the comparison between infant speech and adult speech. The basic acoustic parameters are determined as formants and the analysis of speech patterns. The basic acoustic parameters are extracted using Praat software and the speech analysis platform written by MATLAB. As shown in Fig. 2, this platform is written by Matlab under the Windows platform to play and analyze the collected signals in a targeted way, as well as extract and save relevant parameters [4].

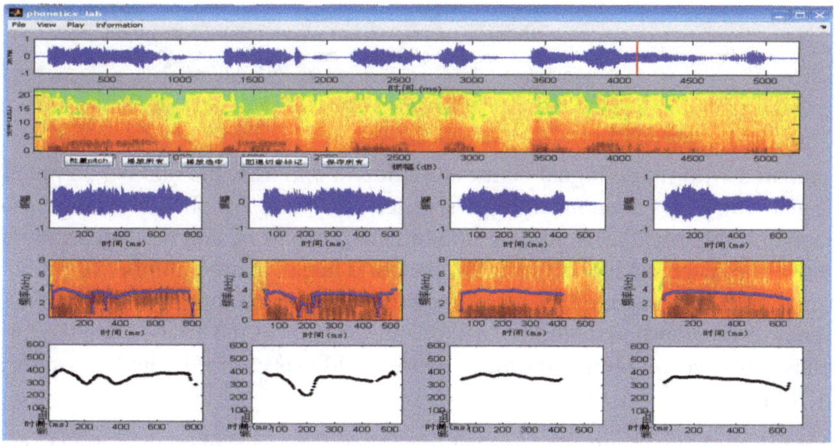

Fig. 2. Infant voice analysis platform interface

Its main functions include: wav file reading; Signal is marked; Signal segmentation; Automatic saving of segmentation files and reading of marked voice files; Calculate and save the relevant parameters; Batch processing of wav files, etc. The flow chart of platform parameter analysis is shown in Fig. 3.

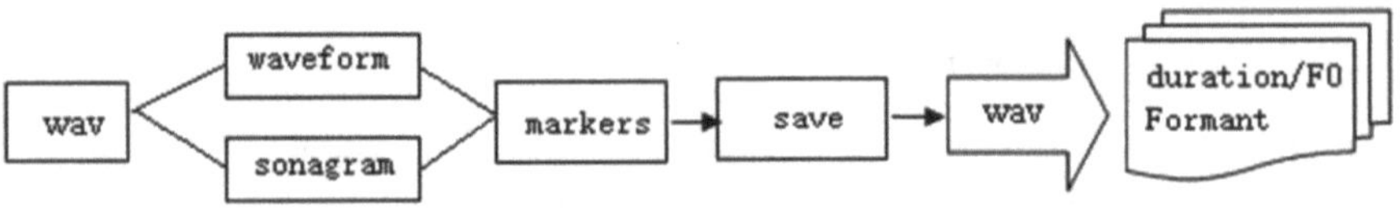

Fig. 3. Flow chart of infant speech analysis platform

3 A Study on the Phonetic Stage Development of Bilingual Children

Studies on the phonetic development of bilingual children are also analyzed from five stages: biological alarm stage (0–8 weeks), Coo learning stage (9–20 weeks), babbling stage (21–50 weeks), and language stage is further divided into phonetic development stage (1-1-year old and half month) and language stage (1–1/2 years old and 2 years old) [5].

0–8 Weeks: At this stage, the child's main vocal expression is crying, which conveys physiological and psychological needs through crying.

9 to 20 weeks: At this stage, the child gradually began to produce phonemes similar to vowels and consonants. Through listening and cutting the phonemes of the recorded samples at this stage, the statistical results are as follows (Table 1).

Table 1. Phoneme statistics of infants from 9 to 20 weeks

monophoneme	diphthong	phoneme of consonant
/a/, /u/, /ə/, /e/	/eɪ/, /aɪ/	/k/, /g/, /s/, /n/, /b/

Compared with monolingual Chinese Mandarin children, the number of phonemes in bilingual children's pronunciation is smaller, and there are phonemes in consonant pronunciation that Mandarin children do not have, such as /s/ and /k/. According to the analysis of the recorded speech samples, the child began a large number of pronunciation exercises at this stage, mostly repeated monophones and diphthong phonemes, and sometimes repeated pronunciation of a phoneme as many as ten times. Then the formant frequency of the vowel phonemes made by the child was counted, and the results were shown in Fig. 4.

21–50 Weeks: At this stage, the number of phonemes produced by the subject are keep increasing, the repeated pronunciation of phonemes is more frequent than in the previous

Table 2. Pronunciation statistics of infants during 21–50 weeks

monophoneme	diphthong	phoneme of consonant	Chinese monosyllabic	Syllable combination	Phonemic combination
/a/, /u/, /ə/, /e/, /o/, /i/	/ eɪ / /aɪ/	/k/, /g/, /z//h/, /n/ /b/, /s/, /h/	/nei/, /gei/ /ge/, /ye/ /ba/, /bie/ /wa/	/bu-ge/, /bu-gei/, /wawa/ /bu-bu/	/ei-ai/, /ai-ai/, /ai-s/ /a-i/, /a-z/, / ə - ai /

stage, and she begin to produce a small number of ordinary Chinese phonetic syllables. The specific statistical results are shown in Table 2.

Compared with the phonetic pronunciation of the previous stage, in this stage, the vowel phonemes /o/ and /i/, and the consonant phonemes /z/ and /h/ were increased, the diphthong phonemes did not change, and the subject can produce 7 Chinese ordinary phonetic syllables. In addition, compared with Chinese monolingual children, there are fewer pronunciations of single syllables in Chinese, fewer pronunciations of overlapping words in two-syllable words, and no fricative pronunciation in Mandarin Chinese children at this stage, the formant statistics of this stage are shown in Fig. 4.

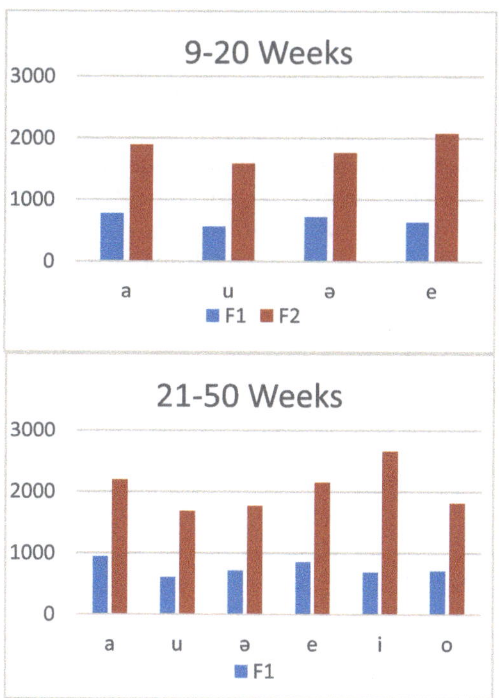

Fig. 4. Vowel formant statistics for infants aged 9 to 20 weeks and 21 to 50 weeks

1 to 1 and a Half Years Old: The child enters the local preschool in the UK after the age of 1, the time she stays at preschool three days a week from 8 a.m. to 6 p.m. In this stage, the child still listen and communicate in Chinese at home, thus the language input is Mandarin Chinese and English, but the use of English began to increase, and the children's language environment in school is only English [6].

Among which Mandarin Chinese is still the main input The phonetic pronunciation of the child at this stage is quite different from that of the three children in Mandarin Chinese. As far as the acquisition of Mandarin Chinese is concerned, it is still a practice of a small amount of syllable pronunciation and syllable repetition. As far as English acquisition is concerned, the child at this stage does not pronounce many English words, but can understand many Mandarin Chinese and English words. The following table summarizes the phonetic statistics of children's Mandarin, English words and English pronunciation at this stage (Table 3).

Table 3. Pronunciation statistics of infants during 1-1/2 years old

Mandarin vocabulary	English Vocabulary	English-like pronunciation
Father, mother, baby, hug, bubble, grandma, bear, brother, sister	Up, bubble, mummy, daddy, bye bye, no	/dæp/, /bæp/, /dɔ:t/, /bʌk/, /es/, /bʌp /

1 ½ to 2 Years Old: In this stage, the number of days the child is sent to the preschool increases by five day per week. The time outside the preschool is mainly based on the communication of Mandarin Chinese. At this stage, the pronunciation of Mandarin Chinese of the child enters the learning stage of monosyllabic pronunciation, and the pronunciation of three-character words is still simple. As the pronunciation of English, the number of English words increases, and the listening and understanding of English

声母	b	p	m	f	d	t	n	l	g	k	h	j	q	x	r	z	c	s	zh	sh	ch	y	w
a	✓	✓	✓	✓	✓	✓	✓	✓	✓														
o	✓																						
e					✓	✓	✓	✓	✓	✓	✓								✓				
i	✓	✓	✓		✓	✓	✓					✓	✓	✓		✓	✓	✓	✓	✓			
u	✓	✓				✓														✓			✓
ü																						✓	
ai	✓	✓								✓													
ei	✓																						
ao	✓		✓					✓															
ou									✓	✓	✓									✓	✓		
ia																							
iao			✓			✓								✓									
iou																							
an																							✓
en			✓																				
iang												✓											
iong														✓									

Fig. 5. Speech sound statistics during the period of 1 to 1 ½ years old

are enhanced. In the following figure, we have calculated the pronunciation of ordinary Chinese phonetic syllables of the child at this stage [7] (Fig. 5).

At this stage, children can pronounce 21 initial consonants of Mandarin Chinese, while the two variants of the rhotic vowels have not appeared yet. The biggest leap in the language at this stage is the expression of the sentence, the sentence pattern most often spoken is declarative and interrogative, long sentences have intonation changes, some sentences will appear the reversal of the main predicate, and the expression of the short sentence is more rhythmic.

4 Systematic Pronunciation Errors of Mandarin Vowel in Bilingual Children

This part only examines the pronunciation of the seven first-order vowels in Mandarin Chinese, so there is no vowel deletion, but only vowel substitution and vowel offset. Vowel replacement is the main pronunciative tendency of infants in the lower age group below two years old. At one and a half years old and two years old, the primary pronunciative tendency of the front vowel /i/ is not round enough and the vowel /ɿ/ was substituted; At the age of one and a half years and two years old, the main pronunciation tendency of the high round vowel /u/ is also not round enough, as shown in Fig. 6 (front unrounded vowel) and Fig. 7 (back high round lip vowel) [8] .

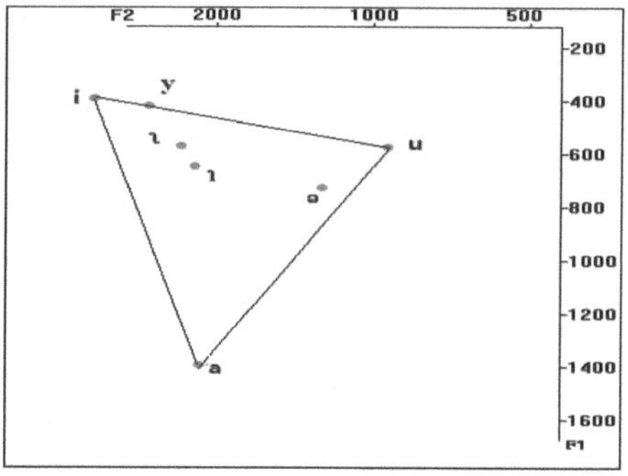

Fig. 6. Front Unrounded Vowels Pattern of 1 to 1 ½ years old

At the age of one and a half years and two years old, the two lingual apex vowels /ɿ, ʅ/ were mainly replaced by the anterior high lingual vowel /i/, and the phoneme was preposition, as shown in Fig. 8 and Fig. 9.

By the age of two, the vowel /y/ is pronounced centrally, similar to the central high-rounded vowel /u/, as shown in Fig. 10 and Fig. 11.

Therefore, it can be considered that children cannot coordinate the relationship between round lips and correct tongue position before the age of two. At the same time, due to the restrictions of physiological conditions, if children pay attention to the correct tongue position during pronunciation, they will not have round lips in place, and if they pay attention to round lips, they will have inadequate tongue position, thus the

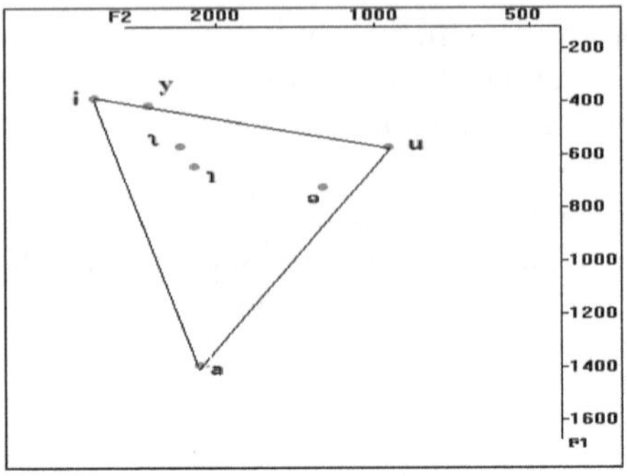

Fig. 7. Front Unrounded Vowels Pattern of 2 years old

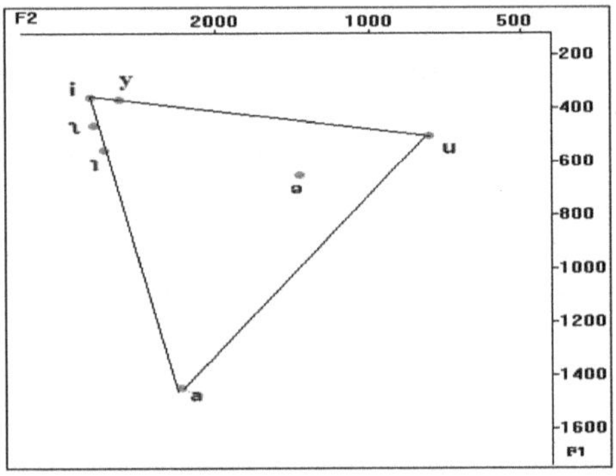

Fig. 8. Back High Rounded Vowels Pattern of 1 to 1 ½ years old

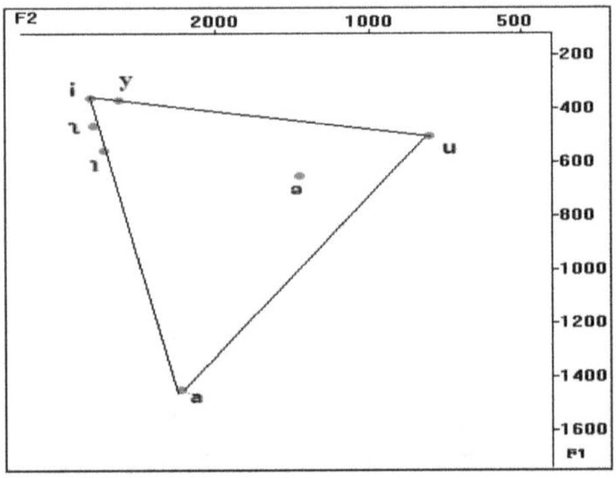

Fig. 9. Back High Rounded Vowels Pattern of 2 years old

whole pronunciation mechanism cannot be coordinated. When children pronounce the /u/ vowel, if they pay attention to keeping the tongue in front, the degree of rounded lip is often not enough, and the vowel /i/ is pronounced. If they pay attention to the round

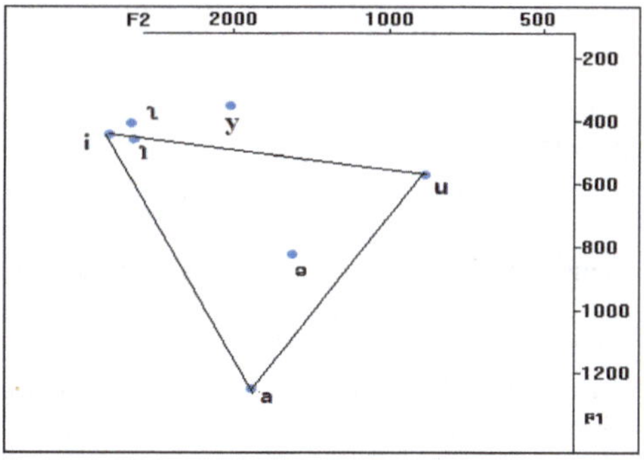

Fig. 10. Central High Rounded Vowels Pattern of 1 to 1 ½ years old

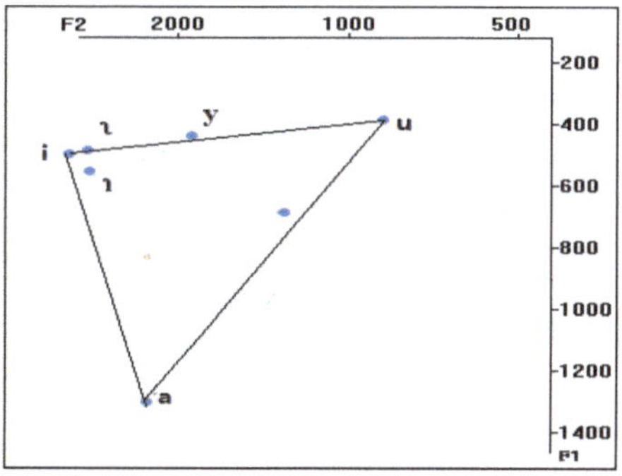

Fig. 11. Central High Rounded Vowels Pattern of 2 years old

lips, then it's easy to place the tongue in the front, resulting in a central tongue position, similar to a high round lip vowel /ə/.

5 Conclusion

This paper mainly analyzes the changes in the phonetic development of a bilingual child, including the research on the systematization of phonetic pronunciation of bilingual children, and summarizes the systematization of phonetic pronunciation errors of bilingual children. Compared with the phonetic development of the Chinese monolingual children, it is found that there is a certain lag in the linguistic development of bilingual children compared with that of monolingual children of the same age. In phonetic pronunciation, bilingual children cannot coordinate the relationship between round lip and tongue position before the age of two years old, and there are characteristics of insufficient round lip and tongue position in pronunciation.

Acknowledgement. This paper was subsidized by Central Universities Fundamental Research Funds for the project: The Ideological and Political Construction and Practice of College English Courses in Ethnic Universities under the guidance of the Consciousness of the Chinese Nation Community (31920230109).

References

1. Eimas, P.D., Siqueland, E.R., Jusczyk, P., Vigorito, J.: Speech perception in infants. Science **22:171**(3968), 303–306 (1971)
2. Benedict, H.: Early lexical development: comprehension and production. J. Child Lang. **6**(02), 183–200 (1979)
3. Dore, J.: Transitional phenomena in early language acquisition. J. Child Lang. **3**, 13–28 (1976)
4. Swingley, D., Aslin, R.N.: Lexical competition in young children's world learning. Cogn. Psychol. **54**, 99–132 (2006)
5. Fenson, L., Dale, P.S., Reznick, J.S., et al.: The MacArthur Communicative Development Inventories: User's Guide and Technical Manual. Singular Publishing Group, San Diego (1993)
6. Hart, B., Risley, T.R.: The Social World of Children Learning to Talk. Paul H.Brookes Publishing Co., Baltimore (1999)
7. Lees: Crying and behavior pattern in breast-and formula-fed Infants. Early Hum. **58**, 133–140 (2000)
8. Bachorowski, J.A.S.: The acoustic features of human laughter. J. Acoust. Soc. Am. **3**, 1582–1597 (2001)

Early Warning Monitoring System for Fresh Food Safety Based on K-means Clustering Algorithm

Qianqian Chen(✉)

Department of Food Science and Technology, University of California, Davis, Davis 95616, USA
chelsea_cqq1@163.com

Abstract. In recent years, with the increase of people's income and living standard, people's demand for quality of life has become higher and higher. With the continuous development of social economy and network technology, etc., people's demand for food has increased dramatically, which has led to the increase of per capita consumption level of Chinese residents. Therefore, this has to a certain extent driven the rapid growth of the domestic catering industry and related industries, but due to factors such as consumer eating habits, food safety awareness is not strong, making some fresh products have serious hidden problems. Some of the food products contain harmful substances, which are harmful to human health and even endanger life and property, causing great economic loss and social panic to people and affecting normal life order. The K-means algorithm is of great significance for early warning and monitoring of food safety. Based on the K-means algorithm and combined with the safety warning and monitoring system of fresh products, this paper analyzes it and establishes the corresponding index system.

Keywords: Early Warning Monitoring · Food Safety · K-means Clustering Algorithm · System Design · System Test

1 Introduction

With the improvement of people's living standards, food safety has become more and more concerned by consumers, and food safety has become more and more important to consumers. As people's incomes continue to rise, they are demanding higher levels of food, and as a result of rapid economic development and a significant increase in the income of urban and rural residents, food expenditures are rising sharply [1]. Along with the accelerated pace of people's lives, resulting in more and more garbage in the city, these problems affect the normal consumption activities of the people and the development of health level, and this also seriously threatens the health safety of consumers. Therefore, in order to solve these problems, there is a need to establish a safety early warning monitoring system for the fresh food category, it is necessary to effectively monitor the food, so as to guarantee people's healthy life [2].

Y. Pei et al. (Eds.): IC 2024, LNEE 1214, pp. 275–284, 2024.
https://doi.org/10.1007/978-981-97-4193-9_29

2 Study on the Current Situation of Fresh Food Supply Chain and Prioritization of Food Safety Indicators

2.1 Fresh Food Supply Chain Status and Deficiencies

(1) The complexity of the supply chain at the source. In the process of getting food from the "field" to the table, some companies will do whatever they can to maximize their own profits. Firstly, when food is still in the ground, pesticides and chemical fertilizers are used to increase food production, and the massive abuse of pesticides and chemical fertilizers can contaminate the soil and air, thus making the food contaminated and affecting human health [3]. Therefore, it is necessary to conduct timely food testing. Secondly, food is prone to spoilage during transportation, and consumers are often not well informed in this regard. Therefore, an efficient, safe and sustainable monitoring system is necessary and meaningful.

(2) The cold chain of fresh supply chain is broken. In the process of transporting cold storage, the fresh products need to be monitored throughout the process, which requires their temperature to be maintained within a certain range. Fresh products are highly susceptible to other factors during transportation, which can lead to spoilage, so real-time monitoring of temperature is required. In order to ensure food quality and safety production must use insulation technology to avoid, air humidity changes cause food spoilage problems. At the same time, the refrigeration equipment can also be adjusted so that the refrigeration compressor to achieve optimal operation, thereby improving the fresh product storage and transportation capacity and reducing costs. The current rate of cold chain transportation in China in aquatic products, meat. Fruits and vegetables are 69%, 57% and 35% respectively, and the average loss rate

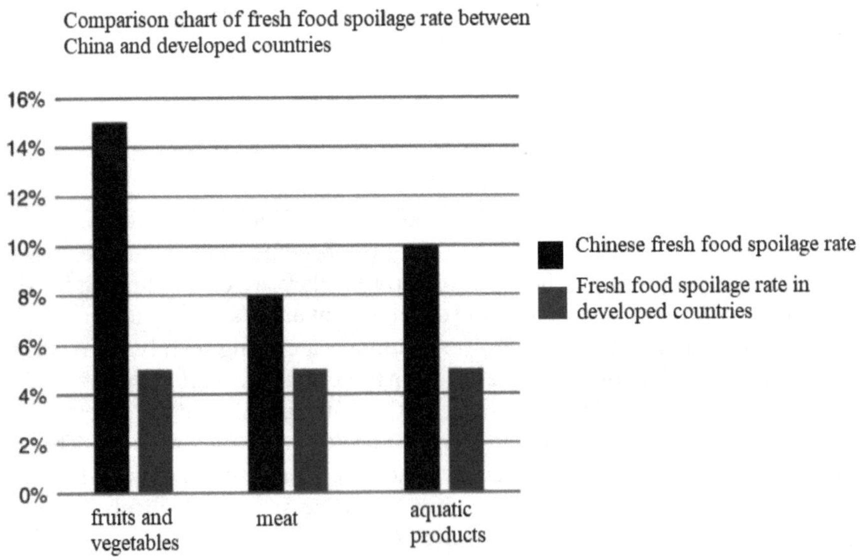

Fig. 1. Comparison of fresh produce spoilage rates between China and Europe and America

of fresh produce in China is more than 10%, which is 2–3 times of that of European and American countries, as shown in Fig. 1 [4].

(3) Information asymmetry in food sales. In the process of food sales, consumers may make purchases due to abnormal factors, such as: counterfeiting, adulteration of deceptive ingredients, etc. Therefore, there is a need for a system that can monitor the safety of fresh products with early warning and feedback information. However, due to their own consumption habits and other reasons cannot accurately identify whether the product is safe, for example: food packaging materials are not qualified, in the process of transportation, product quality varies, resulting in consumers found after the purchase does not match the actual. Some merchants will make fraudulent sales to consumers in order to make huge profits, so that some people are blinded, which will easily cause information leakage and misleading problems. Therefore, the establishment of a fresh food safety early warning and monitoring system is of great significance to improve the development of China's agricultural market and consumer health, which is also an important measure to promote the development of our national economy [5].

2.2 The Need of Fresh Food Safety Index Prioritization Study

(1) The food supply chain is lengthy and has many influencing factors. At present, the level of food supply chain management in China is not yet high, and there are many problems, such as confusing organization and interlocking blocks. The low level of food safety management leads to nutritional waste of consumers, which affects people's lives. In addition, there are many other factors that affect the quality of fresh product supply chain operation. Due to the lack of effective supervision resulting in excessive pesticide residues in food, etc., some consumers become suspicious of food safety and make consumer purchases. In addition, the companies themselves do not monitor food safety risks early enough, leading to hidden dangers in their production processes, which threaten the entire food sector and human health. Therefore, it is particularly important to conduct early warning monitoring of food safety.

(2) The limited resources of society and the efficiency of the work required. With the rapid development of the economy, people's living standard is also improving, and they are paying more and more attention to food safety and quality issues. Therefore, in order to meet consumers' requirements for fresh, tasty, nutritious, healthy and green food, various new e-commerce platforms are emerging. However, due to the limited social resources and the fact that some Internet users do not have enough awareness and ability to identify good and bad food products, some unscrupulous elements take advantage of these loopholes, and people often choose to be silent and shield the buzzwords on the Internet after a food safety incident. Therefore, it is urgent to establish an early warning and monitoring system for fresh food e-commerce [6].

(3) As we all know, in today's competitive social environment, the limited social resources and the efficiency and quality of our work require us to have a focus, in which the safety of food is the most important issue. As people's living standards continue to improve, they have higher and more stringent requirements for food quality. Therefore, in order to meet the needs of consumers and social development as well as the relevant national laws and regulations, we should study the

early warning and detection system that can quickly identify and deal with the food safety loopholes in various fresh food e-commerce, so as to ensure that the supply of agricultural products is not contaminated or spoiled and affect human health.

3 Food Quality Safety Early Warning System Based on K-means Clustering Algorithm

3.1 Model Introduction

(1) K-Means clustering algorithm. k-means algorithm is a decision tree based machine learning method. It uses a computer as the input device for the process of transforming a sample set into a certain individual according to some specific rules, and then training and statistical analysis to draw conclusions and apply them. The algorithm predicts the possible hazards in the target area by judging the set of similar attribute values (i.e., boundaries) in the data. It is based on decision trees to perform classification by judging the potential risk values (i.e., thresholds) in the sample set before determining the target area. In terms of data expressions, suppose the class is divided into (C1, C2, …Ck), then our objective is to minimize the squared error E, expressed as shown in Eq. (1). The principle of the k-Means clustering algorithm (as in Fig. 2) is to represent the similarity of sets and clusters based on the magnitude of the distance between each influencing factor in the coordinate system to a given sample point, obtained from a survey study, and to calculate the probability of the presence of danger in each cluster in the set, i.e., to determine which class the sample belongs to [7].

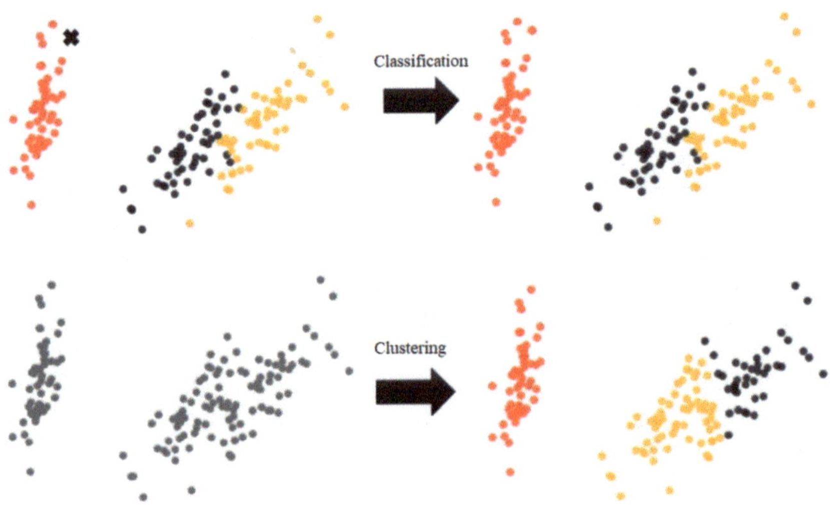

Fig. 2. Principle of k-Means clustering algorithm

$$E = \sum i = 1k \sum x \in Ci \|x - \mu i\|_2^2 \tag{1}$$

Where μi is the mean vector of the cluster Ci, sometimes called the center of mass, and the expression is shown in Eq. (2).

$$i = 1|Ci| \sum x \in Cix \tag{2}$$

The basic steps of k-means algorithm are as follows:

Firstly, choose the N objects you want from the expectation value as the initial point and detect them by k-means algorithm to get the early warning value, then determine whether the fresh food is a safety hazard.

Secondly, the distance from each influencing factor to the expectation point is calculated to divide, by dividing each factor into multiple indicators and making a comprehensive assessment. Then, the prediction of food safety problems in the fresh food category is made based on the obtained results and the pre-defined warning threshold.

Thirdly, each expectation point is obtained from the survey study, and after the classification and prognosis of food, it is then transformed into an early warning value, so that precautionary measures can be developed accordingly to the problems that may occur in each area.

Fourthly, the standard measurement function is calculated until the maximum number of iterations is reached, then it is stopped, otherwise, the operation continues [8].

(2) Quadrant rule. The four quadrants rule of time management is an event processing priority time theory that classifies events according to the degree of urgency and importance, it is to classify the time of occurrence and the subsequent moment that may cause serious consequences, and then according to different types can be divided into unexpected events, general uncontrollable situations and major accidents. The key to the four-quadrant rule of time management is how to get the cause of the division of the degree of impact of each factor, and then according to the cause analysis and early warning indicators, so as to determine the causative factors and the degree of impact, and finally the corresponding risk assessment.

3.2 Food Quality and Safety Early Warning System

(1) The general framework of indicators. In this paper, we use K-Means clustering algorithm and four-quadrant rule to divide the food quality and safety early warning system into target layer, index layer, and indicator layer. The target layer is the most important, core and valuable part of the food safety early warning and monitoring system, which determines whether the whole system can be carried out smoothly. The indicator layer plays a pivotal role in the entire detection process and is used to analyze the food safety early warning system by establishing the indicator layer model. The index layer is among the indicators of the monitored object, which reflects the quality and safety status of the food under test. The index layer is used to reflect the influence of each factor in the index layer on the overall target safety degree by setting different weights, and to determine the weight of each indicator in terms of the overall target safety degree according to this. Firstly, the food supply chain safety supervision system is divided into food production, food processing,

food transportation and food safety supervision according to food circulation, and the indicator layer model is established. Then risk identification is carried out for each food production link, and its possible safety problems are analyzed. Finally, the K-means clustering algorithm combined with the entropy weight function is used to quantify and standardize the information contained in each module through the fuzzy comprehensive evaluation method to obtain a relatively objective and accurate result, which provides a reliable basis for the subsequent fresh food safety early warning monitoring system. This improves the accuracy and credibility of the indicator layer model [9]. According to the description, the food quality and safety early warning system is shown in Fig. 3.

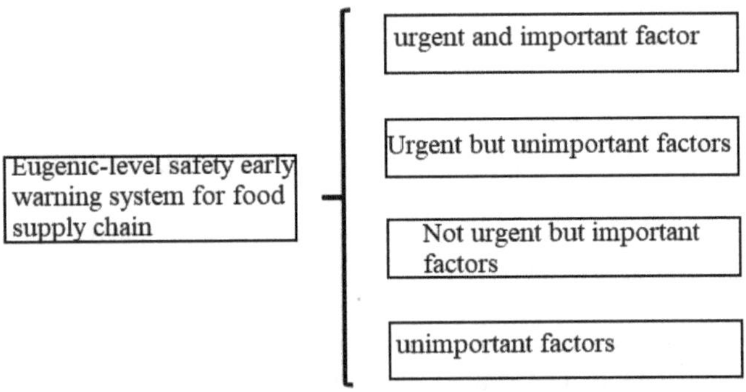

Fig. 3. Food quality and safety early warning system

(2) The establishment of indicators. The construction of indicators is after determining the risk early warning analysis model, the selected samples should be screened first, and then the corresponding evaluation values should be selected according to the different feedback from users of fresh food e-commerce platform or the amount of consumer purchase demand. Fresh food has the characteristics of perishability, wide geographical distribution and its own defects [10]. Therefore, it is necessary to establish an indicator that accurately reflects the current state and development trend and can respond in a timely manner. Based on the results of the mentorship, the relevant factors affecting food safety are shown in Table 1.

(3) Calculation results of indicators

Group A: P7, P8, P12, P16, P17; Group B: P2, P3, P4, P5, P9, P13, P14, P15; Group C: P1, P6, P9, P11, P13; Group D: P10, P11, P16. The center of mass of Group A, B, C, D was calculated separately as shown in Eq. (3).

$$A : K1 = ((1 + 2 + 2 + 3 + 0)/5,$$
$$(2 + 1 + 2 + 2 + 2)/5) = (1.6, 1.8)$$
$$B : K2 = ((2 + 1 + 2 + 1 + 3 + 3 + 0 + 0)/8,$$
$$(5 + 5 + 4 + 4 + 4 + 5 + 5 + 4)/8) = (1.5, 4.5)$$

Table 1. Relevant factors affecting food safety

Target layer	Exponential Layer	Indicator layer
Food Supply Chain Priority Safety Early Warning System	Food production	Irrigation Water Quality Index Ambient air index Soil Environmental Indicators Pesticide detection indicators Pest Control Indicators
	Food processing	Raw material input qualification index Qualification index of processing equipment Processing technology index Processing Staff Hygiene Indicators
	Food Transportation Indicators	Packaging technical indicators Storage Specifications Transport technical indicators Specifications for transportation and handling
	Food safety supervision	Information accuracy index Testing Institution Construction Indicators Market Supervision Indicators Food Safety Hazard Rate Indicators

$$
\begin{aligned}
C: K3 &= ((5+4+3+4+3)/5, \\
&\quad (4+4+4+3+5)/5) = (3.8, 4) \\
D: K4 &= ((5+4+3)/3, (2+3+2)/3) = (4, 2.3)
\end{aligned} \tag{3}
$$

4 System Test

4.1 Test Environment

Through the selection of the test environment, we need to determine a suitable test method, and according to the time required by different algorithms to configure, so as to determine a suitable security early warning monitoring method for different environments. According to the basic requirements of system operation, the software and hardware environment suitable for B/S structure is selected for system test, and the test results are analyzed [11, 12]. Finally, we need to adjust the operating environment, hardware conditions and software design of the system. The specific test environment is shown in Table 2.

Table 2. Test environment

name		Server side	client
Hardware configuration	CPU	Intel i5	Intel i7
	memory	4GMemory	16G Memory
	Hard disk	750G HD	250G HD
	network	10/100M LAN	10/100M LAN
Software configuration	Operating system	CentOS6.5	Windows 7
	Database system	MySQL5.0	
	WEB server	Apache	

4.2 Functional Test Cases

A total of 1000 test cases were designed and used in this system test. On the basis of test cases, combined with K-means algorithm, the fresh food safety warning system was designed and implemented. This paper mainly takes registration function test cases and login function test cases as examples, specifically as follows:

(1) Register for functional testing

In the registration of the function, we need to carry out the design of the test steps, mainly to the system used to test the database and hardware equipment. For specific use, see Table 3:

Table 3. Register functional test cases

Use case name	User registration	
Use case number	Login_01	
role	Ordinary user	
Use-case specification	Realize user registration	
preconditions	Using an unregistered cell phone number	
Basic event stream	Participant action	System response
	After browsing the site, users click the "Sign Up Now" button	The system queries the database and saves the user's information to the database

(2) Log in to the function test

Login function is a basic test of the system; mainly detect whether the user registration information is correct, as well as the user login name and password. For specific use, see Table 4:

(3) Test Results

In this paper, 1000 test cases of the system operation test, the test results show that all aspects of the system meet the design requirements, which basically meet the

Table 4. Login function test cases

Use case name	landing	
Use case number	Login_02	
role	Registered user	
Use-case specification	Realize user registration	
preconditions	Using an unregistered cell phone number	
Basic event stream	Participant action	System response
	1, the user enters the basic information, that is, user name and password, click the login button 2. Select Third-Party Login	1. The system searches the database to check whether the user data exists in the database. If yes, the system home page is displayed. If no, an error message is displayed 2. Conduct third-party verification, and then conduct system verification. If successful, enter the main page of the system; if unsuccessful, give an error message

relevant national regulations and standards, can carry out real-time monitoring of food, and can timely process the relevant information. This has improved the level of food safety warning to some extent. Specific test results are shown in Table 5:

Table 5. Test results

	Functional test
Total number of test cases	1000
Total number of actual test cases	1000
Test conclusion	Pass the test

The test results show that the system not only realizes the real-time monitoring of fresh food, but also realizes the cluster analysis, and gives the corresponding warning according to the data mining results.

5 Conclusion

To sum up, China's economic development and people's living standards continue to improve, and people are paying more and more attention to food safety. However, with the accelerated process of social informatization and the rapid development and widespread use of Internet technology, the way of food consumption has changed dramatically. The lack of consumer awareness of fresh produce products and the lack of enforcement by regulatory authorities have created some negative problems, which affect the next new stage in its healthy growth and sustainable development - i.e., the distrust and shopless business practices in this segment of online shopping. Therefore, it is imperative to improve the safety and security of agricultural products. In this paper, we calculate the weight of each index of fresh food quality and safety based on K-Means clustering algorithm, and build an early warning monitoring system of fresh agricultural products quality based on K-means clustering algorithm in combination with other food safety indexes. By optimizing the operation environment, hardware facilities and software configuration of the system, it can better monitor the food safety indicators of fresh food category, which provides a guarantee for improving consumers to purchase fresh and delicious agricultural products.

References

1. Awad, F.H., Hamad, M.M.: Improved k-means clustering algorithm for big data based on distributed smartphoneneural engine processor. Electronics **11**(6), 883 (2022)
2. Sofia, I., Claudia, C., Gilsenan, M.B.: European Food Safety Authority open access tools to estimate dietary exposure to food chemicals. Environ. Int. **149**(2), 106357 (2021)
3. Cao, W., Li, C., Li, Z., et al.: Research on low-resistance grounding fault line selection based on VMD with PE and K-means clustering algorithm. Math. Probl. Eng. **2022**, 20–22 (2022)
4. Son, W., Cho, I.: Analysis of trends in mega-sized container ships using the k-means clustering algorithm. Appl. Sci. **12**(4), 2115 (2022)
5. Francesco, M., Filippo, O.S., Renato, R.: Early warning monitoring of exothermic batch reactions: a kinetic-free approach. Chem. Eng. J. **418**(1), 129371 (2021)
6. Fang, Y., Zhang, Q.: The construction and case analysis of crisis early warning monitoring and evaluation system of public health events based on collaborative management of government departments. Basic Clin. Pharmacol. Toxicol. **127** (2020)
7. Zhang, F., Yang, X., Yang, H., Zhang, J.: Food safety control: effects of storage time on the change of nutrients of eggs. In: Proceedings of the 2020 Euro-Asia Conference on Environment and CSR: Finance and Business Management, vol. 32 (2020). https://doi.org/10.26914/c.cnkihy. 2020(8):11–15
8. Ariya, C., Kazuharu, N., Kanako, Y., et al.: Comparison of tongue characteristics classified according to ultrasonographic features using a K-Means clustering algorithm. Diagnostics **12**(2), 264 (2022)
9. Cater, M., Gravois, R., Guerra Gaitan, G.G., Xu, W.: Pregnant women's confidence and perceptions on practices related to food safety: a study in Louisiana. Food Control **113–115**(C), 107175 (2020)
10. Jiang, H., Wang, Y., Deng, J., Xu, M.: A gas pipeline anti-damage early warning monitoring system. J. Phys. Conf. Ser. **2033**(1), 012184 (2021)
11. Minh, H.-L., Thanh, S.-T., Magd, A.W., et al.: A new metaheuristic optimization based on K-means clustering algorithm and its application to structural damage identification. Knowl.-Based Syst. **251**(1), 109189 (2022)
12. Qiu, W.: Application of software data analysis model based on K-means clustering algorithm. Secur. Commun. Netw. **2022**(2), 4505814 (2022)

Research and Application of Multi Field Information Evolution and Data Mining in Progressive Landslide

Dongping Hu, Jinlan Tan(✉), and Zhimei Wang

School of Architecture and Civil Engineering, Chongqing Metropolitan College of Science and Technology, Chongqing, China
lanlanwp@cqcst.edu.cn

Abstract. Due to the low rainfall intensity, the water level of the reservoir rises, and the rise of the groundwater level of the landslide is affected by the rise of the water level of the reservoir. Therefore, the sliding shell immersed in the shell bears the hydrostatic pressure orthogonal to the sliding surface. The landslide is generally stable with no obvious deformation characteristics. However, with the increase of rainfall intensity and the decrease of reservoir water level, the groundwater level in the landslide is far lower than the reservoir water level, so the landslide is always affected by hydrodynamic pressure. After the reservoir water level drops for a period of time, the sliding surface is parallel to the convex slope. Heavy rainfall greatly reduces the stability and deformation characteristics of the landslide. For progressive landslide, the overall instability probability cannot reflect the development of internal failure of landslide. The geological characteristics of the planned landslide are introduced. By comparing the application of flexible protection and traditional landslide control methods, it is concluded that flexible protection and management are characterized by rapid construction, economy, safety and environmental protection. The feasibility and superiority of information evolution and data mining are verified according to the specific construction situation.

Keywords: Data Mining · Information Evolution · Landslide · Multi Field · Progressive

1 Introduction

According to the failure characteristics in the process of landslide development, landslide can be divided into two types: sudden instability and gradual instability. The sudden deformation and failure time of unstable landslide is very short, especially in the case of rainstorm, sudden drop of reservoir water level, earthquake and other sudden changes in slope tension state. The deformation development time of progressive failure slip is relatively long, which indicates that the failure process is obviously progressive. Before the landslide, the shear strength of the soil mass gradually decreases, leading to the gradual destruction of the landslide. At this time, the shear strength of local landslide is

Y. Pei et al. (Eds.): IC 2024, LNEE 1214, pp. 285–293, 2024.
https://doi.org/10.1007/978-981-97-4193-9_30

fully exerted, but the stability of the static equilibrium state of the entire landslide can only be guaranteed by the gravity of the landslide itself, precipitation and other external factors. Change of reservoir water level, etc.

2 Mechanical Model of Progressive Failure Landslide

Through in-depth analysis of groundwater sources and water intake channels, the different stages of the degradation process are divided into fragmentation stages, the limits for obtaining different precursor information are established, and the early warning standards for water resources are established. Through on-site testing, business monitoring and early warning are realized. In terms of water monitoring and early warning technology, a coal mine water inflow monitoring system based on Bragg fiber grating technology and radio sensor has been established, which can realize real-time monitoring and remote early warning of data, and contribute to the development of geological tunnel disaster early warning system. The progressive failure probability of landslide is calculated according to the safety limit of the belt. The tape safety limit can be derived from the boundary state equation as a random variable. Safe area of reinforcement - the difference between anti slip force and sliding force of the slider, including various forces on the slider. Based on the random model of progressive failure of slope proposed by Wang Jiasen, the following mechanical analysis is carried out. The tension of each reinforcement is divided into two parts in the direction of parallel sliding and vertical sliding. As shown in Fig. 1, the sliding force of strip i is Formula 1:

$$S_i = (W_{i1} + W_{i2})\sin\alpha + P_{wi}\cos(\theta_i - \alpha_i) + F_i - 1\cos(a_{i-1} - \alpha_i) \tag{1}$$

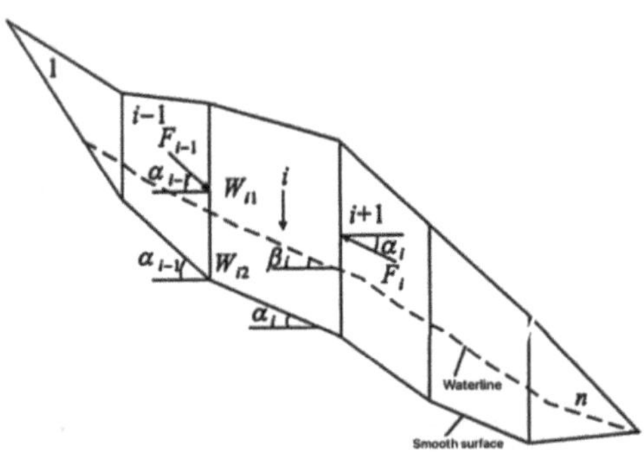

Fig. 1. Overall Slice Stress Diagram

The anti-sliding force of the blocked strip is Formula 2:

$$R_i = N_i\tan\varphi_i + cili \tag{2}$$

The force of the sliding block on the sliding surface is Formula 3:

$$N_i = W_{i1} + W_{i2}\cos\alpha_i + p_{wi}\sin(\theta_i - \alpha_i) + F_i - 1\sin(a_i + 1) \tag{3}$$

3 Geological Characteristics of Landslide Slope

The landslide deformation has many evolutionary characteristics, and the multi field information provides a basis for correctly diagnosing the landslide development process. The existing sediment prediction and control measures in mountain areas are often based on the migration map or characteristic parameters of a certain period, without considering or ignoring many development processes of sediment in mountain areas. Landslide is a complex open system composed of solid, liquid and gas. Its deformation and sliding have many characteristics of physical field. The K8 + 720 ~ K8 + 950 landslide is mainly composed of weathered carbonate deposits, and the upper part is composed of Quaternary slope deposits. The permeability of landslide soil is high, and sediment filtration is easy to form local groundwater. Carbonate fluid has certain expansion and contraction characteristics. [1] Due to expansion, dissolution and water absorption softening, the strength decreases rapidly, and the carbonate peat behind the slope is exposed to the atmosphere. When the landslide is gradually stabilized under multiple wet cycles, the decomposition and softening of the surface create favorable conditions for the infiltration of sediment. On the slope, the large catchment area distributed on the slope is covered by sand on the high permeability residual slope. After dehydration, local ponding occurs in the low carbon sedimentary weathering layer; Due to the erosion of slope bottom caused by the weathering of the carbon dioxide deposition distribution area, the contraction activity expands after the excavation unloading, which is easy to form a softening zone, leading to the gradual flattening of the landslide. However, due to the management of this bid section, the excavation and spoil are too much, resulting in a substantial increase in the building area, water and soil loss and ecological damage. The retaining wall at the toe of the slope can only bear the sliding of the toe of the slope, and the slope skeleton cannot bear the pressure on the top of the slope; The upper part of the slope surface is covered with ditches, which can only intercept the surface runoff behind the slope surface and will not affect the filtration of sediment generated by groundwater. In smooth landslide, this kind of landslide is mainly caused by the expansion, disintegration and softening of sediment caused by rainwater infiltration after the landslide. The treatment effect is poor, because traditional procedures cannot deal with the impact of trauma [2, 3].

In the process of landslide monitoring information processing, the classical data mining algorithm is introduced, and a process of obtaining information of different parts of the landslide is proposed. The hidden mode of monitoring data can be extracted, and a multi-dimensional integrated time series association model of landslide information is established. A multi-level database of landslide development process will be established to determine the external factors of landslide. Data collection is a category of landslide monitoring and data processing. At present, data mining algorithms can intelligently and automatically extract the sea level monitoring data in the hidden model, so as to obtain the w-value monitoring data in a form that is easy to understand [4–6]. The process of data mining mainly includes data collection, data preprocessing, qualitative processing of multi-level information and extraction of relevant rules. Two step clustering algorithm and association rules are used for data mining. The correlation between the development of landslide deformation and external inducing factors is demonstrated, and a multi-dimensional time correlation sequence is established. This program aims to establish the correction and evolution process of multi domain values of information valves in the process of data mining. The process of data mining is of great significance: data collection, data preprocessing, deformation development and qualitative change, and decision tree model. Data mining process is based on two-step clustering algorithm and decision tree algorithm, which can generate a lot of information between landslide deformation development and external stimulus. The stability of landslide is divided into five levels according to the probability of landslide failure, as shown in Table 1.

Table 1. Stability of Landslides with Different Failure Probability

<0.05	0.05–0.3	0.3–0.6	0.6–0.9	>0.9
Stable	basically	Stable poor	Basically unstable	unstable

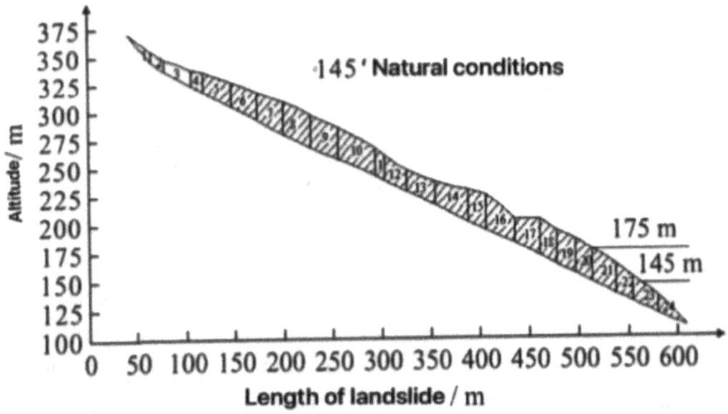

Fig. 2. Stability State Diagram of Strips

Figure 2 shows the initial development process of progressive landslide, which depends on the probability and stability level of strip landslide. Figure 2 shows that most of the landslides are in a sub stable state below the natural state of the reservoir in front of the 145M dam; When the water level of the reservoir reaches 1755 m, the front edge area remains unstable and the rear edge is basically stable; In 145 m + 50 years of rainfall, an unstable hotbed appeared at the front edge of the landslide. At present, the edge zone is unstable and damaged. When the former edge gradually extends to the rear edge, a tensile landslide is formed [7–9].

The results show that Rule 1 is the relevant criterion for the monitoring points of the slope at the initial deformation stage, including the monthly average water level (high water level) of the reservoir. This can be interpreted as that if the monthly average water level of the reservoir is high (160.14 m · h − 174.74 m), the slope monitoring point will enter the initial deformation stage. Rules 2–3 are relevant standards for early monitoring points of slope deformation, including the monthly maximum of high-intensity daily rainfall. This may be due to the initial deformation of the slope monitoring point when the maximum monthly rainfall in the landslide area is high (rain lamp).

4 Multi Field Information Evolution Characteristics of Landslide Model

The geological evolution of landslide covers the whole landslide cycle from conception to extinction. The study of landslide evolution process is very important to understand the geological transformation, prediction trend, prediction and technical management measures of landslide in different periods. At present, the main methods to study the landslide development process can be summarized as engineering geological inference based on on-site monitoring data and qualitative analysis of geological phenomena, and numerical analysis methods to establish an ideal mathematical model for quantitative calculation. Numerical modeling method based on abstract geological model and physical model of landslide. As an effective method to study the deformation and failure mechanism of debris flow, the physical modeling method of debris flow can simulate and reproduce the entire debris flow development process that is difficult to observe in the indoor field in the short term. Therefore, the physical method of landslide modeling has great advantages in the research of landslide development. The method of testing landslide physical model includes measuring landslide physical model, including extension, elongation, ground pressure, displacement, crack, deformation and failure form. With the development of science and technology, the testing technology of landslide physical model has developed rapidly. It is helpful to improve the level of landslide physical model testing by using new testing methods. In particular, since the 21st century, researchers have made some progress in studying observation methods related to the testing of landslide physical models, as described below. In 2005, in the physical model test of debris flow by Roschenj et al., Two new technologies are introduced: the Y-ray method for measuring soil moisture and the automatic detection of grid point displacement, which greatly improve the reliability of landslide and other soft soil quality tests. In the same year, Zhang Junfeng et al. Water pressure sensor, landslide track sensor and other measuring instruments were installed on the downhill, and water flow simulation tests were conducted on the inlet and outlet valves and the right cork box on the drawer. Use observation and measuring instruments to obtain test data. In 2010, Liu Dongyang et al. Use the automatic goniometer to measure the inclination of the bottom of the landslide physical test box, and record the whole process with the camera and camera. They accurately determined and analyzed the evolution characteristics of the landslide, re analyzed the experimental phenomena, and improved the accuracy of the test results. In 2011, Wang Yongbo et al. proposed a multifactor model for large-scale physical landslide, including a deformation measurement module for measuring the process of tilt deformation, and a unit for measuring the water pressure and water content in inclined wells. Although modern measurement methods have promoted the development of physical landslide modeling technology to a certain extent, it must be pointed out that the above methods or physical landslide testing system have limitations or defects.

The landslide evolution process has the characteristics of displacement field, stress field and temperature field. The landslide evolution process research based on the landslide physical test model can reflect various short-term evolution characteristics, but the traditional measurement methods can only conduct preliminary research on the landslide evolution process from two aspects of displacement field and stress field. Ignoring other important phenomena related to landslide evolution.

The traditional displacement measurement method cannot record the process of landslide failure, crack propagation and model deformation in detail.

The traditional digital camera method is inefficient in capturing images. Image processing is a complex task, which requires a lot of manpower, material resources and time to process data.

Traditional digital camera technology can only obtain two-dimensional graphics of landslide surface, and cannot establish three-dimensional model of landslide physical model change process. It is not conducive to the visual observation and analysis of the deformation characteristics of each stage of the landslide physical model evolution.

The traditional digital camera cannot collect the motion data of the tidal slide zone. The lack of data in tidal landslide area affects the analysis of the influence of water level on landslide evolution.

Due to the size of the displacement sensor, the measured displacement data cannot accurately reflect the absolute displacement of the slope, and it is difficult to accurately measure the absolute displacement of each point of the landslide model.

In landslide physical model test, the location of landslide and landslide surface is usually determined according to the prototype of landslide, but for landslide (or rock), no method is proposed to determine the location of landslide surface.

When monitoring the ground pressure change, the installation position of the pressure vessel should be carefully monitored, but the traditional laying method is difficult to ensure the installation accuracy.

Deformation sliding has many characteristics in information development. In different evolutionary stages, there are certain differences in several characteristics of information. Based on the deformation development stages of the front W section, the different characteristics of information development in each stage are studied. The different field data in this paper mainly include slope deformation characteristics, slope pressure and slope temperature. Using particle image velocimetry, the velocity vector of landslide control section is obtained under the condition of deformation of landslide model. The development characteristics of slope deformation and the deformation vector of slope control points are studied [10–12].

The basic condition of landslide is that there is sliding space in front of the slope and cutting surfaces on both sides. The main factors inducing landslide are: earthquake, rainfall, snow melting, surface water scouring and soaking; Unreasonable human engineering activities, such as slope toe excavation, blasting, reservoir water storage (discharge), mining, etc.; In addition, tsunami, storm surge, freezing and thawing can also induce landslides. Simultaneous activity rule of landslide - some landslides will be active immediately after being affected by induced factors, such as strong earthquake, rainstorm, tsunami and storm surge; Lagging - some landslides occur slightly later than the time of induced factors, such as after rainfall, snow melting and human activities, and this lag law is most obvious in rainfall induced landslides. The spatial distribution of landslides is mainly related to geological factors and climate. Generally, the following areas are prone to landslides: bank slopes of rivers, lakes and seas; Canyon area with large terrain elevation difference; Slope sections of mountain areas, railways, highways and engineering buildings; In geological structural zones, such as fault zones and seismic zones; Rainstorm prone area or abnormal heavy rainfall area.

The ideal processing effect is obtained, as shown in Fig. 3, which can be obtained from the velocity vector; At the beginning of deformation, the slope velocity vector is a unified model, and the slope velocity vector represents the main sliding direction in the same direction. The slope velocity vector is mainly concentrated on the trailing edge of the slope. This panel indicates that the slope deformation is mainly concentrated on the trailing edge of the hull, rather than the middle and front. Point cloud data on the slope shows that there are no visible cracks on the slope. In the stage of uniform deformation, the horizontal component of slope velocity vector increases significantly. At present, the velocity vector is concentrated in the slope center; The cloud map data of the slope surface shows that there are periodic cracks in the vertical direction of the main landslide. By comparing the slope velocity vector of the crack location, it can be seen that the crack location is the main focus of the slope velocity vector. In the accelerated deformation stage, the horizontal component of the slope velocity vector increases continuously [13]. The velocity vector on the slope surface is mainly concentrated at the foot of the slope; Further penetration of intermittent cracks in the early slope; By comparing the velocity vector of the slope crack location, it can be seen that the velocity vector of the occurrence stage is also concentrated in the slope crack location [14].

Fig. 3. Effect of Treatment on Rear Slope

It can be concluded that the velocity vector of the slope at the initial deformation stage presents a uniform model at the trailing edge of the slope; At the stage of uniform deformation, the velocity vector of the slope body concentrates on the crack position of the slope body; With the increase of load and deformation, the velocity vector concentrates on the slope toe.

5 Multi Field Information Evolution Characteristics of Physical Model of Landslide Anti-sliding System

When sliding forward along the front of the body, the front edge is disturbed by peristalsis compensation. With the increase of the number of landslides, the tension and deformation of the slope are transferred to the front edge of the slope. The strength of the landslide cannot slide further and can only coordinate with the W-shaped continuous deformation at the rear edge of the slope. The rock sand shell is loaded under the action of the horizontal average counter pressure (vertical sliding direction), and due to the bending of the sand shell, transverse expansion cracks occur. The results show that the influence of slope instability index on the overall instability is decreasing. Under visual stability, the slope displacement field changes from a stable sequence to a more stable sequence, and then from a more stable order to an unstable disaster mode.

6 Conclusion

The local failure probability analysis method based on landslide can better understand the location of the maximum initial failure probability of landslide under different working conditions, and predict the initial failure location and deformation model. The main observation post (or control post) is located to help monitor and control landslide. At present, with the development of component monitoring, data acquisition, transmission and storage technology, landslide monitoring data shows an explosive growth trend. The complexity of the relationship between a large number of monitoring data and multi-level monitoring data increases the technical difficulty of landslide monitoring data analysis. Traditional qualitative and quantitative data processing methods are often insufficient. The data processing requirements and applications of ocean volume monitoring pose challenges to existing theories and lead to the deepening of data collection.

By analyzing the whole process of data mining and summarizing the association criteria, the following main conclusions and conclusions are drawn:

1) The data of ZG93 landslide monitoring point are analyzed and verified. When the flood season starts from June to September, heavy rainfall will affect the deformation and displacement of the landslide. The influence of water level on landslide displacement is mainly reflected in the large deformation caused by the first connection point and the beginning of the connection point in June 2007. With the decrease of reservoir water level, the displacement and deformation of landslide increase synchronously with the increase of rainfall intensity.
2) Combined with the data mining process, under the multi field coupling model of landslide, the relevant criteria of influencing factors and landslide displacement and deformation are established. In general, 87 landslide multi field information correlation standards have been generated, among which 10 standards with high correlation have been selected for analysis. The analysis results of relevant criteria show that the strong rainfall event and the high intensity maximum effective continuous rainfall are the main factors leading to the deformation of slope monitoring points.

Acknowledgments. This work was financially supported by the Science and Technology Research Program of Chongqing Municipal Education Commission (KJQN202202501): Under Double-Carbon Background Flue-Cured Tobacco Room "Optical Storage, Direct Use and Flexible Use" Energy-Saving Technology Application Research; the school-level natural science project of Chongqing Metropolitan College of Science and Technology (XKY202110): Application of "Optical Storage, Direct Use and Flexible Use" Technology and Research on Airflow Organization in Flue-Cured Tobacco Room under Carbon Neutral Strategy.

References

1. Sun, L., Li, C., Shen, F.: Two-surface progressive failure mechanism and stability quantitative evaluation of water-induced weakening retrogressive landslides: case study for clay landslides, China. Bull. Eng. Geol. Env. **81**(9), 382 (2022)
2. Zou, Z., et al.: Progressive evaluation of buckling landslide stability considering strain softening behavior of the slip zone. Int. J. Numer. Anal. Meth. Geomech. **46**(14), 2680–2697 (2022)
3. Zuo, S., Zhao, L., Deng, D., Han, Z., Zhao, B., Zhao, Z.: Back analysis of shear strength parameters for progressive landslides: case study of the Caifengyan landslide, China. Bull. Eng. Geol. Env. **81**(1), 1–13 (2021)
4. Pei, X., Cui, S., Zhu, L., Wang, H., Luo, L., Zhang, X.: Sanxicun landslide: an investigation of progressive failure of a gentle bedding slope. Nat. Hazards **111**(1), 1–28 (2021)
5. Earth Surface Research; Researchers from Swiss Federal Institute of Technology Detail Findings in Earth Surface Research (From toppling to sliding: progressive evolution of the Moosfluh landslide, Switzerland). Sci. Lett. (2020)
6. Glueer, F., Loew, S., Manconi, A., Aaron, J.: From toppling to sliding: progressive evolution of the Moosfluh landslide, Switzerland. J. Geophys. Res. Earth Surf. **124**(12), 2899–2919 (2019)
7. Science - Earth Science; Studies from Hong Kong University of Science and Technology Add New Findings in the Area of Earth Science (Progressive failure of buildings under landslide impact). Sci. Lett. (2019)
8. Luo, H.Y., Zhang, L.L., Zhang, L.M.: Progressive failure of buildings under landslide impact. Landslides **16**(7), 1327–1340 (2019)
9. Earth Surface Research; New Findings on Earth Surface Research from Oregon State University Summarized (Quantifying the sensitivity of progressive landslide movements to failure geometry, undercutting processes and hydrological changes). Sci. Lett. (2019)
10. Leshchinsky, B., et al.: Quantifying the sensitivity of progressive landslide movements to failure geometry, undercutting processes and hydrological changes. J. Geophys. Res. Earth Surf. **124**(2), 616–638 (2019)
11. Kumar, V., Gupta, V., Jamir, I.: Hazard evaluation of progressive Pawari landslide zone, Satluj valley, Himachal Pradesh, India. Nat. Hazards **93**(2), 1029–1047 (2018)
12. Zhang, W., Wang, D., Randolph, M.F., Puzrin, A.M.: From progressive to catastrophic failure in submarine landslides with curvilinear slope geometries. Géotechnique **67**(12), 1104–1119 (2017)
13. Yerro, A., Pinyol, N.M., Alonso, E.E.: Internal progressive failure in deep-seated landslides. Rock Mech. Rock Eng. **49**(6), 2317–2332 (2016)
14. Bernander, S., et al.: Downhill progressive landslides in long natural slopes: triggering agents and landslide phases modeled with a finite difference method. Can. Geotech. J. **53**(10), 1565–1582 (2016)

Research on the Application of Mobile English Learning Platform from the Perspective of Big Data

Yan Wei(✉)

School of Foreign Languages, Wuhan Business University, Wuhan, Hubei, China
antheawy@163.com

Abstract. The continuous innovation of mobile communication technology, the development of Internet technology and the renewal of learning concepts have created a new form of learning through mobile learning. Mobile platform brings great convenience to people's life. This paper mainly studies the application of mobile English learning platform from the perspective of big data(BD). The paper first introduces the changes in English teaching in the era of BD, then states the needs of the development of mobile learning platform. Next, it analyzes the mobile English learning platform which is designed by using PHP language and MVC pattern. The platform consists of four functional modules: login module, learning resource module, online discussion module and personal center module. In order to verify the effectiveness of the mobile English learning platform, a questionnaire survey was conducted among freshmen. The results show that the platform can help students learn English effectively, but it needs to be improved in some aspects.

Keywords: Big Data · English Learning · Mobile Platform · MVC Mode · PHP Language

1 Introduction

With the rapid development of information technology and network technology, intelligent communication technology has also been widely popularized. An increasing number of people use various forms of intelligent terminals for communication and learning. Mobile Internet, BD and cloud platforms have been applied in all areas of life [1]. "Internet + education" has realized the development and innovation of the education industry, and promoted the development of education towards artificial intelligence, digitalization and networking. In recent years, with the rapid development of wireless communication technology and the popularization of mobile terminals like smart phones and tablet personal computers, learning methods such as mobile learning, micro learning and ubiquitous learning have quickly entered people's daily life. People can learn conveniently, quickly and succinctly through mobile terminals, thus these kinds of learning methods have become more and more popular. In the era of information, the vigorous development of wireless mobile communication technology has truly realized the dissemination of information anytime and anywhere, and laid a solid foundation for the development of

Y. Pei et al. (Eds.): IC 2024, LNEE 1214, pp. 294–300, 2024.
https://doi.org/10.1007/978-981-97-4193-9_31

mobile learning. Microlearning is a combination of digital learning and mobile learning, which has the characteristics of socialization and situational learning [2]. Traditional learning resources are developed in a chapter or a lesson, and their resource design mode is different from the resource mode of micro learning, which does not conform to the design characteristics of micro resources, such as small modules, small resource granularity and small knowledge points [3].

Nowadays, lots of the foreign researches focus on mobile learning. In the application of mobile learning in colleges and universities, Kingston University has used the Short Messaging Service on campus to provide learning evaluation services for many years. After the course is completed, students can feed back the experience and effect of using to teachers, so as to promote the cooperation between teachers and students [4, 5]. University of Sheffield Harlem in the UK and University of Helsinki in Finland are using projects such as UniWap which are devices supporting collaborative mobile learning and using mobile devices as learning and management tools to try to carry out mobile learning with the purpose of enabling college students to adopt better personalized learning [6]. M-learning is a mobile learning project funded by the European Union, which is established for young people with literacy and cognitive disabilities. The goal of the project is to create a learning environment and resources for this special group [7].

College English is a compulsory course for college students. However, due to the limited time in class, it is difficult to complete the learning of all knowledge. Therefore, students' autonomous learning after class can be used as a supplement to classroom teaching. But the time for students' autonomous learning after class is relatively scattered, and their ability of autonomous control is not that strong. Large-scale teaching resource is difficult to attract students to achieve autonomous and centralized learning. Therefore, the research on mobile English platform based on BD has certain practical value and significance.

2 Construction of Mobile LP in the Era of BD

2.1 The Necessity of Innovation in English Learning Methods in the Age of BD

The teaching method of college English course is not unchanging. It will follow the changes of the times, and the changes of college students' thinking and behavior. It will experience an infinite cycle process from cognition to practice, and then to practice. In this process, it will continue to innovate and develop. In China, college English teachers constantly make use of the advanced technology actively in different historical periods to innovate teaching methods. For example, from the initial use of blackboards and chalk to the use of Internet technology in the 1990s, and to the application of BD technology today. BD has led a new technological revolution. The arrival of this new technological revolution has profoundly affected and changed many fields of human social life, and has also had a profound impact on the teaching methods of college English in China [8, 9]. As we all know, the post-90s and post-00s college students born in the era of BD have broad horizons, sharp thinking and unique personality. BD analysis technology can also be used to carry out teaching activities by combining online and offline methods so as to make teaching methods more diversified, personalized and accurate, thus give vitality to college English classes.

How to improve teaching effectiveness is a problem that every college English teacher will encounter and need to solve in the teaching process. The level of teaching effectiveness is not only related to the evaluation of teachers' teaching ability, but also of great significance to measure whether students really learn something. The poor effectiveness of traditional college English teaching is partly due to the restriction of technical conditions, which limits college English teachers' selection and use of teaching methods. In the era of small data, in most cases, college English teachers use the same teaching methods to teach, which makes college students lose their enthusiasm for learning, and then leads to the unsatisfactory teaching effect of college English. However, the era of BD has provided rich teaching resources for college English courses. English teachers in college are no longer limited to teaching fixed books, but can also use rich teaching resources and innovative teaching methods to carry out teaching activities, which can effectively avoid the phenomenon of "one size fits all" and "spoon-feeding education". At the same time, BD era also provides convenience for college English teachers to have a more comprehensive understanding of college students' thinking. Teachers can use BD technology to analyze students' data on major network platforms, combine hot topics of common concern with theoretical teaching, and select highly targeted teaching methods to organize accurate teaching [10]. In this way, on the basis of understanding the real demands of students, not only does he realize the frequent communication between teachers and students, makes the majority of students have a better understanding of college English classes, but also improves the teaching effectiveness.

2.2 Design of the Mobile LP

The purpose of mobile LP is mainly to provide teaching assistance services for teachers and students participating in offline classrooms. The use of this system assists students in their after-class learning, helps teachers grasp the real-time learning situation of students conveniently, so as to ensure uninterrupted online and offline learning of students. In order to facilitate the use of teachers and students, the platform combines the content of textbooks with supplementary materials to ensure students can easily use various functions provided by the system to assist their learning. At the same time, teachers and students can log in through browsers or APPs to make sure they can log in and perform various operations conveniently in and after class.

The design of mobile LP mainly adopts the way of development, which is a design pattern in software engineering and can divide the main code in software development into three parts: model, view, and controller. Among them, M (model) is mainly responsible for processing part of the functions of the data logic in the code, the operation of the database and the realization of interaction logic; V (view) processes the part including the display of data and the design of page style, that is, processing everything that can be seen; C (controller) processes in the corresponding model layer according to the actions performed by the user in the view layer, and returns the result, so that the user can see an interactive operation. The system is based on B/S structure, using the back-end technology, PHP language, CI framework, front-end technology HTML, JavaScript, Ajax, and database technology. The display of pages and interactive functions of users are achieved by JavaScript, Vue, JS, jQuery, and Sass.

As shown in Fig. 1, the mobile LP is mainly divided into teacher module and student module. The teacher module performs the function of teaching management. The student module can be divided into the following four functions: classroom tests (including online tests and historical review); assignments (including downloading templates and uploading assignments); sign-in; other functions (including courseware materials and leaving a message).

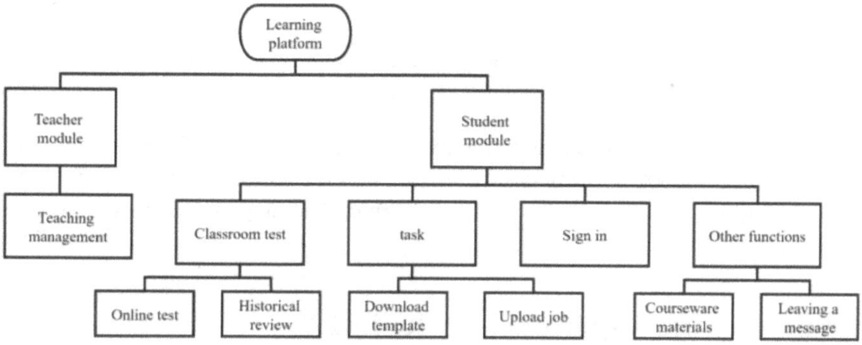

Fig. 1. Modules of the LP

3 Application of Mobile LP

3.1 Application of the Modules

1) Teachers' activities: Teachers' use of the platform is mainly divided into two aspects: uploading materials and supervision. Teachers should sort out and select learning materials and test resources needed by students, upload learning resources on the mobile LP, specify the tasks that students need to complete, and check the learning progress of students in real time. When students complete the test, teachers can check their answers to understand students' mastery of the knowledge. Teachers will summarize students' feedback both on the mobile LP and in the classroom, solve students' puzzles, and focus on the explanation of the key and difficult points to deepen students' understanding. Finally, teachers can assign after-class tasks on the LP to test students' mastery of the knowledge.
2) Students' activities: By logging in to the mobile LP, students browse learning resources in order according to the tasks assigned by the teacher, get to know the difficult points of the part, take notes, and upload homework as needed. At the same time, students can put forward the problems they have encountered in the learning process on the message board. At the end of the course, students can take a test to examine how well they have mastered the knowledge. At the same time, students can review previous tests as needed to facilitate their review. Learning through the mobile platform can help students do a good job in preview, thus keep up with teacher's thoughts, and facilitate the development of classroom activities.

3.2 Data Collection

In order to understand and master students' feelings and views on the use of the mobile LP, this paper conducted a questionnaire survey on the freshmen who used the mobile LP for one semester. Questionnaires were distributed to 150 freshmen and 142 valid questionnaires were collected. In order to ensure the validity and reliability of the questionnaire, SPSS is used to analyze the reliability of the questionnaire data. The following formula is used to analyze the validity of the questionnaire:

$$t = \frac{\overline{X} - \mu}{\frac{\sigma_X}{\sqrt{n-1}}} \tag{1}$$

In the formula, t is the deviation statistic between the mean of the sample and the population; $\overline{X}$ is the average number of samples; μ is the overall average; σ_X is standard deviation of the sample; n is the size of the sample. The questionnaire is proved to be valid after verification.

4 Analysis of the Application of Mobile LP

4.1 Effect of Mobile LP

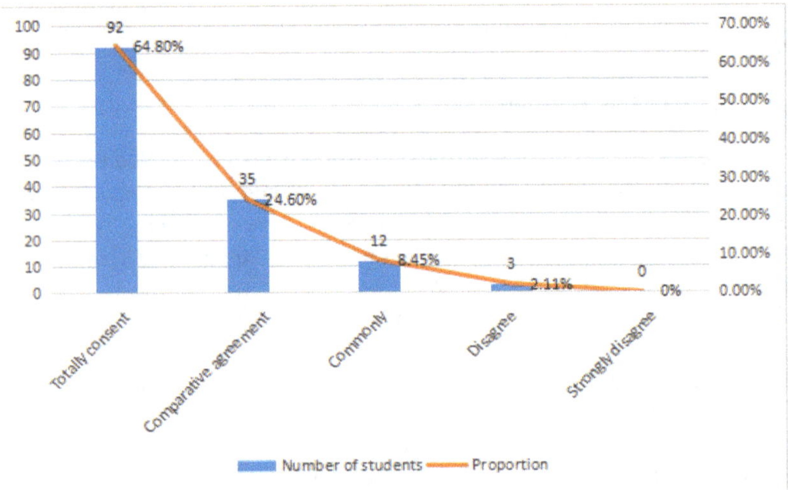

Fig. 2. Effect of Mobile LP on Students' English Learning

As shown in Fig. 2, 64.8% of the students think using this mobile LP is quite helpful to their English learning, and 24.6% of the students think the LP is helpful. These students generally believe that through the use of the platform, they can use the fragmented time more effectively, and learn vocabulary, sentence patterns and other knowledge points related to the text anytime and anywhere. To them, the learning of vocabulary and sentence patterns is within their ability and they can handle that easily. In class, with good preparation, teachers and students can have more time to practice speaking, critical thinking, etc., which can make the class more vivid and arouse students' in-depth thinking at the same time. The tests arranged by teachers on the platform also make it easier for students to check their learning effects and make them aware of their own deficiencies in learning. The supplement of extracurricular knowledge can also further broaden students' horizon and enhance their interest in learning. Only 2.11% of the students think the mobile LP has no obvious help. The overall result indicates the mobile LP has a good effect on helping freshmen make adequate preview, break through difficult learning points, and consolidate knowledge points.

4.2 Aspects for the Improvement of Mobile LP

Although most students have a positive attitude towards the impact of mobile LP, when asked what aspects of the mobile platform need to be improved, the students put forward many suggestions for improvement as shown in Fig. 3. Among them, 53.5% of students believe that the aspect of teaching needs to be improved. The explanation of the knowledge points can be more diversified, and the learning content needs to be updated in time to keep up with the development of the world. At the same time, it is necessary to further refine the teaching content to meet the needs of students at different levels. 19.7% of students want to improve the function of the platform. They think the web design of the platform can be more interesting and vivid. In addition, the interface of operation needs to be more friendly and easy to operate. 16.2% of students believe that this mobile LP needs to be improved in terms of innovation. In addition, nearly 10.6% of students believe there are other aspects that are needed to be improved, such as adding the function of automatic grading of essays, etc.

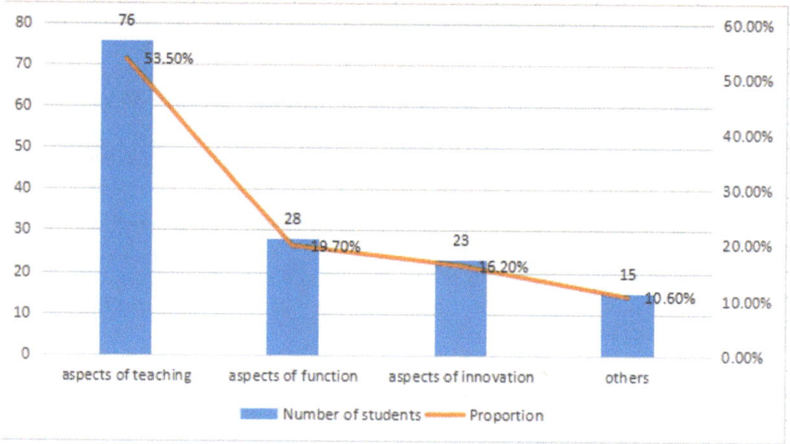

Fig. 3. Statistics of Improvement Suggestions

5 Conclusions

Through the research results of the paper, it is obvious that the use of mobile LP for auxiliary teaching has promoted the innovation of teaching mode and teaching method, and realized the integration of online and offline mixed teaching, provided students with new ways to acquire knowledge, and created good learning environment. By autonomous and exploratory learning through mobile LP, students' interest in learning will be continuously improved. In the process of improving their learning habits, their knowledge will be expanded and it will lay a good foundation for them to make better use of their spare time for fragmented learning in the future. The research on mobile LP assisted teaching is a long-term process. Theoretical support and research alone cannot give full play to its due value. Therefore, researchers and teachers need to explore and practice actively to improve the design of the mobile LP, thus to create better learning environment and learning resources for teachers and students.

Acknowledgment. This research is financially supported by "Research Center of Applied Linguistics in Wuhan Business University".

References

1. Lytridis, C., Tsinakos, A.: MLearn: a mobile LP. J. Eng. Sci. Technol. Rev. **10**, 81–83 (2017). https://doi.org/10.25103/jestr.104.11
2. Akinyokun, O., Oluwamodupe, A.: Augmented Reality (AR) - based human digestive system mobile LP. Int. J. Comput. App. **174**, 975–987 (2021)
3. Punnarumol, T., Sataworn, C.: Multi-agents platform for mobile learning using objective distance based personalisation method. Int. J. Mob. Learn. Organ. **12**, 293 (2018). https://doi.org/10.1504/IJMLO.2018.10011934
4. Sievertsen, N., Carreira, E.M.: Apoc social: a mobile interactive and social lp for collaborative solving of advanced problems in organic chemistry. CHIMIA Int. J. Chem. **72**, 43–47 (2018). https://doi.org/10.2533/chimia.2018.43
5. Talib, C.A., Aliyu, H., Malik, A.: Sakai: a mobile LP. Int. J. Interact. Mob. Technol. **13**, 95 (2019)
6. Merayo, N., et al.: AIM-mobile LP to enhance the teaching-learning process using smartphones. Comput. App. Eng. Educ. **26**, 1753–1768 (2018). https://doi.org/10.1002/cae.21979
7. Chaka, J.G., Govender, I.: Implementation of mobile learning using a social network platform: Facebook. Probl. Educ. 21st Century **78**, 24–47 (2020)
8. Novikov, M.: Educational quests, quizzes and games of the platform learn is in the system of mobile learning methods. Rus. Digit. Libr. J. **23**, 109–118 (2020). https://doi.org/10.26907/1562-5419-2020-23-1-2-109-118
9. Alghazzawi, D.M., Hasan, S.H., Aldabbagh, G., et al.: Development of platform independent mobile learning tool in Saudi Universities. Sustainability **13**, 1–2 (2021)

Analysis of Risk Assessment Model for Accounting Resource Sharing Management Based on Genetic Algorithm

Yue Li(✉)

Xi'an Mingde Institute of Technology, Xi'an 710124, Shaanxi, China
yyytdl@163.com

Abstract. The survival and development of enterprises are inseparable from accounting information, and sharing economy is a new business model based on computer network to realize resource allocation optimization through information technology under the market economy, which not only improves the economic efficiency of enterprises, but also provides an efficient, convenient and fair competition environment for the whole society. Accounting information sharing management refers to the integration and optimization of relevant resources among various departments through network technology, so that the resources within the enterprise can be shared to the greatest extent, and it can also reduce the cost of accounting information processing and improve the working efficiency. In this new era, the construction of accounting business processes, management control and performance evaluation are essential topics to improve the efficiency and level of the sharing society. In this paper, we analyze the management of accounting resources sharing by genetic algorithm and construct an evaluation index system of accounting resources sharing based on genetic algorithm.

Keywords: Genetic Algorithm · Accounting Resources · Shared Management · Risk Assessment · Model Analysis

1 Introduction

With the development of the Internet, the competition among enterprises no longer simply relies on material resources, but more on intangible assets such as talents, information and technology, which can bring great economic benefits to enterprises, and accounting business and information technology are important ways for enterprises to obtain intangible assets. Therefore, how to effectively manage the shared financial information becomes an urgent problem to be solved [1]. In order to make the accounting process more efficient and convenient and reduce repetitive workload, it is necessary to improve financial management efficiency, achieve cost reduction and risk prevention measures to achieve better control, thus establishing a collaborative relationship between enterprises under the sharing economy model. This paper uses a genetic algorithm to optimize and improve the internal financial risk of enterprises, and constructs an ARSM (Accounting Resource Sharing Management) model based on it, so that enterprises can be more efficient in financial risk control [2].

Y. Pei et al. (Eds.): IC 2024, LNEE 1214, pp. 301–309, 2024.
https://doi.org/10.1007/978-981-97-4193-9_32

2 Theoretical Research on Basic Methods

2.1 Genetic Algorithm

Genetic algorithm starts from an individual and gradually expands the optimal solution space by constantly adjusting its fitness. In nature, every organism is unique, stable and orderly. The genetic mechanism is composed of a group of groups with similar genes and multiple intergenerational structures, which interact with each other to complete the information exchange process and produce new results. And a theoretical method of optimal evolution in response to the characteristic parameters expressed by these populations, which is mainly achieved by constantly adjusting its fitness.

The basic genetic algorithm can be expressed as formula (1):

$$SGA = (C, E, P_0, M, \Phi, \psi, \Gamma) \tag{1}$$

2.2 ARSM Risk Assessment Theory

With the development of the times, ARSM has become a trend. How to evaluate it scientifically and reasonably and use it effectively has become a topic of concern for researchers. At present, two approaches are mainly used at home and abroad: one is based on game theory analysis. This approach calculates the financial performance of enterprises at different times by establishing mathematical models. Although this method can evaluate enterprises, it lacks scientificity, and it cannot guarantee decision makers to make correct judgments. The second is the hierarchical qualitative-quantitative approach (AHP), in which each sub-objective function value is derived by quantifying financial indicators and then determining the weight of each indicator in the whole system according to the size of each sub-objective function value [3]. Although this method can evaluate enterprises scientifically and reasonably, it cannot guarantee its accuracy and is difficult to meet the development requirements. Therefore, this paper proposes a risk assessment model. Genetic algorithm can reduce the probability of decision makers to make wrong judgment to a certain extent by simulating the natural evolution process to search for the optimal solution [4].

3 Risk Assessment Model Design of ARSM Based on Genetic Algorithm

3.1 Genetic Algorithm

Genetic algorithm is a kind of optimization problem solving by random search, which has obvious advantages in solving large-scale, nonlinear and multivariate complex system problems. It is able to combine individual characteristics of information in the data with each other and the population so that it can effectively improve the global optimal solution. This approach allows the use of genetic algorithms for a large number of recurring and discontinuous problems or when there is a large scale. For some simple and easily understood models, the use of genetic algorithms can be difficult: the operation process is complex, the computational effort is large, etc. [5]. Therefore, genetic algorithms often have limitations in solving large-scale, nonlinear and multivariate complex system problems.

3.2 Selection of Evaluation Indicators

The performance assessment of ARSM needs to take into account its use by different levels of users, as well as the needs of the enterprise for the type of services provided by that information user within the same time period. Based on the above analysis, it is clear that accounting resource sharing in China is a dynamic process. Since the level of socio-economic development, technological application capability and people's living environment factors affect accounting, the risk assessment is mainly affected by human factors, technological factors. The different needs of information users in business management activities lead to the variability among ARSM systems [6].

3.3 Weighting of Evaluation Indicators

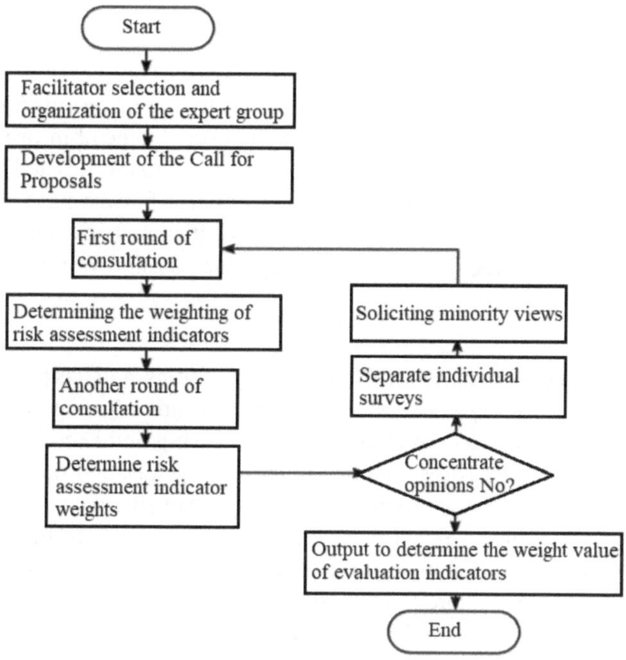

Fig. 1. Delphi method process

Indicator weighting refers to the extent to which a certain objective or criterion is used to measure the influence of that factor on the evaluation results, and assessment methods are limited by their own characteristics, scope of application, and subjective judgment. Therefore, different researchers have chosen to different methods. The Delphi method is a subjective judgment method, and after determining the index weights, each data is analyzed to obtain the comprehensive evaluation results of each ARSM risk [7]. The Delphi method process is shown in Fig. 1.

3.4 Quantitative Risk Assessment Model Construction

When building a risk assessment model, the first step is to establish a mathematical foundation that is scientifically sound, comprehensive, effective, and easy to operate and control. In order to obtain the relevant information, it is necessary to first determine the required primary data through certain methods. For example, interviews, expert interviews, literature searches and field surveys can be used to collect relevant information and then analyze and evaluate the shared management of accounting resources based on these primary data [8].

A quantitative model of ARSM risk evaluation is built, and the model is evaluated based on genetic algorithm to construct an ARSM risk evaluation index system, and then fuzzy comprehensive judgment method and hierarchical analysis method are applied to calculate the weights of each element.

The basic risk assessment model of ARSM is constructed, and the expression is shown in formula (2):

$$R = F(V, T, C) \tag{2}$$

The sharing of accounting resources refers to the sharing of various resources owned by an enterprise and external stakeholders through the exchange of information within a certain period of time to enhance their own value creation capabilities and social benefits. Share of accounting resources is the sharing of information between an enterprise and its external stakeholders for a certain period of time to enhance its own interests. In terms of risk as defined by ISO/IEC, Eq. (1) can be expressed as shown in formula (3):

$$R = F(Pt, Pv, V) \tag{3}$$

Enterprises face a variety of risks in their daily operations, and it is particularly important to manage these issues effectively. Accounting resource sharing is a new thing that arises and develops in the mode of information processing and exchange, and provides the technical support required for the corresponding services and the establishment of trust between accounting service providers. The purpose of sharing accounting resources is to solve complex problems faced by companies and to manage them effectively in order to control risks and maximize benefits. The more effective the risk control measures are, the less risky the shared accounting resources management is. The effective degree of risk control measures is calculated as shown in formula (4):

$$Sm = 1 - \frac{Nv}{NR} \tag{4}$$

This paper uses Poisson distribution quantitative risk assessment index system to assess the risk of ARSM platform and proposes corresponding countermeasures, which provides a theoretical basis for enterprises to build an efficient and standardized accounting resource sharing model. The Poisson distribution sets the random variable x, which takes the values of [0, +∞], and its computational representation is shown in formula (5):

$$P(x = k) = \frac{e^{-\lambda}\lambda^{x}}{Sm} \tag{5}$$

where, λ is a constant and the Poisson distribution formula can be converted to a normal distribution formula when its value reaches a maximum, as shown in formula (6).

$$P(x = n) = \frac{1}{\sqrt{2\pi\lambda}} e \frac{(n-\lambda)2}{2\lambda} \tag{6}$$

When the entry is equal to 10, the Poisson distribution curve is close to the normal distribution curve, as shown in Fig. 2.

To sum up, this paper uses Poisson distribution to quantify the establishment of risk assessment index system, builds an improved index system based on accounting resource sharing, and uses analytic hierarchy process to evaluate its risks, and then realizes the ARSM risk assessment model, which is shown in formula (7). By analyzing the data before and after improvement, a mathematical model based on the improved index system is established. Based on the evaluation results, feasible suggestions are put forward which are easy to be quantified and popularized [9].

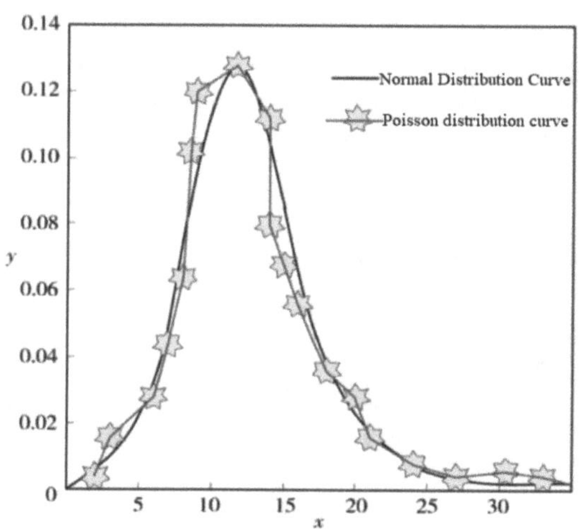

Fig. 2. Poisson distribution curve

$$Q = \begin{cases} \sum_{i=1}^{k} \left(V \times \frac{e^{-\lambda}\lambda^x}{Sm} \times Pv \right), k < 10 \\ \sum_{i=1}^{k} \left(V \times \frac{1}{\sqrt{2\pi\lambda}} e^{\frac{(n-\lambda)2}{2\lambda}} \times Pv \right), k \geq 10 \end{cases} \tag{7}$$

4 Experimental Results and Analysis

Through the above analysis, we can find that the genetic algorithm based on solving the ARSM model has high generality, adaptability and high efficiency. However, there are also some shortcomings, such as limited data access and a single way of coding information, which can lead to large deviations of the calculation results from the actual situation.

In order to verify the evaluation performance of this paper's risk assessment model for ARSM, we designed a simulation comparison experiment, in which the logistic-based financial risk assessment model for manufacturing enterprises, the FAHP-based financial risk assessment model for SMEs and this paper's assessment model were used to evaluate the performance of the model by evaluation time and evaluation accuracy.

4.1 Experimental Environment

In order to realize the shared management of accounting resources, we need to conduct simulation experiments. Since China is currently in the economic transition period, all aspects are not mature and sound enough. Therefore, the main purpose of this paper is to establish an accounting resource sharing system to reduce enterprise transaction costs, improve efficiency and reduce information asymmetry, as well as to provide reference value for other regions and to contribute to their rapid development. In this paper, we conducted simulation and comparison experiments to find out the advantages of the accounting resource sharing system, and then obtained its advantages in different application environments through comparison and analysis. The risk assessment model for efficient management in different application environments is then obtained through comparative analysis, and the feasibility of the system for efficient management in different application environments is verified through examples [10].

4.2 Analysis of Experimental Results

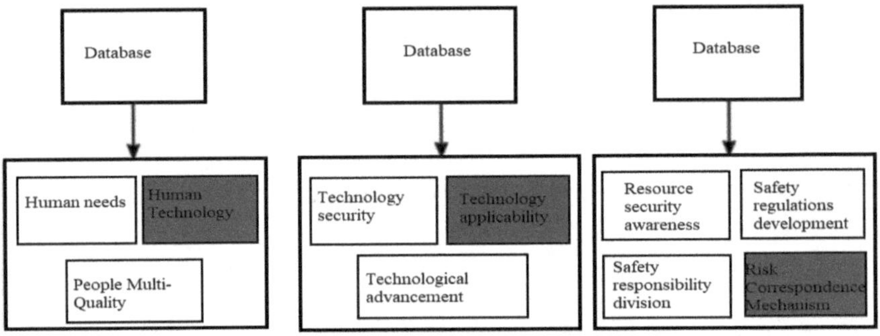

Fig. 3. ARSM risk assessment indicators

The sample data is processed by Matlab software, and quantitative analysis is carried out. The relevant models are used to build the accounting resource sharing decision support platform for information interaction, business process interaction and system interconnection between users and enterprises. Figure 3 shows the risk assessment indexes of ARSM, with the gray part being the high-risk assessment indexes. The information within the enterprise is made public by building the ARSM platform. At the same time, the evaluation methods such as fuzzy comprehensive analysis and hierarchical analysis are used to calculate the weights and affiliation values of the factors in the system and find out their probability matrices [11]. Then, the two concepts of decision vector and entropy power in the genetic algorithm solution model are combined to determine the optimal solution of the objective function chosen by the decision maker under each scenario, and the different risk levels of the shared management platform in selecting accounting resources at different stages.

Through the above analysis, it is found that under the ARSM mode, the enterprise management can allocate and utilize the system reasonably and effectively. But with the complexity and uncertainty of accounting resources sharing, enterprise management needs to make reasonable and effective planning when making decision-making. At present, our country is not perfect for the system. First of all, the service quality provided by the accounting resource sharing platform is not high, which leads to the use efficiency of accounting resources is not high. Secondly, the management of the enterprise does not manage and utilize the shared service platform reasonably and effectively, so that there are many uncertain factors in the operation process of the system, which increases the decision-making risk. Based on this, this experiment mainly analyzes the model adopted in this paper, so that the improved model puts forward corresponding suggestions for the ARSM platform, and compares the model adopted in this paper with the risk division model based on Logistic and FAHP. The comparison results are shown in Table 1:

Table 1. Comparison of evaluation time results

Number of iterations	Logistic-based evaluation model	Valuation model based on FAHP method	This paper evaluates the model
20	23.00	30.20	10.20
40	26.02	32.50	10.00
60	25.12	33.60	9.15
80	20.10	35.61	8.23
100	20.41	36.78	8.00
120	19.58	37.22	8.40
140	23.58	38.12	7.12
160	18.77	39.14	6.45
180	17.46	40.15	5.00
200	20.15	42.85	6.14

The ARSM model based on FAHP method can well reflect the problems existing in the financial risk of enterprises. Based on the Logistic regression model, the improvement scheme of the ARSM platform is proposed, and it has certain reference significance for the subsequent enterprises to use this system for risk assessment. However, the disadvantages of the above two models are that they cannot quantify the risks of enterprise ARSM, and there are some defects in the construction process of this model. The evaluation model proposed in this paper can make up the deficiency effectively, can quantify the risks of enterprises in the process of ARSM, and can provide some reference for the future establishment of the appropriate accounting resource sharing system in our country [12]. In order to verify the effectiveness of the method in this paper, the accuracy and error rates of the above three models are compared according to the test samples, and the comparison results are shown in Table 2.

Table 2. Evaluation results of the three models

Model	Number of test samples	Number of misjudgments	Evaluation error rate/%
Logistic-based evaluation model	200	9	48
Evaluation model based on FAHP method	200	17	52
Evaluation model of this paper	200	1	2

From Table 2, it can be seen that the accuracy rate of ARSM risk evaluation of the evaluation model gradually increases with the number of iterations and has an obvious increasing trend. Based on this, this paper proposes an improved genetic algorithm time series model, which is a mathematical method to establish the ARSM risk evaluation index system based on probability theory. In this system, randomly generated sample data can be used as a reference basis for evaluation, and then this information is compared with historical results to obtain corresponding conclusions and analyze the degree of impact and reasons caused by factors such as regularity and variation on the enterprise.

5 Conclusion

To sum up, ARSM mode is an inevitable trend of modern enterprise development, and its generation and existence have brought a huge impact on traditional financial work. In this context, how to effectively build an accounting resource information platform to optimize the internal structure of enterprises, improve the efficiency of capital use, reduce costs, and achieve sustainable business development has become an urgent problem to be solved and needs to be studied. Based on this, this paper takes ARSM as the research object, and builds a risk evaluation index system for ARSM, so that enterprises can reasonably and effectively avoid and control its possible adverse effects in financial management, realize the optimal allocation of resource information platform, improve the efficiency of capital use and reduce the cost of enterprises. This will enable the enterprises to avoid and control the possible adverse effects, optimize the allocation of resource information platform, improve the efficiency of capital use and reduce the business risks.

Acknowledgment. Educational and Teaching Reform Research Project of Xi'an MingDe Institute of Technology, Project number: JG2022YB04, Project name: Research and Practice of

Multi-blended Teaching Mode of Ideological and political Education integrated into Financial Management courses.

References

1. Deng, X., Zhang, D.: Trajectory optimization of interceptor missile based on hybrid genetic algorithm. In: Proceedings of 2021 2nd International Conference on Electronics, Communications and Information Technology (CECIT 2021), pp. 1292–1296 (2021)
2. Xiao, P., Xu, Q., Cao, W.J., et al.: Risk prediction for deep venous thrombosis after total knee arthroplasty based on modified Caprini risk assessment model: a confirmatory study. Zhongguo gu shang = China journal of orthopaedics and traumatology. **3**, 32–35 (2022)
3. Wu, H., Wen, Y.: Pricing strategy of software products based on cloud accounting resource sharing platform. In: Fifth International Conference on Economic and Business Management (FEBM 2020), pp. 23–25 (2020)
4. Deng, X., Zhang, D.: Trajectory optimization of interceptor missile based on hybrid genetic algorithm. In: Proceedings of 2021 2nd International Conference on Electronics, Communications and Information Technology (CECIT 2021), pp. 1292–1296 (2021)
5. Anna, M.: Ship model identification with genetic algorithm tuning. Appl. Sci. **11**(12), 5504 (2021)
6. Rajiv, A., Patrick, R., White, W.B., et al.: Individual patient risk assessment and cost–benefit analysis of patiromer in AMBER – Authors' reply. The Lancet **396–400**, 311–312 (2020)
7. Berg, R., Maja, N., Bruun, M.T., et al.: Detecting mismatched donor HLA types from allograft biopsies – an easily applicable tool for improved individualized risk assessment. Human Immunol. **81–88**, 337–349 (2020)
8. Wu, H.P., Li, H., Sun, X.L.: Evolutionary game for enterprise cloud accounting resource sharing behavior based on the cloud sharing platform. IAENG Int. J. Appl. Math. **49–51**, 1–8 (2021)
9. Song, X.: Feature extraction of online English learning behavior based on genetic algorithm. In: Proceedings of 2021 International Conference on Intelligent Transportation, Big Data & Smart City (ICITBS 2021) Part III, pp. 225–228 (2021)
10. Dong, W., Tse, T.Y.E., Mak, L.I., et al.: Non-laboratory-based risk assessment model for case detection of diabetes mellitus and pre-diabetes in primary care. J. Diabetes Investig. **13**(8), 1374–1386 (2022)
11. Xiaosheng, W., Wei, L., Haiying, G., et al.: A risk assessment model of WSMC projects under emergencies. Water Supply **2**, 20–22 (2022)
12. Chen, R., Lu, Y., Wang, Y.: Genetic algorithm application on the risk assessment for ARSM. Math. Probl. Eng. **20–22**, 1–10 (2022)

Intelligent Analysis of University Apartment Management from the Perspective of Face Recognition Technology

Haisheng Wu(✉)

Student Affairs Office, Nantong Institute of Technology, Nantong, Jiangsu, China
wuhaisheng669@163.com

Abstract. The use of face recognition technology can greatly improve the effectiveness of the management of college students' apartments in view of the current situation that there are many security risks in school apartments, the process data is difficult to trace, the identity verification of off-campus personnel is difficult, it is difficult to check bedtime without returning home at night, and the epidemic prevention and control data is not accurate. Face recognition technology in the confidentiality, security, stability and other advantages, is currently in and out of the apartment for students informationization, intellectualized ideal way of identity, so this article will focus on the importance of university apartment for personnel training and management frequently encountered problems, through the facial recognition technology to implement the intelligent of the apartment management. Using the power of science and technology can improve the efficiency and accuracy of the apartment administrator's attendance, reduce the human cost of manual attendance, and thus improve the effectiveness of the entire apartment management.

Keywords: Apartment Management · Face Recognition Technology · Intelligent · Promotion Of Information Technology · Security

1 Introduction

With the reform and development of education system, the importance of student apartment to the healthy growth of college students is becoming more and more prominent. The Ministry of Education proposed in Several Opinions on Further Strengthening the Management of Student Apartments in Colleges and Universities that student apartments are an important place for students' daily life and study, and an important front for ideological and political work and quality education in addition to classroom study. In recent years, in colleges and universities in the promotion of fraud, theft, personal injury and other cases frequently appear in the news reports, school security problems are becoming more and more prominent, colleges and universities in the safety management is becoming more and more difficult. Apartment should be reassured students and parents the most the most safe place, but the reality is counterproductive, student

Y. Pei et al. (Eds.): IC 2024, LNEE 1214, pp. 310–320, 2024.
https://doi.org/10.1007/978-981-97-4193-9_33

apartments become the university security incidents in fertile soil, which makes the traditional rely on manpower to apartment management model has been unable to adapt to today's education reform and the development of The Times, by increasing the technology to assist in daily apartment management is imminent [1]. At present, a lot of school attendance, knowing, service, or by artificial forms, these methods have larger gaps and potential safety problems, the human eye could not tell the precise nature, the human brain memory limit on duty, human's high-cost, these factors have restricted the university apartment safety prevention, thus necessity, relying on scientific and technological means to enhance the level of safety At the same time, using the power of science and technology can also improve the efficiency and accuracy of apartment manager attendance, reduce the labor cost of manual attendance, so as to improve the effectiveness of the whole apartment management.

2 Importance of Apartment Management

The student apartment is the "home" for students to study in other places. Most students spend time in the student apartment during their study. Student apartment should be the ideal training place to improve the comprehensive ability of college students. To improve the comprehensive ability of college students, it is not enough to only rely on the classroom, we should look for space outside the classroom. The cultural construction of college students' apartment has a direct impact on the construction effect of school spirit and style of study. Strengthening the cultural construction of the apartment plays a positive role in promoting the comprehensive quality of college students [2].

2.1 Student Apartment is the Educational Front to Strengthen the Ideological and Moral Construction of College Students

With the reform and development of the university system, university apartment in college students is more and more important role in the process of development, student apartment is the place where they live, not only it is the important place that they learning and communication, the quality of education students apartment management, cultural construction, the college students' learning and health of body and mind, thought personal character had a great influence. The management department of the school apartment needs to change the simple management concept, in line with the principles of education, management and service, and deeply realize its own positioning [3]. It is not only the manager and server of the apartment, but also the student educator. It follows the law of education and strengthens the awareness of education. In the daily management and service, the ideological and political education and management of students are integrated into the management of student apartments, adhering to the purpose of "management education, service education, environment education".

2.2 Direct Needs for Student Development

The student apartment at the university stage is the first place for most students to live independently, which is an important transition stage from family to society. College students spend most of their time in student apartments. The life of students in the apartment runs through the construction period of their knowledge system and the shaping period of their values. In the student apartment, students should not only learn knowledge, but also learn to adapt to collective life, deal with interpersonal relationships, develop healthy work and rest, and make full preparations for the society.

2.3 The Objective Needs of Building Virtue and Cultivating People

To establish morality and cultivate people is to integrate ideological and moral education, cultural knowledge education and social practice education into each link to form a long-term mechanism of teaching, scientific research, practice, service, management and culture. The student apartment undertakes the basic function of providing students with living and learning space, and plays an important role in the work of moral education. A good apartment atmosphere is conducive to creating a good learning and living environment for students, so as to form good living habits and learning conditions, and then cultivate students' correct outlook on life and values, which will produce an educational effect imperceptibly.

3 Demand Analysis of Face Recognition System in Apartment Management

3.1 Face Recognition System

At the entrance and exit of the school apartment, students' passage is controlled by face recognition channel gate or intelligent check-in is carried out by face capture device and back-end deep learning server. This method can not only reduce the security risks brought by the entry of outsiders to the apartment to the greatest extent, but also enhance the efficiency and reliability of the attendance of the apartment administrator.

Face recognition system with face recognition technology as the core, is a new biometrics technology, is today's international science and technology field of advanced technology. Face recognition system is the use of visible light to obtain face image information, and different from fingerprint recognition needs to use electronic pressure sensor to collect fingerprints, face comparison equipment can capture the face comparison analysis, generate comparison events, and reported to the platform. The platform can receive real-time face comparison events and process them.Through the face monitoring system, apartment managers can effectively monitor and control strangers, high-frequency personnel and key personnel, so as to facilitate school management more effectively. The irreplaceable and unreplicable human face ensures the accuracy of students' access information and solves the defects of apartment management.

Face recognition system realizes electronic processing of visiting information. All the visiting information of legal visitors and temporary visitors, including the identity of visitors, visit time, visit place and the information of visitors, are recorded in the system. According to the above functions, the system is divided into three functional modules, including system management module, identity identification module and user information management module [4]. The functional structure of the system is shown in Fig. 1.

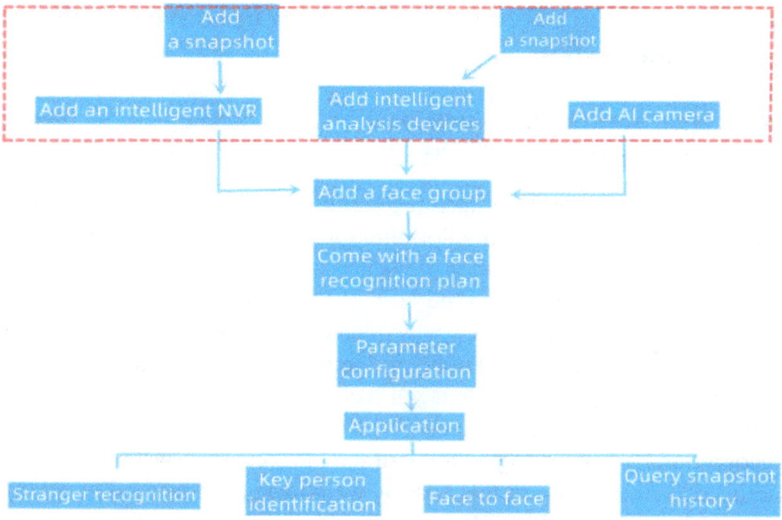

Fig. 1. Function diagram of face recognition system

3.2 Demand Analysis of Access Control System for Face Recognition Technology

With the continuous expansion of college enrollment and the openness of the campus, the existing college apartment management model is relatively backward, and some students have relatively weak sense of self-protection, leading to more and more prominent problems of school security in recent years, and more and more difficult in the security management of colleges and universities. A large number of people come in and out of the student apartment every day, and there is a large flow of people. Illegal elements enter the girls' dormitory, posing a threat to the girls' personal safety; Some illegal sales personnel enter and leave the dormitory, which affects students' rest. Apartment management personnel judge the incoming and outgoing personnel by human eyes, which is prone to miscalculation or omission. Manual registration is cumbersome and inefficient.

In order to solve this kind of security problem in the apartment, the general practice of colleges and universities is that after closing the building door, the management personnel should check and confirm each person who enters the dormitory building or manually check the dormitory building, and register the persons who are late or not. Some schools also try to purchase and install all-in-one card access control system and fingerprint access control system. Simple card swiping access control system and fingerprint system will also cause problems such as forgetting to bring the card, campus card lending and fake fingerprint cover. Therefore, the school has invested a lot of personnel and funds but failed to achieve the expected management effect. The introduction of face recognition access control system can not only improve the travel efficiency of the dormitory, but also achieve intelligent statistics on the return of the apartment, reducing the management pressure of managers. In addition, it can strengthen the security management and control ability of the dormitory, prevent unauthorized personnel, strangers and blacklisted personnel from entering the dormitory outside the school, and solve the problem of swiping cards and checking attendance on behalf of others.

4 Application of Face Recognition System in University Apartment Management System

4.1 Student Apartment is the Educational Front to Strengthen the Ideological and Moral Construction of College Students

In order to explain the necessity of face recognition system, we have made a questionnaire, and the respondents are mainly dormitory administrators and members of college students' self-management Committee of Nantong Institute of Technology. Meanwhile, the survey is designed according to the current situation of the school. For example: "What is your attitude towards the new way of entering the apartment by" brushing your face "?" ", "Is face recognition system helpful to your work? ", "How do you feel about your experience with the apartment's facial recognition system? Does it support facial recognition technology into the apartment?" ", "What do you think are the disadvantages of face recognition? ", "What do you think of the face brushing system, and what needs to be improved?" And so on. In this survey, we adopted the method of questionnaire distribution, and a total of 158 valid questionnaires were recovered. This questionnaire is mainly for the management of the front line, from the questionnaire survey data shows that the management of the application of face recognition system urgency and necessity (Table 1).

Table 1. Questionnaire survey on the use of face recognition system

The serial number	Face recognition system is helpful for the work	Whether facial recognition technology is supported to enter the apartment	How do you feel about the dormitory face recognition system
1	96% think it helps	95% student support	96% of the students felt good
2	3% said they were unsure	4% are on the fence	Three percent felt average
3	1% said it was not helpful	1% is not supported	1% felt bad

4.2 Student Behavior Big Data Analysis Case of Student Apartment

Face recognition system will record every student in and out of the apartment, through the behavior of the student apartment big data analysis, the state of the students in the apartment in a timely and comprehensive understanding, master the ideological status of special students, security risks timely warning, so that the student apartment management more scientific and effective. At the same time, the regular of apartment administrators in the form of questionnaire survey (Table 2), hope from the process of investigation found the apartment management problems, analyze the problems and put forward the corresponding countermeasures, in order to improve the quality of apartment management service, better for college students to provide a comfortable and quiet environment of study and life. So that students have a more positive attitude into learning, to promote the development of high quality of school education.

Table 2. Student Satisfaction Survey with Dormitory Manager - January 2022

The serial number	Students' understanding of the management system	Dormitory administrator safety precautions are in place	Are you satisfied with your work as a dormitory administrator
1	78.49% of the students fully understand	85.47% of students think it is very good	88.37% of the students think they are very satisfied
2	18.02% of the students know better	13.37% of the students think it is better	9.88% of the students think they are satisfied
3	3.49% of the students do not understand	1.16% of the students thought it was not good	1.75% of the students said they were dissatisfied

5 Optimization of University Apartment Management by Face Recognition Technology

5.1 Access Control or Attendance Management

Face recognition for student apartment can effectively identify the school staff and students, to prevent outside personnel into the student apartment to promote or fraud on students, to give timely and effective early warning of special circumstances [5]. At the same time, brake machine has brush face + card (standby) recognition, using pendulum brake, brake machine in case of an emergency on the one hand, through a unified control management personnel opened machine, on the other hand each brake machine installed the emergency door open, students can pass, in an emergency to maintain the students' personal and property security provides more and more effective guarantee. Face recognition system can also monitor and attendance management of all personnel in and out, can accurately and completely record the dormitory administrator and counselors in and out, so as to help the management department dormitory administrator attendance and counselors into the apartment statistics.

5.2 Face Monitoring Application

Through time, capture point condition such as query personnel identification records and capture records, or search by search face to face, and can look at the track and capture the information such as the original image, in order to track human activity traces, path analysis and event analysis, is helpful to scientific assessment of safety risk and make corresponding measures, in order to prevent crime and incident investigation [6].

5.3 Daily Management of Student Apartment

The face recognition system can complete the docking with the data center library and the student work system in the apartment module, realize the automatic synchronization of the apartment check-in information (generally at 0 o 'clock every day for automatic synchronization), set the focus on students, adjust the apartment, check out and other functions [7]. The system supports query records such as the total number of people who have not returned to bed, the real-time statistics of the number of people who have not returned to bed in each building today, the ranking of the number of people who have not returned to bed, and the information of students who have returned to bed in real time according to the building. It can realize the real-time push of students' attendance information to the management personnel and counselors, so as to know the students' situation in real time. At the same time, data statistics can be carried out on special students who have not returned for consecutive days, who have not gone out for consecutive days, who have returned late for many times, and who live alone in dormitories, so as to facilitate accurate management and assistance by student managers, so as to achieve efficient and accurate dormitory management.

5.4 Integration of Face Recognition into Daily Management

When face recognition technology is applied to daily teaching, there are a large number of students in college classes, especially in public classes, and it is easy for students to skip classes or substitute classes. In order to avoid this situation, schools can use face recognition technology to identify students in the classroom. In the classroom of the school, there will be monitoring equipment. Relevant personnel can use the monitoring equipment to collect the students' faces, input the collected information into the face recognition system, and verify their identity. During the collection, attention should be paid to that relevant personnel should conduct multiple collections to avoid truancy. Before class, the teacher can let students use their mobile phones to log in to relevant software for face recognition. In this way, students who are absent from class can be accurately found. At present, the progress of science and technology in China has led to the development of face recognition technology. Now, face recognition technology has begun to develop in a dynamic direction. Gradually, relevant departments have begun to develop dynamic face recognition technology. This kind of technology is more flexible than traditional face recognition technology. It can recognize faces according to students' behavior in class. During recognition, it can analyze students' listening status and listening time, and send the collected information to teachers as a whole. Teachers can evaluate students according to their listening status. Applying this kind of technology to teaching management can reduce teachers' teaching and working pressure, and teachers can systematically evaluate students according to the information sorted out. It is difficult to eliminate truancy in class by roll call, card swiping and other methods. Because face recognition is based on biological recognition and unique, it cannot be replaced by others, so it has obvious effect in preventing truancy.

Colleges and universities hold many large-scale examinations every year, especially the national unified examination. In the course of the examination, it is easy to substitute and cheat Before the students enter the examination room, the invigilator will check the students' information, and the examinees can enter the examination room only after passing the inspection. However, the invigilator does not know the students, so there will be loopholes in this way. The face recognition technology is applied to the examination management. When the examination enters the examination, the existing face recognition is used to ensure that the facial information in the examination is consistent with the information on the ID card. During the examination, the camera in the examination room can clearly and accurately capture the examinee's examination dynamics. Once the examinee is found whispering, the relevant person in charge will notify the teacher and give corresponding treatment. To a certain extent, face recognition technology can prevent the phenomenon of substitute examination, at the same time, it can also serve as a warning to examinees. Using this technology, we can strengthen students' moral awareness, let examinees establish correct moral concepts, and eliminate those bad ideas.

"Face recognition book borrowing" is to solve the problem that students are easy to lose their library cards by using the results of the combination of new biometric technology and library services, based on the relatively single and fixed consideration of campus readers. By saving personal information to the face recognition system in advance, readers can borrow and return books and query relevant information immediately without carrying their library cards. At the same time, it can record, count and analyze the library borrowing information of teachers and students, operate the database according to different permissions, and provide access control record query, data backup, data printing and output and other functions.

5.5 Face Recognition Orientation Management

Using face recognition systems, new report just need to brush "face", can complete id, shooting scene faces, comparing two validation, admissions information can be either to experience the new AI era brings sense of science and technology, and can work for the college freshman report added safety check, presenting an impostor in college. In the face recognition process, freshmen place their ID cards on the face recognition equipment for one second to quickly complete the identification comparison, which improves the efficiency of on-site freshmen identity verification and reduces the verification pressure of welcome personnel [8]. In the links of health screening and nucleic acid test certification, students who need to submit nucleic acid test certification are screened out in advance through the big data of the health check-in platform of our school for 14 consecutive days in advance, which reduces the pressure of the welcome scene.

5.6 Epidemic Management in Apartments

The movement of people in student apartments is relatively dense, but the risk of virus transmission is still not to be underestimated, which requires constant vigilance. Daily self-monitoring of students is inseparable from the measurement of body temperature. Since the normalization of epidemic prevention and control, the integrated temperature measurement gate, which integrates temperature measurement, face recognition, monitoring and protection, has become more important. The intelligent temperature access control device installed in the school's apartment can also identify people [9]. Once they have abnormal body temperature, the system will immediately send a voice prompt to forbid them to go through. School manual temperature measurement and intelligent temperature entrance guard gate system has formed a "civil air defense + dimension is" prevention and control of double insurance, was also able to identify temperature measurement, check the health code, persons together, identify in epidemic prevention and control under the normalized persons only takes a few seconds, implements the management of personnel entering and leaving, measuring temperature, health inquiry "out of contact, the second recognition". At the same time, it can also query the statistics of temperature measurement of each building every day, including the total number of temperature measurement, whether there is any abnormality in temperature measurement, etc., and one person in the apartment can query the temperature measurement records and changes of the personnel in 30 days.

5.7 Face Recognition Student Registration Verification

It is urgent for colleges and universities to check the status of students and ensure the consistency of students, certificates and status. However, the huge data makes the screening work particularly difficult, and there are certain loopholes in the manual screening [10]. The review of the admission qualification of freshmen through face recognition can greatly improve efficiency and reduce the error rate. Through the "enrollment photos", "on-site photo collection", "ID photo" and other portrait photo information matching, can accurately verify the identity information of each new student, to ensure the unity of witness, thus greatly improving the efficiency and accuracy of the review.

6 Existing Problems and Optimized Solutions

The face recognition rate is not only related to the photo quality and algorithm of the face database, but also to the adaptability of the hardware. For example, there is a problem with the face information acquisition equipment. A school uses a self-service image acquisition machine for verification of human identification, and hardware cameras such as the acquisition machine use wide dynamic technology, which can adjust the brightness parameters of infrared and white light compensation, so as to solve the problems of backlight and weak light in the acquisition. However, the light compensation function of the collector can only manually adjust the parameters, and cannot automatically adjust the light compensation in the adaptive environment. In addition, the speed of sending system photos to front-end devices is very slow, so it is considered to upgrade the distribution mode of face photos, so that photos can be modeled on the platform, and modeling data can be synchronized and distributed to Facepad, which not only reduces the time of platform distribution, but also helps to ensure the security of face data and improve the stability of the system.

For the current normalization of epidemic prevention and control, developing or upgrading the information platform for epidemic prevention and control has become the new development needs of education informatization. The construction of school access control transformation is considered to be connected with the quarantine center, so as to realize the sharing of epidemic prevention information, and carry out self-service inspection of health codes and body temperature detection for incoming and outgoing personnel. The results of the non-contact infrared temperature measurement module currently used are greatly affected by the surrounding environment, especially in high temperature or low temperature seasons.

7 Conclusions

Face recognition access control system is no longer a simple and narrow application, but provides a basis for the integration of multiple application systems and the needs of all parties. When each application system accesses the face authentication service, you can swipe your face to sign in, enter and exit the campus, dormitory and other places, and enter the examination room. Help colleges and universities truly realize "integrated and intelligent" management, build a safe and smart campus, and let teachers and students have a better life and service experience on campus.

Student apartment is the place where students come and go most frequently. The use of face recognition technology actively promotes the intellectualization of student apartment management service. It can fundamentally break through the defects of traditional apartment management mode, so as to achieve the comprehensive supervision and management of all students living on campus. This paper proposes the introduction of face recognition technology, it and access control system organic combination, and through the effective recognition of face, can achieve more refined management of students, reduce and eliminate security risks, effectively protect the safety of students' life and property, promote the continuous development of the school.

References

1. Luisa, S., Derya, T., Marcelo, V., et al.: When students are house-poor: urban universities, student marginality, and the hidden curriculum of student housing. Cities **124**, 103572 (2022)
2. Prajapat, N., Turner, C., Tiwari, A., et al.: Real-time discrete event simulation: a framework for an intelligent expert system approach utilising decision trees. Int. J. Adv. Manuf. Technol. **110**(11–12), 2893–2911 (2020)
3. Alaa, B., Charkari, N.M.: Complex network-based feature extraction in RGB-D humanaction recognition. J. Visual Commun. Image Represent. **82**, 103371 (2021)
4. Arya, S., Avinandan, B., Kumar, S.P., et al.: 3D human action recognition: through the eyes of researchers. Expert Syst. App. **193**, 116424 (2022)
5. Ayush, B., Sakshee, P., Saurabh, K., et al.: Empowering knowledge distillation via open set recognition for robust 3D point cloud classification. Pattern Recogn. Lett. **151**, 172–179 (2021)
6. William, D., Fernanda, A., Rafael, P., et al.: Cross-dataset emotion recognition from facial expressions through convolutional neural networks. J. Visual Commun. Image Represent. **82**, 103395 (2022)
7. Padmanabhan, S., Radhika, K.R.: Optimal feature selection-based biometric key management for identity management system: Emotion oriented facial biometric system. J. Visual Commun. Image Represent. **74**, 103002 (2021)
8. Tan, G., et al.: Microbial community and metabolite dynamics during Soy Sauce Koji making. Front. Microbiol. **13**, 841529 (2022)

9. Sotomayor, G., Hampel, H., Vázquez, R.F.: Water quality assessment with emphasis in parameter optimisation using pattern recognition methods and genetic algorithm. Water Res. **130**, 353–362 (2018)
10. Rohit, R., Abhinav, T., Himanshu, U., Devendra, S.: Algorithmic analysis of automatic attendance system using facial recognition: a revolutionary approach for future education. Int. J. Decis. Support Syst. Technol. (IJDSST) **14**(1), 1–9 (2022)

Design System of Smart Campus Big Data Management Service Platform Based on FCOS Algorithm

Chenxiao Lv(✉)

Wuhan Business University, Hubei 430056, China
Lvchenxiao_always@163.com

Abstract. With the development and application of information technology, campus network has become a new type of information exchange platform, which is not only one of the indispensable parts of university management, but also plays an important role in the school. The campus big data analysis and processing platform is based on computer technology and information technology, which uses advanced and reliable management software to realize the functions of information collection, storage and retrieval, analyze, process and apply the data, and realize the construction of campus big environment through the related information platform, so as to provide services for the university management. The campus big data management platform built based on FCOS platform has the characteristics of fast and convenient. It can provide efficient and high quality services for university teachers and students, so that it can provide convenient services for school management and meet the needs of teachers and students for campus big data management platform, which has good application prospects. This system designs a large database platform based on and combined with FCOS algorithm fusion technology to provide reliable support for smart campus construction in order to realize information sharing and intelligent management, and thus help students learn modern education knowledge efficiently and conveniently.

Keywords: Big Data Management · FCOS Algorithm · Service Platform · Smart Campus · System Design

1 Introduction

With the rapid development of computer technology, the Internet has become an indispensable part of people's lives, and the amount of information on campus has increased dramatically. In the information era, network management is the trend, and the traditional office mode can no longer meet people's needs. The campus big data management platform is the development of network technology to a certain stage in the information era, and the application of modern information technology in the information environment for the scientific and standardized construction of the school. The combination of smart city and Internet of Things can effectively solve the problems existing in the traditional office mode [1]. At present, most universities in China have realized the construction

Y. Pei et al. (Eds.): IC 2024, LNEE 1214, pp. 321–330, 2024.
https://doi.org/10.1007/978-981-97-4193-9_34

and put into use the multimedia resource sharing platform such as digital library and electronic reading room. However, due to the limited scale and capital investment of traditional schools as well as the lack of professionals, their development is restricted or even stagnant, while the huge amount of campus information makes it impossible to process many data at the same time, which leads to the quality and value of data being greatly damaged. Based on this, this paper proposes a campus big data management platform design based on FCOS algorithm [2].

2 FCOS Algorithm

FCOS algorithm is a data management method based on distributed storage structure. The technology is computer-based, and it ensures the speed and accuracy of transaction processing in information flow by compressing a large number of files, which can improve the efficiency of information storage in data warehouse. The algorithm is computer-based, combining traditional methods with modern technologies to analyze large amounts of big data, and using relevant computing tools such as Python to implement a campus big database management system based on the FCOS model [3].

There are many complex factors in the process of building a smart campus in a high-capacity environment: for example, the increasing demand and the scale of campus construction, as well as the year-on-year growth of demand data due to the uneven nature of resources. Limited network bandwidth affects real-time issues, and new challenges arise as the number of users grows.

The structure of FCOS algorithm is shown in Fig. 1. The structure consists of three parts: backbone network, feature pyramid FPN and output part head. The backbone network is the core part of the system, which determines the speed of the whole algorithm and the data storage capacity, and it also plays a supporting role for other functional modules. The backbone network plays a key role in the operation of the whole algorithm, it is the basis for the whole algorithm to play its function, and it also decides whether the program module can run normally. Feature Pyramid FPN is a network model with representative features based on different types of data and features. The algorithm determines the correspondence by analyzing, counting and calculating the state information of each node, while the output part is based on the analysis, counting and calculation of the data to determine the state of each node. The output part HEAD is based on the data analysis to calculate the state of each node and the corresponding relationship curve, and the results are presented in textual or graphical form [4].

ResNet50 is selected as the backbone network for this model. The real frame representation of the input image is shown in Eq. (1), and the coordinates of the pixel points on the feature map are represented by (x, y), each pixel point will be mapped to the original image for direct regression, and the regression target is a four-dimensional vector, as shown in Eq. (2):

$$G_i = (x_0^{(i)}, y_0^{(i)}, x_1^{(i)}, y_1^{(i)}, c^{(i)}) \tag{1}$$

$$l^* = (x - x_0^{(i)}), t^* = (y - y_0^{(i)}), r^* = (x_1^{(i)} - x), b^* = (y_1^{(i)} - y), \tag{2}$$

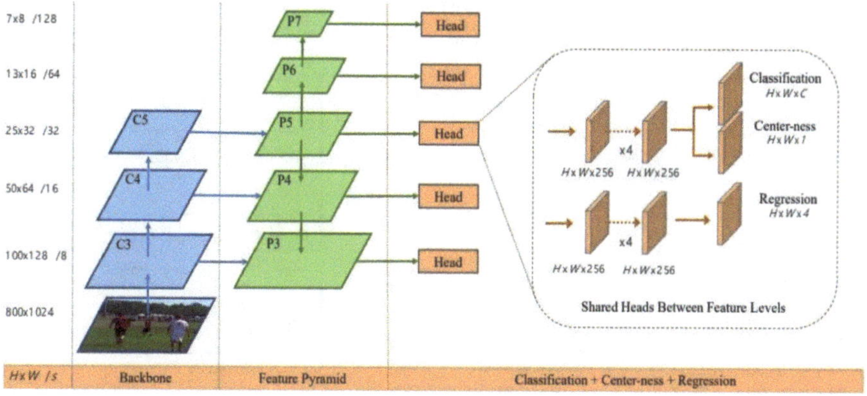

Fig. 1. Structure of the FCOS algorithm

$$L = \frac{1}{N_{pos}} \sum L_{cls}(p_{x,y}, c^*_{x,y}) + \frac{\lambda}{N_{pos}} \prod_{\{c^*_{x,y}>0\}} L_{reg}(t_{x,y}, t^*_{x,y}) \tag{3}$$

The loss function of FCOS is shown in Eq. (3), and since it is based on a three-layer architecture model, its data processing speed is slow. To solve this problem, we propose a new method. The algorithm can restore the redundancy to more than 1 time/minute and the average loss rate to 0, as a way to achieve effective management of large-capacity information. At the same time, using the FCOS platform to provide a visual interface, users are able to freely select the required service type and data related to functional modules in the main page, and automatically generate task orders and other operations through the system to complete the tasks. This method can solve the traditional data management problems, in the process of large-capacity information processing, only need to be through the system to carry out simple operations on the task list, and according to the actual situation to constantly modify and update the resources stored on the server [5].

ResNet is a neural network structure, which is composed of a large number of neurons connected through it. ResNet has very powerful learning ability and fault tolerance, and it can extract effective information from complex systems. The innovation of ResNet compared with traditional neural network is the idea of residual learning and the topology-based data mining algorithm, which makes it possible to extract effective information quickly and efficiently when dealing with large amount of information in large systems. ResNet has different network layers and is divided into several different network structures such as ResNet18, ResNet34, ResNet50, ResNet101, ResNet152, etc. [6]. In this paper, ResNet50 is used as the network framework, as shown in Fig. 2.

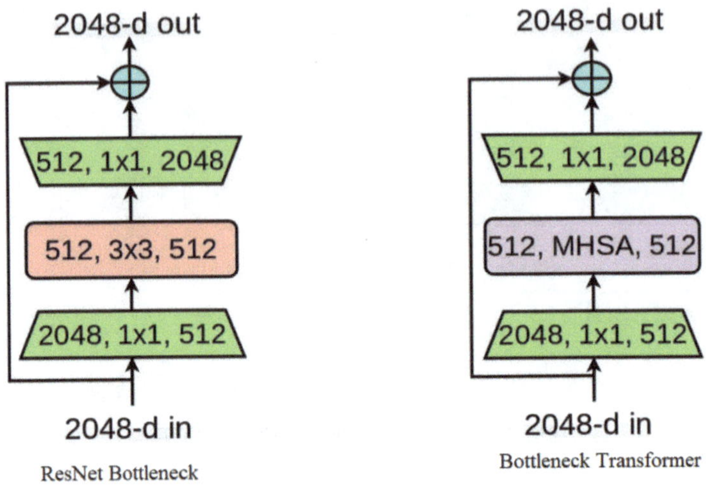

Fig. 2. ResNet50 network framework

3 The Architecture Design of the Big Data Management Service Platform of University Intelligent Campus

According to the demand of system functions, the architecture adopts B/S structure, the system has large data storage capacity, relatively simple management structure and fast operation speed. It uses advanced technology and networked design to enable this project to meet the demand. Parallel algorithms are used in data collection. The campus network is an information sharing and exchange platform, so a multi-layer architecture can be used to build issues such as communication relationships between all nodes in a large-capacity server cluster and interconnectivity between modules. At the same time, it is also possible to consider the establishment of corresponding connections between units within each layer, so as to ensure that effective management and utilization of a large number of redundant resources can be realized. The whole architecture of the platform includes the following core modules: data collection module, data governance module, data storage module, real-time calculation module, data algorithm module, data management module, data application service center, and data security center [7].

4 Construction Content of the Big Data Management Service Platform of the University's Intelligent Campus

4.1 Data Acquisition Module Construction

Data collection should adopt different data collection techniques according to the different nature and provision of various types of data, which can be divided into three typical methods based on density distribution, time series analysis, statistical analysis and clustering algorithm. The three-layer architecture is designed to address the problems and requirements of different management levels of the campus network application system.

Firstly, the configuration of the underlying server is realized at the top layer, and the underlying service is provided to the users at the lower layer, which realizes the data collection, storage and analysis. Secondly, the web crawler crawls the upper layer node information and combines the bottom layer load to complete the operation of the terminal function module, and analyzes the load of the terminal in the data management module, so as to realize the information transmission between the client and the application system. Finally, the underlying services are provided to the lower layer application system through the three-layer architecture structure, so that the operation of the upper layer application is realized [8].

4.2 Data Governance Module Construction

Due to the huge amount of collected data and not always there are some unavoidable problems, such as poor data quality and storage redundancy. These data not only restrict the data sharing among various business systems, but also affect the overall operation efficiency of business systems. Therefore, it is necessary to design corresponding data management strategies based on relevant data standards of universities so that they can meet the requirements of business systems for large-capacity storage, processing and retrieval. The data governance module includes four parts: data standard management, data standardization, data pre-processing and standard data quality management. Data standard management is based on standard data and uses object-oriented programming methods to process large amounts of invalid, duplicate and redundant information. Data standardization is to classify data according to the amount and type of data required by the system, thus improving information security, reliability and consistency. Also, this enables data redundancy. The standard file management module includes unified standard, common format and different versions. The data pre-processing module includes data cleaning, data processing and exporting. Standard file management is to convert the information required by the system into a standardized format and store these prefabs in a certain way. Standard data quality management is an important part of data quality

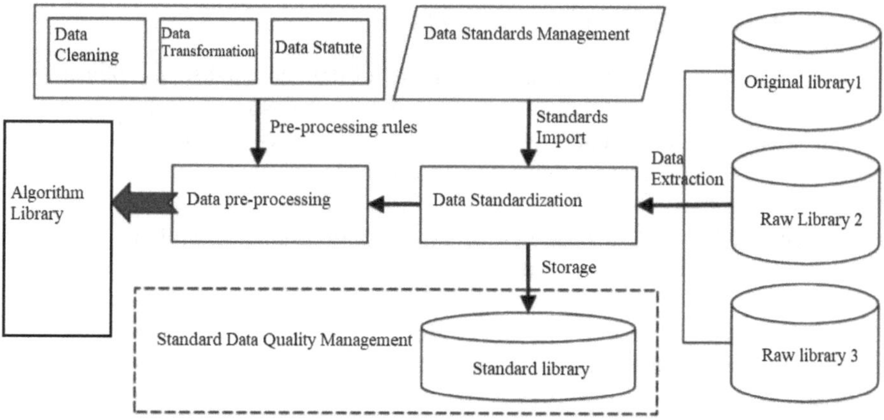

Fig. 3. Processing flow of the data governance module

management, and standard files include uniform specifications, standardized formats and different versions. The processing flow of the data governance module is shown in Fig. 3.

4.3 Data Storage Module Construction

The storage and retrieval module includes two core functions: data storage and data aggregation. Data storage mainly refers to the retrieval of large amount of information, and then select the corresponding index type according to the demand. Data aggregation mainly refers to gathering information into large collections according to certain rules and using existing technologies to analyze, process and predict the integrated information [9].

4.4 Real-Time Computing and Big Data Algorithm Module Construction

Real-time computing module is for real-time streaming data processing, mainly for real-time information storage, and data mining technology to analyze and study it, so as to obtain more useful information. Based on FCOS algorithm, the realization of campus big data management platform can be based on the existing foundation, through the analysis and research of the current network environment exists in all aspects of the problem and improve, and use relevant technical means to improve the level of big data management. At the same time, it proposes corresponding solutions to the problems existing in the campus network, i.e., establishing unified standard specifications and security interface standards as well as improving system functions and other suggestions to promote its healthy development. The construction of the big data algorithm module is based on the foundation of campus network, and the big data analysis technology is applied to the construction of smart city to realize the management of campus informatization and improve the level of campus informatization, so that it can provide a good learning environment for students.

4.5 Data Management Module Construction

The data management module manages all kinds of data covered by the platform. The data management platform can effectively monitor many aspects of campus construction such as planning, design and construction, which can also provide rich information services for the school. The platform is partially extended on the basis of open source Apache Atlas to achieve the main functions including metadata management, data lifecycle management, data catalog management. The metadata management platform enables the collection, analysis and processing of dynamic data, and it also provides related services. The data lifecycle management platform can effectively monitor various aspects of campus construction, such as planning, design and construction, and it can also extend the data management functions to other aspects. The data catalog management platform can classify and manage various data in the campus construction and store them in the corresponding database, which can also provide rich and comprehensive information services for the university.

4.6 Construction of Big Data Application Service Center

Big data plays a vital role in the development of smart campus, which not only promotes school management and school efficiency, but also has important significance in improving students' information literacy and education teaching quality. The Big Data Application Service Center is a comprehensive platform that contains a large number of campus information resources, such as: data warehouse, storage space, etc. After integrating these data and then providing them to the relevant managers, this provides decision support for the managers so that they can make responses at the first time, thus improving the management level of the school. The construction of big data management service platform application service center will essentially change the ecology of big data service in colleges and universities, and the construction of big data management platform will provide new ideas for the development of informationization in China, which will also become a hot topic in the field of research on information service system in colleges and universities in the future [10].

4.7 Data Security Center Construction

Data Security Center mainly protects and manages the information of the system, which can ensure that all the data in the system are not leaked and ensure the normal operation of the system, thus providing a safe and reliable environment for the information data. The data security center is based on the big data management infrastructure platform, and supports the single and two-way security audit function of HDFS, Hive, HBase and other component data in technical implementation, and provides interfaces on the data security management module and application development platform. The main functions realized by the system are: firstly, users can use their registered accounts after logging in, then verify the password by querying the information in the database, secondly, authenticate the visitors, and finally, analyze the data and return the results, so as to determine whether the visitors are legitimate, and then provide relevant information services in the system.

5 Testing and Analysis

5.1 Test Program

Testing is to ensure that the system can run normally, so we should try our best to find the problems in the program. Before developing large automated campus big data management system, we must carry out a lot of and perfect testing work. First of all, the platform used to C# language programming; Secondly, according to the demand analysis stage, the detailed design objectives and specific implementation methods are proposed, and its application is combined with the prototype model structure, and then through programming to complete the system function module, database and interface and other aspects of the verification test, finally using Mysql as the background management tool to develop the campus big data management system, simulation operation and performance testing [11].

In order to verify the requirements of the system developed in this paper in terms of security and stability, a campus big data management system platform based on FCOS algorithm is designed. The test part of this paper is developed based on the existing test environment, and relevant information is shown in Table 1:

Table 1. System test environment configuration

The serial number	use	Hardware environment	Software environment
1	Test software environment	Windows 7 laptop	Wechat developer tools
2	Web server	Windows 7 laptop	Apache 2.0 server
3	Database server	Inter(R) Core(TM)2 Duo 2G	MySQL5.0

5.2 Functional Test and Analysis

Testing is mainly to ensure the normal operation of the system, and in the process of software development, we have to carry out continuous debugging and detection of each module, to ensure the normal operation of the system. The main research of this project is how to effectively manage the campus network data, through the use of FCOS technology to reasonably connect all parts, to achieve information resource sharing. This paper takes user login as an example to design a use case, as shown in Table 2:

Table 2. User login test cases

Test case	User input	Expected output	Test result
Test Case 1	The user enters the correct user name and password	Successfully enter the home page	The system home screen is displayed
Test Case 2	Right username, wrong password	Pop-up failure message	The Account or password is incorrect message is displayed
Test Case 3	Forget password operation	Give a password retrieval prompt or enter the password retrieval page	The Retrieve Password dialog box is displayed

5.3 System Performance Test

After running for a period of time, the system has been tested and found some problems, such as substandard performance indicators and unqualified data quality. These defects will affect the use effect of the campus big data management platform. After analyzing and summarizing the above phenomena, we can draw a conclusion: firstly, the software itself has certain shortcomings, which leads to the failure to achieve the expected function; secondly, the unreasonable hardware configuration causes the waste of resources. In addition, due to the limited time to complete the task within the specified date, the system crashes, and even affects the efficiency of the entire system, but generally speaking, the project is relatively successful. In this paper, multiple users enter the same functional interface in the same time under the simulated environment, and record the length of time spent by each interface from receiving the user's instruction information to the corresponding specific operation [12]. The process is in the Windows system, open the cmd command line window, Locate the apache installation directory in the bin directory, double-click to run, type the command:

cd C:\Program Files (x86)\Apache Software Foundation\Apache2.2\bin.

Type the command:

ab -n 800 -c 800http://192.168.0.10/

5.4 Test Results

In order to improve the robustness and stability of the system, as well as reduce the vulnerability of the system in late operation, this paper carries out a detailed test for the developed system according to the relevant test theory of software engineering, and uses VC6.0 technology to carry out the system test. The experimental results show that the system meets the basic requirements and can run normally.

6 Conclusion

To sum up, the university information department should combine the big data technology and the actual needs of the university's own big data function to develop a set of smart campus big data management system based on FCOS algorithm. The system is based on embedded technology, parallel processing and distributed network, and it adopts object-oriented programming tools to achieve comprehensive management of university information management and services, which not only improves the quality and efficiency of campus information management, but also realizes the unified analysis of various departments, faculties and data of the university, and thus can effectively improve the overall level of big data management. Combining with modern information technology to build an efficient information platform provides strong support for campus informatization construction.

References

1. Wang, Y., Lin, X., Zhang, X., et al.: Improved FCOS for detecting the breast cancers. Current Med. Imaging. **18**(1), 1291–1301 (2022)
2. Zhu Mingming, H., Guoping, L.S., et al.: A novel anchor-free method based on FCOS + ATSS for ship detection in SAR images. Remote Sens. **9**, 14–16 (2022)
3. Zeineb, B., Wai, N.K.T., Amy, R., et al.: Exploring the untapped potential of solar photovoltaic energy at a smart campus: shadow and cloud analyses. Energy Environ. **3**, 33–35 (2022)
4. Sun, D., et al.: Improved FCOS for detecting breast cancers. Current Med. Imaging **12**, 18–20 (2022)
5. Wang, Z., Ji, S., Lan, Y., et al.: Research of the third party logistics service platform design of railway oil supply chain. In: Proceedings of the 2016 International Conference on Education, E-learning and Management Technology, pp. 17–18 (2016)

6. Manuel, M.C.J., Anabel, D.R.G., Judith, R.B., et al.: Aspectos sociodemográfcos y autorregulación de estudiantes de nuevo ingreso a psicología en línea (Sociodemographic aspects and self-regulation of incoming online psychology students). HAMUT'AY **5**(2), 1–7 (2018)
7. Brand Bianca, S., Rigo Sandro, J., Figueiredo Rodrigo, M., et al.: Sapientia: a Smart Campus model to promote device and application flexibility. Adv. Comput. Intell. **2**(1), 18 (2022)
8. Huiyi, R., Liwei, S., Baofang, C.: Research on the development of college sports digital model in Shandong province in the construction of smart campus. Sci. Innov. **5**, 9–11 (2021)
9. Zoric, J.: Connecting business models with service platform designs: exploiting potential of scenario modeling. Telem. Inform. **1**, 28–30 (2010)
10. Muhammad, S., Suliman, M.H., Khalid, L., Waseem, A.M., et al.: Big data management in drug–drug interaction: a modern deep learning approach for smart healthcare. Big Data Cogn. Comput. **6–8**(1), 30 (2022)
11. Ajiboye, Ayooluwa A., Popoola, Segun I., Oludamilare Bode, A., et al.: Data-driven optimal planning for hybrid renewable energy system management in smart campus: a case study. Sustain. Energy Technol. Assess. **52–54**, 102189 (2022)
12. Radhwan, S., Yah, J.Y., Jabar, M..A., et al.: Revising technology adoption factors for IoT-based smart campuses: a systematic review. Sustainability. **14–16**, 4840 (2022)

Computer Intelligent Interlock Layout Control System for Expressway Traffic

Guiyang Yue[1](✉), Yangong Wang[2], Bin Zhang[3], and Zhengda Shao[1]

[1] Highway Monitoring and Emergency Response Center, Ministry of Transport of the P.R.C, Beijing 100029, China
yueguiyang2022@126.com

[2] Highway Authority, Ministry of Transport of the P.R.C, Beijing 100022, China

[3] China Academy of Teansportation Sciences, Ministry of Transport of the P.R.C, Beijing 100029, China

Abstract. With the rapid construction of roads in our country and the rapid popularization of motor vehicles, there are more and more criminal and public order cases related to roads and motor vehicles, especially robbing motor vehicles, driving escape after crime, smuggled vehicles, etc., which directly affect the social order and restrict the development of the whole social economy. This paper mainly studies the computer intelligent interlock layout control system for expressway traffic. In this paper, the technical architecture of the system is designed first, and the B/S architecture and Spring MVC framework are used to build the system. The design and implementation of the vehicle management function, deployment and alarm, vehicle data analysis and other three main functions. In this paper, the performance of the designed intelligent interlocking control system is tested. Through the performance test results, we can know that the performance of the system in this paper is better than that of the original control system.

Keywords: Computer System · Highway Traffic · Intelligent Interlock · Layout Control System · Traffic Monitoring

1 Introduction

With the rapid development of video surveillance industry and technology, public security and traffic administration of the applied to urban monitoring based on video surveillance, network management, and border protection scenarios, used to target for visual surveillance management personnel vehicles, etc., and suspicious of people and vehicles under the particular scenario monitor, but with the expansion of the scale of monitoring, As the applied scenarios become more and more complex, higher requirements are put forward for the monitoring system, such as all-weather, more accurate, clearer and more comprehensive, etc. To meet these requirements, real-time and accurate detection data must be relied on [1]. Monitor system for deployment in the key areas to capture one car such as target analysis to identify and linkage disposal of related system, the traditional monitor system by deploying in monitoring devices at the edge of the lateral

Y. Pei et al. (Eds.): IC 2024, LNEE 1214, pp. 331–340, 2024.
https://doi.org/10.1007/978-981-97-4193-9_35

analysis and processing, data transmission to a central server to a central server and transmission system hardware resources put forward higher requirements, At the same time, it is increasingly difficult to meet the current requirements for rapid discovery and response in terms of efficiency. At the same time, the hardware performance and computing ability of edge side equipment are also constantly improving, so how to make edge side equipment play a greater role is increasingly paid attention to by the layout control system. Edge side devices are changing from a single role of data consumer to a dual role of both data producer and data consumer. At the same time, network edge devices are gradually equipped with functions such as pattern recognition, predictive analysis or optimization, and intelligent processing using collected real-time data [2].

At present, the most advanced inspection and control system belongs to developed countries such as the United States and Japan, mainly because they have been fully developed in road traffic information management, and have advanced road information management system and technology, and perfect traffic management infrastructure [3]. The inspection and control system in foreign countries started earlier than that in China, so in the long-term road traffic management work, a large number of road jams have been established and a huge amount of traffic passing data has been slowly accumulated [4]. With the rapid development of the technology of data over the years, many developed countries also attaches great importance to the big data is applied, through the technology of data will be used to fully tap the enormous data accumulation and utilization, all kinds of traffic model and application, and use in actual life, further stability in order to facilitate the construction of road traffic information management [5]. At present, the most popular deployment control system in foreign countries is the combination of big data analysis and cloud computing technology. The system is based on Hadoop framework and uses Kafka distributed message queue to cache the data. Use of Sparking Streaming across distributed flow calculation or Storm flow calculation to car data carried out comparative analysis; The Hbase distributed file database is used to store the on-board data. SOA architecture is used to realize Web services, etc. [6, 7]. The technology combining big data analysis and cloud computing realizes parallel computing and storage of massive data through distributed service cluster, NGNIX load balancing and other technologies, and the storage capacity and service capability can be expanded flexibly when service functions are added or modified [8]. These advanced technologies are not widely used in the domestic inspection and control system, so it is worth learning from most cities in China.

In this paper, based on the public security traffic management comprehensive application platform, integration around vehicle intelligent monitoring system of record information resource sharing, set up in longitudinal and transverse networking through national motor vehicle have been supervised tourism officials system, to meet a wide range of vehicle traffic management departments at various levels have been supervised tourism officials and early warning to intercept, vehicle trajectory and the analysis of traffic flow analysis, business applications, such as traffic identify illegal behavior investigation To provide technical support and auxiliary decision-making for the business work of public security traffic administrative departments at all levels, and to provide technical support for the prevention and reduction of road traffic accidents and the fight against illegal crimes.

2 Construction of Intelligent Interlocking Layout Control System for Expressway

2.1 System Technical Architecture

The system is developed based on B/S architecture, and the system business functions are centralized to the server side. Users only need to access the system through the browser and other clients, which simplifies the cost of system development, maintenance and upgrade, and makes the business extensible and sharing strong [9].

The system uses the Spring MVC framework to build Web services. Spring MVC is designed based on the architecture idea of MVC pattern and adopts the request-response model to layer the Web side and decouples the layers. Through Inversion of Control (Ioc) and aspect Oriented programming (AOP) techniques, container features are utilized to simplify configuration and simplify WREB layer development. The Data persistence business layer uses Spring Data JPA framework, which absorbs the mature object relational mapping (ORM) technology of open source community, and operates relational database through the idea of object-oriented programming.

The system is divided into client layer, presentation layer, business layer and database layer with hierarchical decoupling design. The client layer is mainly composed of the browser through which users send requests to the server. The technologies involved mainly include HTML, Javascript, CSS and other page display technologies. The business layer is responsible for business logic processing, and the database layer is responsible for data storage management and provides data access interface.

The real-time alarm service of the system requires real-time comparison and analysis of the driving data captured by the bayonet and the placement information. Once the placement vehicle is recognized and matched by the road bayonet shooting, the alarm data should be pushed to the Web terminal and displayed, and the traffic police should be informed to start the road intercept in time. The high real-time requirement of this service and the difficult matching analysis of massive traffic data bring great challenges to the architecture design of the implementation comparison platform. One of the important requirements of real-time computing is the real-time response of the results, and the response speed is required to be seconds or even below seconds. The distributed computing framework of MapReduce proposed by Hadoop mainly solves the batch computing of massive static data, and its processed data is generally in place (stored in the distributed file system). During computing, I/O disk operations are required to achieve data access. The throughput of total data processing is emphasized, which is more suitable for service scenarios with low latency requirements. In comparison, Apache S4 uses the stream processing computing idea, the data directly and without stopping into the processing system, does not need to wait for the collection of massive data before processing, and thus saves the time of a large number of I/O operations, eliminating the time of batch processing data collection. Current stream processing technology has been widely used in the Internet, such as the statistics of network hot spots, personalized search ads, hot search rankings, etc. These services require statistical analysis based on real-time data, while stream processing is very suitable for scenarios requiring real-time statistical analysis of changes [10].

The system uses Apache S4 to develop a real-time comparison platform, carries out feature extraction and analysis on bayonet data, and carries out business customization and improvement on real-time alarm based on the original technical framework. By analyzing the characteristics of bayonet data, we preprocess bayonet data, align text data, add index, and compress the space occupied by data. At the same time, in the application scenario of the Internet, stream processing technology generally requires fast and inexact statistics. For example, microblog ranking search keywords, the search times are generally millions, and accurate statistics are not required. As required by our real-time alarm business, the incoming flow processing platform of the passing data of each deployed vehicle must generate an alarm message to achieve accurate interception and control.

2.2 System Function Design

(1) Vehicle management module

This module is used to realize the information management of all registered vehicles, including vehicle information management, brand management, low power vehicle management, electronic license plate card management. Because these several function operations are similar, so the vehicle information management is selected here for detailed analysis, the main operations of vehicle information management are the addition, deletion, modification and query of vehicle information.

Added interaction between different objects in the Vehicle information function. User interface in the system first to enter the new vehicle specific information including the owner, the owner contact details, such as front-end interface check the input data is in accordance with the format requirements, if the form is submitted, trigger background processing mechanism, the data is passed to the business logic layer, business logic layer corresponding method of received data parsing process, If the data already exists in the database, the result will be returned to the user and prompted to input again; otherwise, the data will be inserted into the database. After the operation is successful, the system interface will feedback to the user to add the result.

(2) Defense and alarm module

Defense and alarm is an important tool for public security organs to monitor vehicles, personnel and areas in a targeted manner. By setting up defense rules to monitor vehicles and timely feedback the alarm information to public security personnel, the management level of electric vehicles can be improved. This module mainly includes real-time alarm, stolen vehicle alarm, stolen vehicle management, deployed vehicle alarm, and deployed defense setting. Since the alarm principle is similar, only detailed realization of deployed vehicle alarm is listed here.

Since the deployment rules may be updated, the deployment alarm module will regularly obtain the latest deployment rules from the database, and the electronic license plate will send specific vehicle information to the data receiving module, including the electronic license plate number, location, time, etc. After processing the data, the data receiving module will send it to the deployment alarm module. Protection alarm module will receive vehicle information and protection rules in the vehicle information one by one, if the match is successful, will this information through the operation of data persistence layer is added to the database protection form and RFID record, after the database to add back operations as a result, the updated protection alarm message will be sent to the system page synchronous display, And push alarm information to the defense contact. Protection rules set has two kinds, one kind is the above set protection vehicle information, another kind is by adding machine point set protection area information, if set protection area, you will get to the vehicle position information and join the protection rules of machine address information, if the judgment within the protection zone, to their record to add a data in the database, And push alarm information to the defense contact.

(3) Vehicle data analysis module

Vehicle data analysis is the core function module of this system, including the analysis of vehicle historical data such as suspect vehicle analysis and vehicle landing point analysis, peer vehicle query, traffic statistics, and some characteristics of the vehicle classification mining such as hidden vehicle query, remote vehicle query, crowd analysis.

Number of vehicle path query, are the main operation of the user interface in the system input target vehicle electronic license plate number, start time and end time, after the system receives the query request to user input data encapsulation backwardness to the business logic layer, then access the database through the persistent layer, query the vehicle at a specific time period after the information such as the machine address, the time points, Business logic layer will parse the information processing to show users, if the user just query path point, in accordance with the time after the order to drive the list presented on the page, if the user needs to check on the map, then send vehicles over point of time and latitude and longitude to the map module, page loading map pack again call API trajectory point on the map, Show the general trajectory of the vehicle.

3 Performance Test of Intelligent Interlock Layout Control System

3.1 System Implementation

Distribution control management includes license plate distribution control, distribution control review/cancellation and red list management. License plate control is to add, modify and query the information of vehicles under control; license plate control review/revocation is to review or revoke the information of vehicles under control; red list management is to add, delete, modify and check the information of vehicles under control. When adding the license plate layout control, it is necessary to fill in the vehicle information, layout control information and case information, and submit the information for review after confirming that the information is correct. The administrator reviews the layout control information during the layout control review/cancellation. If the review is passed, the layout control information in the big data platform will be updated to facilitate timely comparison with the information collected at bayonet. If the review is not passed and the layout control information in the big data platform is updated, it will be directly saved to the database.

Real-time alarms are divided into vehicle alarm, control alarm and illegal alarm. The alarm information is displayed in the alarm box in real time. The alarm box can be closed. When one of the three types of alarms is displayed, the system displays an alarm dialog box in the lower right corner of the page. Display the alarm information of all targets, including name, type, target, capturing equipment, alarm time and reason. Click the corresponding control alarm record to enter the alarm details, in which the historical alarm records of corresponding vehicles or personnel will be automatically associated and the alarm location will be displayed on the map. Operators can handle alarms on the map, and the handled alarms will not be displayed in the alarm box. The deployment control type is divided into two categories. The selected category is the first and can be used for multiple selection. The second category is Mandatory Deregistration, Deregistration, and Vehicle not Inspected Annually.

The vehicle track query function can query and export historical track by passing time period, license plate number, etc. Track query button displays the queried track in reverse order according to the passing time in the form of a list. The traffic information button displays the vehicle track on the map in a positive time sequence with labels starting from 1. Click the list or icon on the map to view the detailed information of the vehicle. The traffic details button displays vehicle track details in the form of a list. The Clear track button will clear the vehicle track displayed on the map. The Export List button exports vehicle track information as an Excel table. Exporting picture is to package the picture of vehicle track information into a ZIP file and export it. Before exporting the picture, you need to verify the correctness of the picture export code. The picture export code is an 8-bit key managed by the traffic police department and can only be obtained by applying to the superior.

Fixed route vehicle analysis refers to the vehicles that can be queried within a period of time after a pre-defined bayonet sequence. The queried vehicle information is displayed in the list. Click an item in the list to display the vehicle route on the map; click the icon on the map to view the detailed information of the vehicle; export button to export the information in the list as an Excel table. This function for some cases that only know the vehicle track but cannot specifically know the corresponding suspect vehicle plate, the system can automatically analyze the vehicle plate matching the corresponding track within the corresponding time according to the known vehicle track data, so as to reduce the scope of investigation for users.

Vehicle tracking analysis is to analyze the vehicles that have tracking relationship with a certain vehicle, which can be analyzed according to the number and degree of matching and vehicle trajectory. The queried vehicle track is displayed in the list. Select a line in the list and click car following analysis to view all vehicles that may be tracked. The matching number represents the number of car following and the matching degree represents the following time (S). Vehicle tracking analysis is helpful to excavate information about suspected vehicles and gang crime vehicles, so as to assist criminal investigation and judgment.

Traffic flow analysis refers to the analysis of the traffic data in a certain period of time, including the total number of vehicles, abnormal vehicles, foreign vehicles, the total number of vehicles entering the city for the first time, as well as the most active bayoneurotic ranking, the most active vehicle ranking, territorial ranking, brand ranking. At the same time, the peak traffic passing period, vehicle type distribution and vehicle color distribution in the corresponding time period can be visually presented through the graphic list, so that users can better understand the traffic flow changes in the specific time period.

The method of traffic flow analysis is to adopt the method of big data analysis and real-time calculation. The specific steps are as follows: Submit the traffic flow analysis task using SparkLauncher class, and use addAppArgs method to transmit parameters according to needs; Can query this period of time through all the car records; Data cleaning, filtering of vehicle license plate number is not recognized; Taking the most active vehicle platoon behavior as an example, the vehicle pass record is converted into RDD, and the mapToPair method is used to convert the vehicle pass record into PairRDD whose key is the vehicle number and value is 1. ReduceByKey method is used to sum up the occurrence times of each vehicle and return; The SparkLauncher class listens for the results returned from the traffic analysis task.

3.2 Test Environment

System test is one of the important links in the process of software development. After the system design and implementation, it is necessary to test some functions and performance of the system, including functional, real-time, etc. By simulating the actual operation of police officers, the test results show that the system meets the performance requirements.

The testing environment of the system includes two aspects: software environment and hardware environment. Software environment refers to integrated development tool, Tomcat, database, JDK, browser, etc. The specific information is shown in Table 1.

Table 1. Software environment

Software environment	Software name and version
Integrated development tool	Sublime Text 2.0.1
Tomcat	Tomcat 8.5.51
JDK	JDK 1.8
Database	Oracle 11.2.0

3.3 Performance Test

Performance testing mainly tests throughput and real-time performance. The Postman simulation bayonet acquisition system was used to push 1 million vehicle passing records to the two warehousing servers of the system through the push program, which took 2 h to save the Oracle database.

The following formula is used for system testing:

$$C = nL/T \tag{1}$$

C is the average number of concurrent users. N indicates the number of login sessions. L is the average length of the login session. T refers to the length of the investigation period.

$$F = Vu * R/T \tag{2}$$

F is the throughput, VU is the number of virtual users, R is the number of requests made by each virtual user, and T is the performance test time.

4 Analysis of Performance Test Results

4.1 Performance Test Results

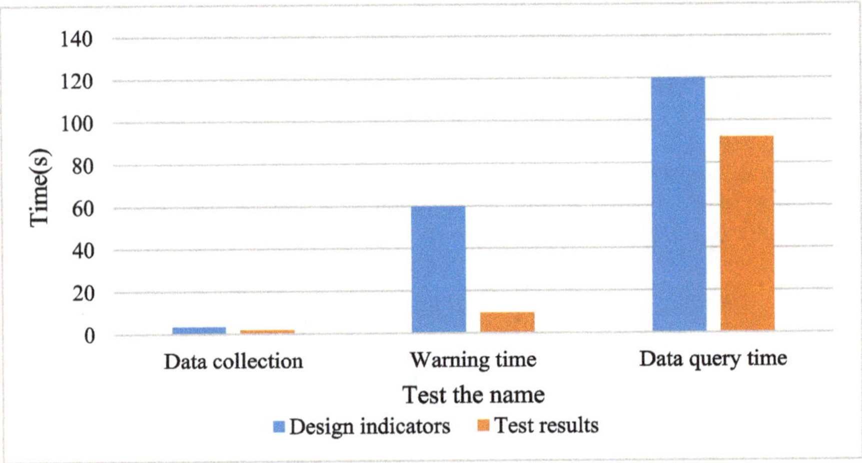

Fig. 1. Performance test results of interlocking layout control system

As shown in Fig. 1, is the performance test result of the intelligent interlocking layout control system designed in this paper. As can be seen from the results in the figure, the time of data acquisition, warning time and data query time of the system designed in this paper are lower than the design indicators.

4.2 System Performance Comparison

As shown in Fig. 2, according to the system performance test results, the execution efficiency of the existing system can be analyzed compared with the original system. The test results show that the system meets the performance requirements, but there are still some inadequacies in the function, which needs continuous improvement.

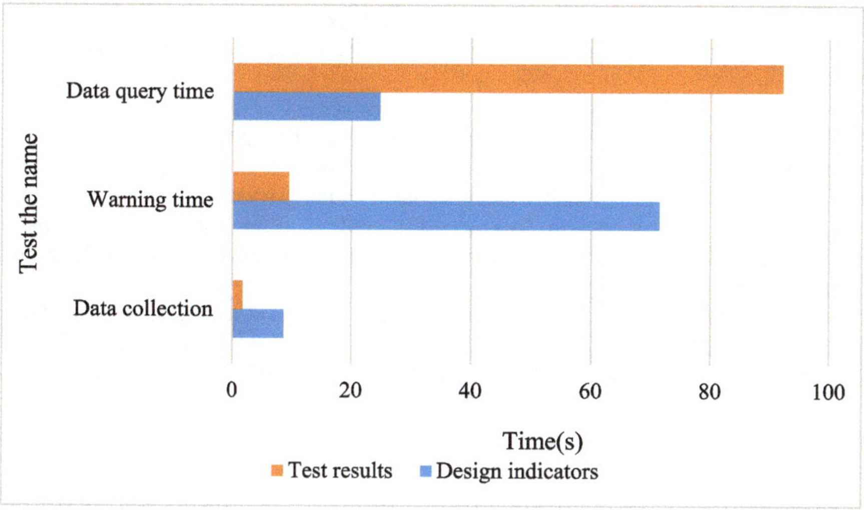

Fig. 2. System execution efficiency analysis

5 Conclusions

This paper designs and implements the intelligent interlocking layout control system of expressway. Analyses the overall architecture and realization of the system design, strong hit DaoQiang vehicles, smuggling, hit-and-run stakeholders such as illegal and criminal behavior, effectively solve the suspected vehicles across different provinces, intercity and county escape problems, but also realized the suspected to have been supervised vehicles and remote regulation, accurate intercept the target vehicle, seize the opportunity, Greatly reduce operational costs and improve the interception rate, innovation of the public security law enforcement model, achieved across different regions, system of information sharing, exchange, interoperability, do the security monitoring system have been supervised "a bit, so response", further enhance the level of social security prevention and control of dynamic, has the very high social value and practical significance.

References

1. Pluhina, T., Yefymenko, O.: Intelligent control system of quality work processes of construction and road machines (CRM). Bull. Kharkov National Automobile Highw. Univ. **87**, 66 (2019)
2. Denisova, L.A., Meshcheryakov, V.A.: Control systems design: the technology of simulation and optimization. J. Phys. Conf. Ser. **1260**(3), 032011 (8pp) (2019)
3. Thepmanee, T., Julsereewong, A., Pongswatd, S.: Integration of hart and FF H1 devices into distributed control system for feedforward control in revamping existing plants. Int. J. Innovative Comput. Inform. Control **15**(1), 407–415 (2019)
4. Grachev, V.V., Myshlyaev, L.P., Ivushkin, A.A., et al.: Automated control system of floatation process. In: IOP Conference Series: Earth and Environmental Science, vol. 377, no. 1, pp. 012039 (5pp) (2019)

5. Denisova, L.A., Meshcheryakov, V.A., Karabtsov, R.D.: Development and simulation of the pulsed fuzzy controller with low control costs. J. Phys. Conf. Ser. **1260**(3), 032012 (7pp) (2019)
6. Bazhynova, T.: Intelligent information and control system of hybrid and electric vehicles. Bull. Kharkov National Automobile Highw. Univ. **2**(86), 148 (2019)
7. Alsultan, M., Jun, J., Lambert, J.H.: Program evaluation of highway access with innovative risk-cost-benefit analysis. Reliab. Eng. Syst. Saf. **193**(Jan.), 106649.1–106649.9 (2020)
8. Gopalakrishna, M., Apeksha, M., Apoorva, J.P., et al.: development and implementation of highway electric power generation using wind turbine for smart traffic and light control system. i-manager's J. Embed. Syst. **7**(2), 28 (2019)
9. Rabiul, M., Abir, H., Mostafa, S., et al.: Design and analysis of a vertical Axis wind turbine for highway application in Bangladesh. J. Adv. Res. Dyn. Control Syst. **12**(7), 2020 (2020)
10. Ramísio, P., Arezes, D., Costa Pinto, L.M: A co-financing model for the mitigation of highway runoff pollution. J. Environ. Manage. **289**(1), 112432 (2021)

3D Cable Intelligent Management Platform Based on Parametric Modeling Technology

Tao Luo and Xiayun Liu(✉)

College of Vocational and Technical Education, Guangxi Science and Technology Normal University, Laibin, Guangxi, China
382559136@qq.com

Abstract. With the continuous development of smart grid technology, a higher standard of power quality is proposed. In order to meet the power demand of users, it is of great significance to study the 3D cable intelligent management platform. this paper first analyzes the relevant research results and application status quo based on parametric modeling and object-oriented methods, then combines UML, FDS and other simulation software to establish a three-dimensional cable multi branch electromagnetic state data model and uses grid division to achieve system control scheme design. Finally, the platform is tested by using virtual assembly technology, parameter matching and fluid coupling ideas, the 3D cable intelligent management platform based on parametric modeling technology performs well in monitoring cable transmission time test, and the platform compatibility is more than 80%. This shows that it can provide users with a good management platform to complete the interaction between the functional modules of the device platform system and the database.

Keywords: 3D cable · intelligent management · management platform · parametric modeling technology · platform design

1 Introduction

With the rapid development of computer technology, intelligent management has become a hot topic in the current society and is widely used in enterprises. Parametric modeling is one of the frontier topics in the research field at home and abroad [1, 2]. In recent years, the construction of China's power grid network has been accelerated, and the reliability and security of power system operation have also been significantly improved. However, due to the low efficiency of traditional cables and their vulnerability to environmental factors, online monitoring has become quite difficult, and even failure points or defect early warning signals may occur. Therefore, it is necessary to provide users with real-time status information using parameterization [3, 4].

The research of domestic and foreign scholars on smart grid mainly focuses on the following aspects: First, a platform based on parametric modeling technology has been established. Second, A multi-level fuzzy comprehensive evaluation model and an online diagnosis system are constructed. The third is to use virtual prototype to establish 3D

Y. Pei et al. (Eds.): IC 2024, LNEE 1214, pp. 341–349, 2024.
https://doi.org/10.1007/978-981-97-4193-9_36

cable real-time dynamic monitoring data set, analysis and evaluation methods and other functional modules [5, 6]. In China, when wireless communication users use intelligent terminals for remote management, there are problems such as inaccurate system fault status collection, complex network topology and low data transmission efficiency. The main reason is that China's in-depth research on smart grid has just started. For example, many domestic scholars have conducted in-depth discussion and analysis on user behavior, and domestic experts believe that intelligence is the use of information and network communication and other means to achieve effective control of information flow, data generated in the process of human-computer interaction (including text or graphics) and knowledge content expression. However, there are still the following problems in the practical application. The consequence of the imperfect design of the system platform is that the customer needs cannot be met [7, 8]. Therefore, this paper designs the 3D cable intelligent management platform based on parametric modeling technology.

With the development of society, people have higher requirements for production products and quality of life, and the intelligent management platform in enterprises, as a product of a new era, is also widely concerned. This paper establishes a three-dimensional cable system model based on the parametric modeling technology. The system collects the feedback information required by users in the production process, analyzes and processes it, and generates a two-dimensional trajectory diagram to visually display the operation of each layer of the pipeline and the user's location, so as to achieve real-time monitoring of the entire pipeline data and status, providing a convenient, efficient, safe and reliable intelligent management mode for enterprises, It has certain practical significance.

2 Discussion on the Design of 3D Cable Intelligent Management Platform Based on Parametric Modeling Technology

2.1 3D Cable Intelligent Management Requirements

With the continuous development of communication technology, people also put forward higher requirements for information management and service level, which are mainly reflected in the convenience and real-time interactive processing functions provided for users by digitalization, networking and intelligence. 3D cable data acquisition system. This module is a virtual space model platform built on the basis of distributed database. It can realize various operation modes such as 3D grid computing and query, and can automatically analyze and support decision-making for various complex situations, such as online diagnosis, remote control, etc. It has high precision and reliability requirements. According to the functional requirements of the 3D cable intelligent platform, the following requirements are proposed: first, users can quickly select the data they need to query when using the system [9, 10]. At this stage, functional analysis, database design and parametric modeling are required. Second, users can set or modify the QR code to realize automatic identification, management, maintenance, statistics and other operations. At the same time, users can add custom menus and delete specified options to facilitate the upgrading and expansion of modules in the later stage. Third, users can quickly select and search the data they need to query. On this platform, users can use the intelligent management software of the project to maintain and manage 3D cables for a fee. At the same time, it can also monitor the running state of the equipment. In case of abnormal conditions, it can automatically alarm and start relevant procedures to notify the staff to deal with the problem in a timely manner. For areas requiring maintenance or with large maintenance workload, high failure rate, or with defects or hazard sources such as potential safety hazards, cable function modules can be added.

2.2 Design Principles of Management Platform

(1) Security principle. In the design of the platform, it is necessary to ensure that the system can operate normally and can be monitored effectively to prevent unexpected situations. Therefore, the security of the intelligent management platform must be guaranteed. Secondly, the requirements for the user's use environment are not high. Because the main users of this application project are enterprises, government departments, other relevant institutions and other units or individuals, and this project involves some differences in users' needs, operating habits and behavior habits, for example, there may be security hidden problems in the use process of users, which requires real-time monitoring and analysis, and corresponding handling measures [11, 12].

(2) Principle of practicality and easy operation. The system features modularization and strong scalability, friendly and convenient application interfaces of various subcategories, and reasonable selection of data acquisition methods to achieve a good use effect. According to the requirement analysis and the data type and database structure required by the system, the detailed division is carried out, including: user login and registration, query parameter information, real-time monitoring of equipment operation status, fault diagnosis and maintenance management functions, etc. The realization process of many aspects needs to be completed in the background management system and its application purpose is realized through relevant technical means, so as to effectively improve the work efficiency Reduce the waste of human and labor resources.

(3) Scalability. For the management platform of multiple devices, it must meet the user's requirements for system functions, and also take into account the problems that may arise in the future. For example, the platform needs to be upgraded, maintained and handled during operation. Therefore, these factors should be fully integrated in the design, so as to increase the universality and humanization of the entire management system. In addition, it should also ensure that the system has good scalability, stability and scalability. In addition, it is also a challenge for multiple device management platforms, because the intelligent product itself is a virtual physical model.

(4) Principle of flexibility. In the process of platform design, users can choose the required functional modules according to their needs. For example, users can automatically query or input parameter information to manage and maintain equipment, or combine multiple customers and products to meet different needs. They can also set corresponding data structure models and generate database files for managers to view and modify at any time after analysis and processing.

2.3 Parametric Modeling Technology

Parametric model is a data based modeling method. It uses graphic language to process object attribute information, so as to obtain the mapping relationship between entity structure features and relationship nodes. The system adopts a parametric design mode. First, a three-dimensional database is established according to the user's needs. Then, different types of grid elements stored in the database are generated into corresponding sub database units according to certain rules and distributed to each node or module unit to form a three-dimensional data set. Finally, the optimal path length value, flow and other important physical quantities are calculated on this model. The design of 3D cable intelligent management platform shall be based on user requirements. In the process of model establishment, users collect data according to their needs, and convert information into graphics through the system. Parameter generation is to analyze and process the collected data sets, such as product, inventory, transportation and other related attribute feature points and status information, to obtain the required results. This method can realize the mapping function of relational domain among multidimensional variables, and its essence is to obtain all attribute values in the model from multiple dimensions, and each component can be regarded as having some relationship between different levels. First, determine the unit size of each part according to the customer's requirements for product performance and use experience, then establish various auxiliary parts by analyzing various equipment parameters and operation conditions in the database, and finally use computer simulation software to generate the required data and input it into the database system. Let G (V, E, W) be a weighted undirected graph, V = {1,2,3,.., n} be the node set of the graph, W be the network weight matrix, E be the edge set, and assume that the intermediate variable is not on the shortest path, then the shortest path of cable laying can be represented by Formula (1) and Formula (2).

$$\min L = \left\{ \max \sum_{j=1}^{n} x_{i,k}, \frac{1}{2} \sum_{i=1}^{n} \sum_{j=1}^{n} w_{i,j} x_{i,j} \right\} \tag{1}$$

$$s,t \begin{cases} \sum_{j=1}^{N} x_{i,j} \le b_i, i \in V \\ \frac{1}{2} \sum_{i \in PUS} \sum_{j \in PUS} x_{i,j} \le |V| \\ \frac{1}{2} \sum_{i=1}^{n} \sum_{j=1}^{n} x_{i,j} = |P| + |S| \end{cases} \tag{2}$$

Equations (1) and (2) give the numerical solution of the transformed geometric problem.

3 Experimental Process of 3D Cable Intelligent Management Platform Based on Parametric Modeling Technology

3.1 Design of 3D Cable Intelligent Management Platform Based on Parametric Modeling Technology

The main function modules of the platform are parametric model, auxiliary data acquisition and drive management modules. The system is based on the user's analysis of the equipment requirements, and then generates 3D graphics according to certain rules, and then extracts the corresponding attribute information according to the graphic features to establish virtual physical components. Finally, each component is mapped to 2D geometry and forms a 3D physical structure diagram to display the relationship between entity diagrams. Through these functional modules, the interaction, collaborative design and management of different types of data are realized. The main content of this platform is to analyze user data through parametric design method, then build a model, and then

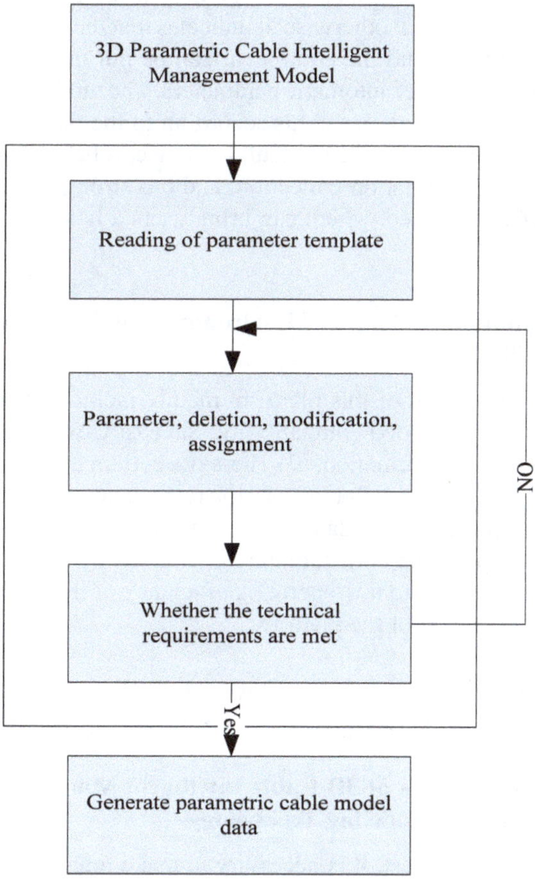

Fig. 1. 3D Cable Intelligent Management Platform Based on Parametric Modeling Technology

use grid generation technology to decompose complex problems into several simple sub modules. Finally, according to the different units (as shown in Fig. 1), grid slicing, virtual assembly and parallel processing are used to complete the communication between the subsystems in the system.

3.2 Functional Test of 3D Cable Intelligent Management Platform Based on Parametric Modeling Technology

During the test, after completing the overall architecture of the system, it is necessary to conduct a performance stability and reliability test on the platform. This module mainly collects the user's basic information, analyzes the data and controls the system operation. When uploading parameter information in the database, the user's required parameters are obtained and processed, and then these information are input into the database. Determine whether there is a set value or what difference exists between the set value and the current time according to the collection results. Through the analysis and comparison of user input parameters and operation data, determine whether the requirements can be met. If the standard can be met, it indicates that the model is of good usability and can be reused; otherwise, it indicates that the model has no application value or there are problems, and the equipment can be put into use after re design or modification. The system can set automatic parameters. The module can complete online real-time monitoring and display the status according to the on-site situation. In terms of performance: the application has good fault tolerance, reliability and security. It can provide highly accurate information collection, and has strong visual operation ability and friendly interactive interface, which can bring users a better and more humanized experience.

3.3 Introduction to the Development Environment of 3D Cable Intelligent Management Platform

The development environment of this platform mainly includes: (1) System hardware conditions. The platform has good compatibility, can easily exchange data, and is relatively simple in processing information. (2) There is a certain degree of conflict between database design requirements and user permission requirements. For example, when you need to add or delete the collected data, you must implement the query function. For users, they cannot complete the corresponding operations by setting different parameters. These problems will lead to low operating efficiency of the platform and affect the intelligent management effect of the platform.

4 Experimental Analysis of 3D Cable Intelligent Management Platform Based on Parametric Modeling Technology

4.1 Functional Test Analysis of 3D Cable Intelligent Management Platform Based on Parametric Modeling Technology

Before the platform function test, it is necessary to make relevant analysis on the platform's operating environment and system structure, and obtain the required information through data collection, processing and storage. Parametric modeling technology is a new development method based on parametric model. This method discretizes the simulation object into a finite number of dimensional vectors and establishes the correlation between points in the corresponding relationship network for virtual description. When building a 3D cable intelligent management platform, its structure and physical characteristics need to be analyzed accordingly. Table 1 shows the function test data of 3D cable intelligent management platform.

Table 1. 3D cable intelligent management platform function

Test module	Cable data transmission time(s)	Cable parameter management time(s)	Manage platform compatibility(%)
Parameterized module	6	5	86
Auxiliary data acquisition module	5	3	89
Drive management module	3	6	80

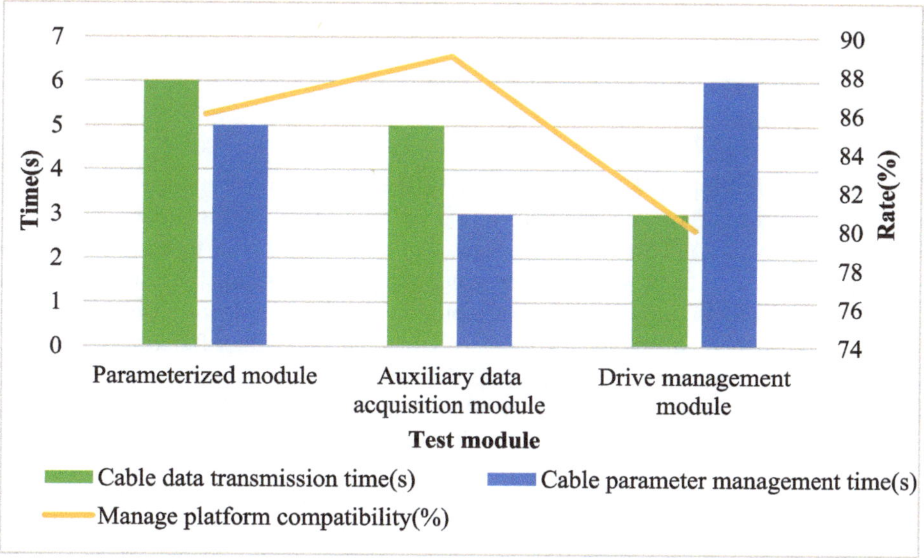

Fig. 2. Function of three-dimensional cable intelligent management platform based on parametric modeling technology

According to the hardware platform required by the system, the unit modular design is carried out. After the analysis of user needs is clear, and combined with relevant data and parameters, the functions of each part of the 3D cable intelligent management platform are established. A complete set of 3D cable intelligent management system is built by using different models, specifications and interface methods. The purpose of functional testing is to judge whether the platform can operate normally and analyze its system performance, including software environment and user requirements. The hardware environment mainly checks whether the selected module can realize the function accurately under the conditions of position, window size and parameter value in the main domain coordinate system. Whether the function test platform for establishing virtual parametric model and online real-time monitoring can operate normally through on-site data collection, and whether its internal state and performance indicators meet user requirements can be determined through simulation of internal components of the system. It can be seen from Fig. 2 that the 3D cable intelligent management platform based on parametric modeling technology performs well in monitoring cable transmission time test, and the platform compatibility is more than 80%. This shows that it can provide users with a good management platform.

When analyzing the requirements of the platform functions, we should make clear what users need to do and how to achieve these functions. According to the above situation, the following suggestions are made: (1) The system should be real-time. For data collection, processing, query, etc., it should have a certain rapid response capability. At the same time, parametric model technology can be used to achieve mutual coordination between modules. Finally, a complete, reasonable and effective information interaction can be established for the relationship between subsystems in the platform through parametric modeling tools, so as to achieve a good interaction between users and devices.

5 Conclusion

With the rapid development of social economy, the field of power management is constantly improving, and the requirements for power quality are getting higher and higher. How to monitor and control the operation status of power grid in real time and accurately is an urgent problem to be solved. This paper proposes to establish a user equipment platform based on parametric model technology to assist intelligent terminals in data analysis and timely feedback to customers and managers and other functional modules. The system architecture is designed using B/S structure and the communication interface and data storage are realized through protocols after the completion of system testing, so as to provide more power management demand services for enterprises.

Acknowledgment. Young and Middle-aged Teachers in Guangxi Universities (2020KY23021); Guangxi Higher Education Undergraduate Teaching Reform Project (2022JGA375); School-level scientific research fund projects (GXKS2022QN006).

References

1. Aso, N., Yanami, H., Ogawa, M.: Automatic parametric modeling technique for structural design standardization. IEEE Access **10**, 81031–81041 (2022)

2. Singhadia, A., Pati, P.S., Singhal, C., Chakrabarti, I.: Efficient HEVC encoding to meet bitrate and PSNR requirements using parametric modeling. Circuits Syst. Signal Process. **41**(8), 4479–4511 (2022)
3. Allerbo, O., Jörnsten, R.: Flexible, non-parametric modeling using regularized neural networks. Comput. Stat. **37**(4), 2029–2047 (2022)
4. El Motaki, S., Yahyaouy, A., Gualous, H.: Modeling the correlation between the workload and the power consumed by a server using stochastic and non-parametric approaches. Softw. Pract. Exp. **52**(10), 2177–2190 (2022)
5. Evers, E., de Jager, B., Oomen, T.: Incorporating prior knowledge in local parametric modeling for frequency response measurements: applied to thermal/mechanical systems. IEEE Trans. Control Syst. Technol. **30**(1), 142–152 (2022)
6. D'Alterio, P., Garibaldi, J.M., Wagner, C.: A constrained parametric approach for modeling uncertain data. IEEE Trans. Fuzzy Syst. **30**(9), 3967–3978 (2022)
7. Rosenberg, L.: Parametric modeling of sea clutter doppler spectra. IEEE Trans. Geosci. Remote Sens. **60**, 1–9 (2022)
8. Ncwane, S., Folly, K.A.: Modeling wind speed using parametric and non-parametric distribution functions. IEEE Access **9**, 104501–104512 (2021)
9. Crescentini, M., et al.: Online EIS and diagnostics on Lithium-Ion batteries by means of low-power integrated sensing and parametric modeling. IEEE Trans. Instrum. Meas. **70**, 1–11 (2021)
10. Santesteban, I., Garces, E., Otaduy, M.A., Casas, D.: SoftSMPL: data-driven modeling of nonlinear soft-tissue dynamics for parametric humans. Comput. Graph. Forum **39**(2), 65–75 (2020)
11. Okyar, F., Guldeniz, O., Atalay, B.: A holistic parametric design attempt towards geometric modeling of the lumbar spine. Comput. Methods Biomech. Biomed. Eng. Imaging Vis. **8**(1), 65–75 (2020)
12. Piraud, M., et al.: Towards quantitative imaging biomarkers of tumor dissemination: a multi-scale parametric modeling of multiple myeloma. Med.Image Anal. **57**, 214–225 (2019)

Informationization Construction of Procurement Management Based on Supply Chain Management Mode

Yu Zhang[1(✉)] and Xiaoxia Zhao[2]

[1] Beijing Normal University, Beijing 100161, China
1305217252@qq.com
[2] North China Electric Power University, Beijing 100000, China

Abstract. As a network structure that connects upstream and downstream enterprises, the supply chain can enhance the competitiveness of enterprises in a dynamically changing market environment. Through supply chain management, core enterprises can establish mutually beneficial and win-win development relationships with other enterprises in the supply chain. However, there are currently issues such as conflicts between the cost and quality of procurement objects, low level of informatization in procurement management, and a lack of procurement risk assessment. Based on the perspective of supply chain management, with the help of Big data technology, this paper analyzes the typical material procurement platform architecture, analyzes the information types and data characteristics collected, and proposes the material procurement management and application methods based on Big data, hoping to provide reference for personnel concerned with procurement information management, and further improve the quality of enterprise procurement management.

Keywords: Supply Chain Management · Big Data · Procurement Management · Information Construction

1 Introduction

At present, Chinese enterprises widely adopt the bidding system for material procurement. This bidding system is an effective system that firmly grasps the needs of economic system reform and marketization, can promote effective competition, save purchasing costs, and improve the comprehensive efficiency of enterprises. However, with the increasing quantity and scale of material procurement in China, the current material procurement and management are facing challenges [1, 2], mainly including: ① the material procurement management system is not perfect, relevant procurement personnel cannot strictly follow standards for procurement, and the supporting reward and punishment and regulatory mechanisms are not transparent and unified; ② The information level of the material procurement platform is not enough. The relevant platforms only collect bidding information, but lack the management of bidding information. The platform has a lot of information, presenting Big data characteristics, and it is difficult to carry out

Y. Pei et al. (Eds.): IC 2024, LNEE 1214, pp. 350–356, 2024.
https://doi.org/10.1007/978-981-97-4193-9_37

further mining and value analysis; ③ There is a lack of corresponding methods for risk assessment in material procurement, and there are certain risks in each link of material procurement. However, many enterprises have not conducted risk assessments for each link, thus unable to fully grasp the quality of the entire material procurement work. These difficulties further constrain the management level of material procurement and the quality improvement of the overall material supply chain.

Supply chain management is a product of the development of the market economy. As an important component of supply chain management, procurement management can successfully transform traditional buying and selling relationships into strategic partners. Procurement management is an important part of the enterprise value chain and can to some extent affect the economic benefits of the enterprise. If information technology construction can be used to optimize the procurement process and mode reasonably, It can maximize the market competitiveness of enterprises. Therefore, it is necessary to analyze the informationization of procurement management centered on the supply chain management model, in order to demonstrate the value that enterprise procurement management should have.

2 Overview of Procurement Management Under Supply Chain Management Mode

In today's market model, competition between enterprises has gradually become competition at the supply chain level, making it difficult for individual enterprises to compare with the supply chain from both resource and other perspectives. Because no enterprise can become the first in all business activities at the same time, in the development stage, enterprises need to collaborate with other upstream and downstream enterprises to create an economic benefit chain, and use the advantages of different industries and resources of different enterprises to strengthen their market competitiveness, thereby enabling all enterprises in the supply chain to achieve mutual benefit and win-win development through joint operation. The supply chain management model emphasizes the complementary advantages between different enterprises, that is, integrating all available resources in the development of enterprises to meet the market competition environment. Since the 21st century, supply chain management has become a top priority for enterprises to strengthen market competitiveness. Supply chain enterprises can strengthen cooperation in information sharing, which will improve the utilization rate of enterprise resources and technology [3].

In the stage of enterprise operation and development, the core work of supply chain management is procurement management, because enterprise procurement is the most direct communication between upstream and downstream supply chains, and the core cost of manufacturing is material procurement cost. Procurement effectiveness not only affects product production costs, but also affects product quality. Therefore, when enterprises operate and develop in the supply chain, Emphasis should be placed on the scientific optimization of item procurement in the supply chain management model, in order to deeply tap into the potential of the supply chain and meet the increasingly diverse market demands. It should be noted that procurement management, as an important part of the supply chain, can establish a bridge for stable cooperation between upstream and downstream enterprises, and procurement is also the starting point of the supply chain. Therefore, in order to achieve seamless connection of the supply chain, supply chain management should be used to further optimize the procurement process and strengthen supply chain cooperation by collaborating with various enterprises [4].

3 Supply Chain Management

Supply chain is a relatively complex system of enterprise cooperation, and enterprises must coordinate their work in order to achieve good performance. Supply chain management can view the supply chain as an organizational integration, comprehensively managing the supply chain from a holistic perspective. The division of labor and development of various enterprises in the supply chain can significantly improve the competitiveness of the collective market.

3.1 Scope of Supply Chain Management

The main scope of supply chain management can be divided into four categories: procurement, planning, logistics, and demand. The supply chain can take integrated production planning as the core, supplemented by various technologies to improve the economic benefits of the supply chain. As a core function, procurement will continue to enhance its value through continuous exploration of its management model.

3.2 Supply Chain Management Application Mode

The application models of supply chain management can be divided into three types based on enterprise needs: manufacturing, retail, and 3PL. Different application models need to face different challenges during supply chain management. The core mechanism of supply chain management lies in competition, cooperation, and coordination among enterprises. As long as enterprises can continuously explore opportunities for cooperation in the supply chain, they can maximize the actual needs of enterprises and even customers. During this period, it is necessary to strengthen cooperation mechanisms in supply chain management, with cooperation mechanisms as the core to maintain the relationship between strategic partners. Then, it is necessary to strengthen decision-making capabilities through information sharing and achieve synchronous development of different enterprises. In addition, the essence of each node enterprise in the supply chain is the relationship between entrustment and agency. As long as it can coordinate upstream and downstream enterprises and utilize problems, supply chain management can play its due role. Therefore, it is necessary to strengthen incentive management in supply chain management and promote enterprise development from multiple perspectives such as incentive goals, targets, and performance evaluation. It should be noted that each enterprise is a stakeholder partner. In order to continuously gain industry advantages, it is necessary to actively analyze competitors, in order to further improve the effectiveness of supply chain management and ensure the effectiveness of subsequent work [5].

4 Informationization Construction of Procurement Management in Supply Chain Management Mode

4.1 Overall Architecture

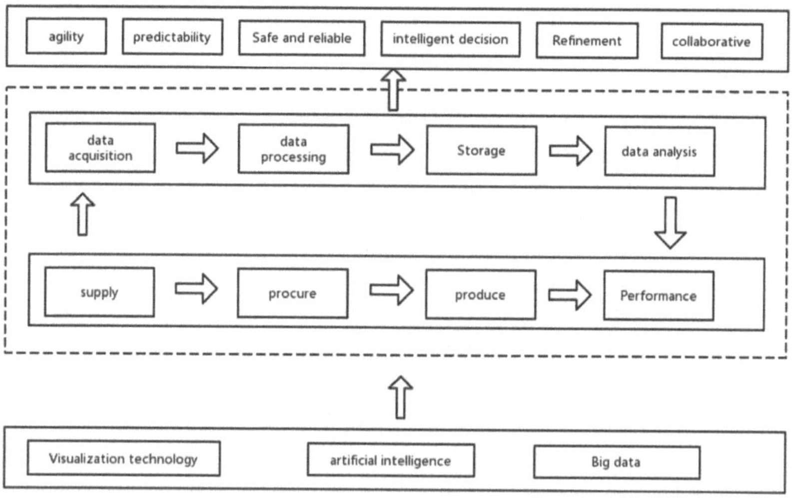

Fig. 1. Schematic diagram of intelligent supply chain architecture

The concept of "smart supply chain" originated in 2009 and has gone through three development stages, namely the online and offline interconnection stage, automation stage, and localization stage. It is the result of the continuous integration of modern supply chain theory and information technology. Its overall architecture is shown in Fig. 1. From the graph, it can be seen that data is the internal driving force of the supply chain. Supported by data, enterprises can achieve refined, collaborative, and intelligent management, accurately predict material procurement needs, and flexibly adjust procurement strategies.

4.2 Technical Support

At the technical level, the intelligent supply chain system based on Big data integrates Cloud storage (data storage), visual display (data display) and intelligent perception (data collection), and its independent decision-making ability depends on artificial intelligence algorithms. For example, there is a phenomenon in traditional supply chains where suppliers and material procurement enterprises are not closely connected, making it difficult to integrate data from both parties. In the smart supply chain, the production data of suppliers can be integrated into the production system of power enterprises, connecting the upstream and downstream chains [6].

4.3 Integration of Material Procurement Information and Data Characteristics

As shown in Fig. 2, the platform's data collection items mainly include basic supplier information, procurement manager information, procurement record information, and external data. Each data item also includes several sub items, such as procurement record information including manufacturer, batch, shipment quantity, arrival time, contract, transaction records, etc. Among these sub items, the data structure and quality level are different, and the data categories are numerous, making processing difficult.

With the continuous expansion of New Infrastructure investment and material construction demand, the number of material procurement is increasing and the number is huge. Various data in the material procurement process are characterized by large scale, multiple types, fast collection speed and low value density, meeting the 4V (Volume, Variety, Velocity, Value) characteristics of Big data [7]. Faced with such a large amount of data and chaotic content of power material procurement information, how to effectively integrate and analyze these data and conduct comprehensive, multi-dimensional, and multi-angle research on suppliers is a technical difficulty in improving procurement quality. At the same time, how to use the Big data technology to connect and integrate the data of the procurement platform with the data of other platforms, cluster and correlate the information of different data characteristics and carry out multi-dimensional analysis, so as to realize the platform's automatic auxiliary decision-making is an urgent task.

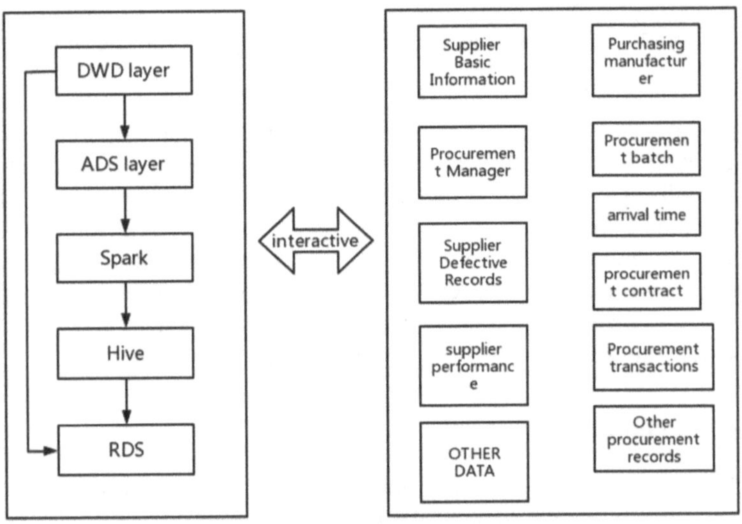

Fig. 2. Material Procurement Platform and Database Framework

4.4 Procurement Management Informatization Strategy Based on Supply Chain Management Mode

This paper proposes a material procurement management and application framework based on Big data, which is no longer limited to data collection, pre-processing and integration, but is to make full use of the platform's internal and external data as well as multi category, multi-source heterogeneous data to conduct data mining, present valuable information in Big data, and assist in procurement decision-making [8, 9].

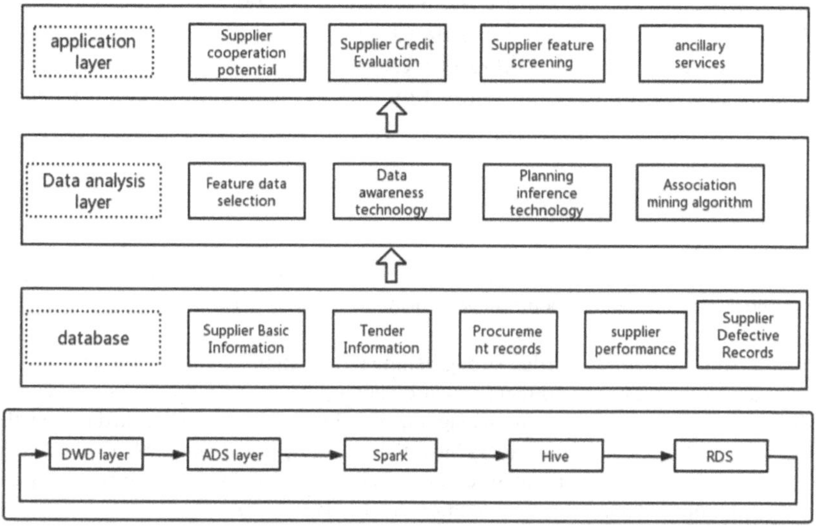

Fig. 3. Material Procurement Management and Application Method Framework

Firstly, the platform integrates its procurement business data into the platform through integration with procurement business, and carries out the data preprocessing, calculation, and storage processes shown in Fig. 3; Secondly, integrate the internal data of procurement business with external business data from platforms such as product quality inspection, lifecycle operation and maintenance, and supplier after-sales service, to break up and reconstruct multiple types of data, and support multidimensional data fusion analysis; Then, in the data analysis layer, the platform needs to build a data analysis model to comprehensively evaluate suppliers by evaluating indicators such as business level, technical level, performance response, and service quality, and establish a procurement decision and risk evaluation model; Finally, the calculation results of the model are output to achieve analysis in various aspects such as supplier profiling, potential assessment, and credit reporting, and to obtain decision results. The described method can analyze Big data in view of the current difficulties faced in the procurement and management of electric power materials, which is mainly reflected in: ① for the data management of procurement personnel and procurement records, directly benchmark the procurement standards, and achieve a supporting and transparent reward and punishment and supervision mechanism; ② Through Big data technology, integrate the internal and external data of the platform for multi-dimensional analysis, and realize the deep informatization and intelligent decision-making of the platform; ③ In data analysis and decision-making, comprehensive evaluation is conducted on supplier performance, poor records, and integration of product quality inspection and maintenance data in the later stage, effectively avoiding procurement risks [10].

5 Conclusion

In view of the problems of nonstandard system, low informatization and risk decision-making in material procurement, this paper proposes a material procurement and management method based on the supply chain management mode with the help of Big data technology, and designs an in-depth Big data mining and analysis framework, which can effectively assist the decision-making process of enterprise material procurement and provide reference for the in-depth innovation and application of material procurement data.

References

1. Yufei, T., Jianyue, F., Yi, S., et al.: Research on the construction and management of an experimental consumable procurement information platform. Exp. Technol. Manage. **38**(04), 271–274 (2021)
2. Reporter from our journal: Further strengthening informationization construction and promoting the development of government procurement – "overall plan for the construction of the national government procurement management trading system" and "basic data specification for government procurement business" were released. China Finance **08**, 57–58 (2013)
3. Xuexi, X.: Exploring the material procurement management of state-owned enterprises under the "Internet plus." China Logist. Procurement **18**, 81–82 (2021)
4. Zhang Bo, Y., Zhiqiang, H.L.: Discussion on methods to improve the management level of material supply and procurement of state-owned enterprises in the new context [J]. China Logist Procurement **3**, 71–72 (2022)
5. Queqiao, D.: Analysis on optimization and improvement of material procurement based on Big data technology. Mod. Trade Ind. **9**, 28–29 (2021)
6. Fang, L.: Research on material procurement management of state owned enterprises from the perspective of supply chain management. Mod. Ind. Econ. Inform. Technol. **9**, 163–164+167 (2022)
7. Jianqiang, H.: Research on Centralized Procurement and Management of Bulk Materials and Materials for Construction Enterprises [J]. Low Carbon World **9**, 173–174 (2020)
8. Jianliang, Z.: New concept of procurement management. Energy Technol. Manage. **02**, 119–120 (2007)
9. Tian, Z.: A brief discussion on the management model of enterprise material procurement in the supply chain environment. Small Medium Sized Enterprise Manage. Technol. **01**, 33–34 (2019)
10. Jiangdong, F., Zhan, G., Yilin, Z., Daqiang, C.: Elements, evolution, and practice of building a modern smart supply chain digital logistics system for power grid. Logist Technol. **40**(02), 102–109 (2021)

Intelligent Monitoring and Management System for Coal Mines Using Internet of Things Technology

Xiaohong Liu, Shuaxia Bai, Jiacheng Yuan, and Huadong Sun(✉)

School of Transportation and Civil Engineering, Shandong Jiaotong University, Jinan 250357, Shandong, China
sunhuadong2004@163.com

Abstract. The underground environment of coal mines is quite unique, and the disasters that occur are also serious. In order to ensure the safety of underground workers, the underground environmental monitoring system of coal mines plays an important role. This article proposes a coal mine underground environmental monitoring system based on fiber optic Internet of Things transmission, taking into account the technological advantages of fiber optic anti-interference long-distance and large capacity transmission. By analyzing the environmental parameters and types of electromagnetic interference in coal mines, four parameters, namely temperature, methane concentration, carbon monoxide concentration, and oxygen concentration were selected for collection. The hardware circuit of the system was designed. Simulation experiments were conducted on the system using a testing platform, and the data transmission function of the system was tested. The average time required for each request was approximately 15.8 ms, verifying the feasibility of the design. The intelligent monitoring method for coal mines in this article helps to improve the safety of coal mining.

Keywords: Coal Industry · IoT Technology · Monitoring Systems · Wireless Transmission

1 Introduction

Coal resources have the characteristics of strong flammability, high portability, low price, and convenient storage [1]. They are an important pillar of the development of light and heavy industries and are closely related to people's lives. In terms of current energy usage, coal is an irreplaceable energy source. Before and after reaching the carbon peak in 2030, the main structure of energy consumption for a long time will still be coal.

At present, the communication methods of the environmental monitoring system in coal mines include wired communication and wireless communication [2]. Most of the coal mines still rely heavily on traditional wired communication methods, using industrial communication networks such as CAN bus and RS485 bus. The related technologies of this type of wired communication method have been developed relatively well in coal mines, but there are still some problems. For example, the interaction rate of

Y. Pei et al. (Eds.): IC 2024, LNEE 1214, pp. 357–365, 2024.
https://doi.org/10.1007/978-981-97-4193-9_38

signals is slow, the transmission distance of ordinary cables is limited, and the cable laying work is more complex, etc. [3]. The underground environmental monitoring of coal mines mainly monitors the gas, dust, temperature, roof pressure, and other conditions in real-time. Among these factors that may cause safety accidents, gas concentration and temperature are the focus of environmental safety monitoring. The main gases detected in coal mines include carbon monoxide, methane, oxygen, and other gases. The concentration of these gases in the coal mine can directly reflect whether the environment is suitable for personnel to work [4]. The fourth stage of development of coal mine monitoring systems began around the 1980s, during which computer network technology also began to rapidly develop. Subsequently, the combination of coal mine monitoring technology, network technology, and computer technology formed the fourth generation of distributed network technology coal mine monitoring systems. Until now, we have entered the fifth stage of development in coal mine monitoring technology. Although a comprehensive fifth generation coal mine monitoring system has not yet been formed, its development direction has been determined, that is, through the integration of database technology, geographic information system, wireless transmission technology, etc., we can achieve a safety warning system that can collect various environmental parameters in coal mines in real-time and provide remote alarms and dynamic tracking of accident hazards [5].

The research work done in this paper is the design and research of the ambient intelligence intelligent monitoring system for coal mine underground based on the optical fiber transmission technology of the Internet of Things. The main content of this article is as follows:

In the first part, the background and significance of the research on the intelligent monitoring and management system for coal mines based on Internet of Things technology.

In the second part, there is a review of relevant concepts, including analysis of the electromagnetic environment in coal mines, system requirements analysis, and fiber optic communication.

In the third part of the coal mine monitoring system design, including the overall circuit design, temperature sensors, etc.

In the fourth part of the system performance testing analysis, the research results and processor capabilities were tested.

In the fifth part, the limitations and prospects of the research on intelligent monitoring and management systems for coal mines based on Internet of Things technology are presented.

2 Overview of Relevant Concepts

2.1 Analysis of Electromagnetic Environment in Coal Mines

The harsh working environment in coal mines determines that the electromagnetic environment in the space they are located in is complex. Most of the underground spaces in coal mines are relatively small or have long channels, which can form complex electromagnetic interference paths. Moreover, the electrical equipment in coal mines is mostly high-power equipment, accompanied by radio frequency transmission electromagnetic fields, and some electromagnetic field characteristics with parameters and propagation characteristics generated by changes in coal rock properties [6]. The electromagnetic interference can be summarized as follows:

(1) Electromagnetic interference of frequency converters

The interference spectrum of frequency converters is relatively wide, with more severe interference in low-frequency components than in high-frequency components, and the output power line interference of frequency converters is the most severe [7].

(2) Radio frequency interference

Wireless devices in coal mines can generate electromagnetic wave components in various frequency bands, which can radiate and cause interference in space [8].

(3) Voltage drop

In addition to electromagnetic interference generated by frequency converters in coal mines, there are also surges, pulse groups, voltage drops, harmonic interference, etc. generated by other electrical equipment due to switch switching, and voltage drops occur very frequently [9].

(4) Power frequency magnetic field

The high-power equipment underground in coal mines is generally concentrated together, and the magnetic field interference generated by the aggregation of some power cables is also relatively serious [10].

2.2 System Requirements Analysis

According to the background requirements of the coal mine underground, the monitoring system must first be able to centrally monitor various environmental parameters in the complex environment of the coal mine underground, and also require real-time data transmission in the complex environment of the coal mine underground. It is required to be able to perform long-distance transmission and normal operation of the system function in the complex environment, ensuring the stability of the system work, real-time data transmission, and reliability of the work. Therefore, the following conditions need to be met [11]:

(1) Multi parameter integrated monitoring

This monitoring system can monitor various environmental parameters underground in coal mines, and can complete the functions of data collection, transmission, and processing of various environmental parameters.

(2) Distributed monitoring

The monitoring system needs to be distributed in the complex environment of coal mines, requiring monitoring of multiple points. Therefore, the application of distributed systems can more comprehensively monitor its environmental parameters.

(3) Long distance transmission

Because the channels underground in coal mines are relatively narrow, it is required that the monitoring system can transmit the collected data remotely.

(4) Accuracy

The testing system has errors in sensor accuracy, hardware circuit noise, data transmission, and calculation. According to China's accuracy standards for environmental monitoring data, the error should be controlled within a reasonable range.

(5) Stability

The stability of the system requires that the functions of each module in the coal mine underground environment can operate normally and stably, and the collected data can be transmitted in real-time [12].

(6) Scalability

The scalability of the system requires the ability to easily add monitoring systems for environmental monitoring of new monitoring points, and the already laid out environmental monitoring system will not be affected, and facilitate the maintenance and replacement of faulty monitoring systems [13].

2.3 Fiber Optic Communication

Fiber optic mainly consists of a fiber core, a cladding layer, and a coating layer. According to the number of modes transmitted in the fiber core, it can be divided into multimode fibers and single-mode fibers. The differences between the two mode fibers are mainly reflected in the types of light sources, fiber core diameter, bandwidth, modal dispersion, and other aspects [14]. In terms of light source category, multimode fiber lasers have higher costs and power consumption compared to LEDs, but have better performance in precise control of light sources. The diameter of single-mode fiber is generally between 8 and 10 um, while the core diameter of multimode fiber is approximately 7–8 times that of single-mode fiber, typically between 50 and 62.5 um. These two classifications are expressed using mathematical expressions, with the step type represented by Eq. (1) and the gradient type represented by Eq. (2).

$$n = \begin{cases} n_1, r \leq a \\ n_2, a < r < b \end{cases} (n_1 > n_2) \tag{1}$$

$$n = \begin{cases} n_1[1 - 2\Delta f(\frac{r}{a})]^{\frac{1}{2}}, r \geq a \\ n_2, a < r < b \end{cases} (n_1 > n_2) \tag{2}$$

According to whether the light waves emitted into the optical fiber undergo total reflection, they are divided into guided waves and radiation waves. The light waves that undergo total reflection become guided waves, and the light waves that undergo refraction become radiation waves. In long-distance transmission, the radiation wave undergoes several refractions and reflections, and the energy in the fiber core is coupled to the external medium. Guided waves, on the other hand, form a total reflection of light waves at the interface between the fiber core and the cladding, with energy mostly present in the fiber core. When the incident angle meets certain conditions, light can form total reflection within the light, thereby achieving long-distance transmission.

At present, the main fibers responsible for long-distance transmission are single-mode fibers, and they are all step type fibers; when multimode fiber is transmitted over a long distance, it has a significant impact on its mode dispersion, and currently cannot undertake the work of long-distance transmission. It is mainly used in some experiments or local area networks [15].

3 Design of Coal Mine Monitoring System

3.1 Overall Circuit Design

The system mainly monitors the environmental temperature, methane concentration, carbon monoxide concentration, and oxygen concentration parameters in coal mines. The signal conditioning and acquisition module of the system mainly includes the sensor circuit and the data acquisition circuit. The sensor circuit is mainly composed of the sensor and its driving circuit, and its function is mainly to process the analog signals of various parameters output by the sensor; The data acquisition circuit is mainly composed of an analog-to-digital conversion chip (ADC) and a main control chip (CPLD). The analog-to-digital conversion chip (ADC) samples the conditioned analog signal and outputs a digital signal. One function of the main control chip (CPLD) is to provide the required timing, and the other function is to encode and process the data, and then transmit it to the fiber optic transmission module.

3.2 Temperature Sensor

The ambient temperature sensor adopts the AD849X series chip from ADI company, with a packaging type of MSOP-8. Its temperature measurement range is between −40 °C and +125 °C, and its voltage output is shown in formula (3):

$$V_{\text{out}} = T_A \times 5mV/^{\circ}C \tag{3}$$

In the formula, Vout is the output voltage, and TA is the ambient temperature at which the sensor is located.

3.3 Oxygen Concentration Collection Module

The oxygen sensor used in this article is the electrochemical sensor ME3-O2 produced by Weisheng Technology. Its working principle is to use oxygen to undergo a reduction reaction on the working electrode to generate a current, which is proportional to the oxygen concentration and has the same characteristics as the carbon monoxide sensor.

3.4 Data Collection and Conversion

The main functions of the data acquisition circuit are the analog-to-digital conversion chip (ADC) sampling and quantifying the output signal of the sensor, and the main control chip (CPLD) providing timing for the ADC and processing the data output by the ADC before sending it to the fiber optic transmission module. The analog-to-digital conversion chip is selected from the ADI company's AD7767 series.

3.5 Main Control Chip

The main control chip selected is the XC9572XL from XILINX company. In this article, a 64 pin chip is used. The XC9572XL is composed of four 54v18 functional blocks and has 1600 available gates internally. The propagation delay under standard test conditions is 5ns. The power supply voltage range is −0.5~+4.0 V, with an input voltage of up to 5.5 V and a storage temperature of −65~150 °C.

3.6 Methane Concentration Sensor

The thermal conductivity detection sensor MP-4 produced by Weisheng Technology is selected in this article. The working principle of this sensor is that the conductivity inside the sensor changes with the concentration of methane gas in the air. The higher the concentration of methane gas, the higher the conductivity of the sensor. Conversely, the lower the concentration of methane gas, the lower the conductivity of the sensor. Moreover, the sensor can replace the change in conductivity in the sensor with a voltage signal corresponding to the concentration of methane gas by configuring some basic circuits. This sensor has high sensitivity to methane gas over a wide concentration range (300–10000 ppm). The heating voltage is generally 5 V DC or AC voltage, and the loop voltage (test voltage in the following text) is less than or equal to a DC voltage with an amplitude of 24 V. The most ideal testing conditions are a temperature of 20 °C±2 °C, a humidity of 55±5% RH, and a DC voltage of 5 V for both the circuit voltage and heating voltage. This sensor has advantages such as strong anti-interference ability, low power consumption, fast response recovery, simple detection circuit, good stability, and long service life.

4 System Performance Testing Analysis

4.1 Analysis of Test Results

Table 1 shows the summary report results of testing the coal mine monitoring system. The pressure test results show that a server throughput (TPS) of 59.3 indicates that 59.3 requests can be processed per second. The average time spent per request is approximately 15.8 ms, basically meeting the system's response speed requirements, able to complete data storage tasks well.

Table 1. System Speed Test

Number of samples	10000
Average time	1458
Minimum time	232
Maximum time	3258
Abnormal rate	0%
Throughput	59.3

4.2 Summary and Analysis of Time Consumption Situation

Figure 1 shows a summary of the request time consumption during the entire testing process. From the graph, it can be seen that the average time consumption is approximately equal to the maximum time consumption of 95% of requests, indicating that for most requests, the system processing speed does not change much. This test indicates that the coal mine monitoring system based on Internet of Things technology can provide normal and reliable data transmission support in most cases, ensuring the smooth operation of the system.

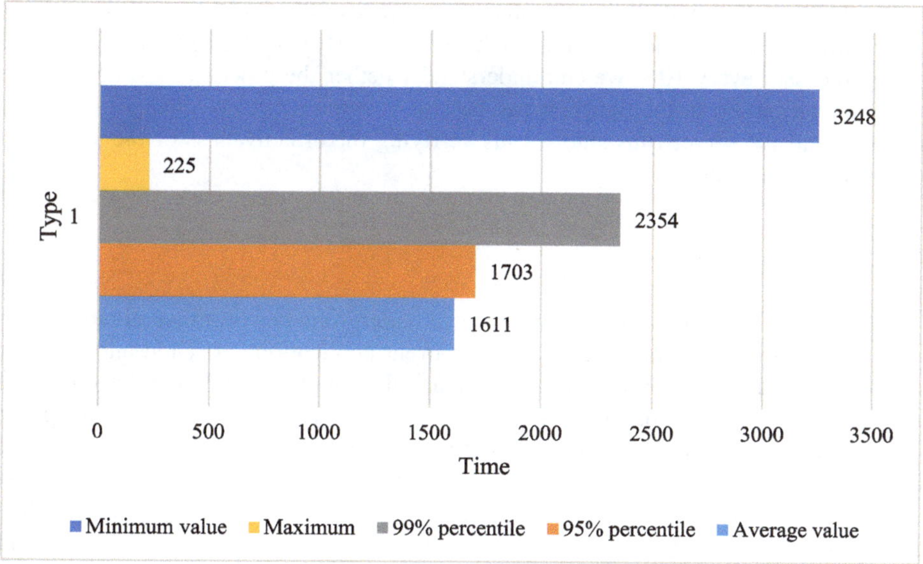

Fig. 1. Time consuming testing

4.3 Processor Achievement Test

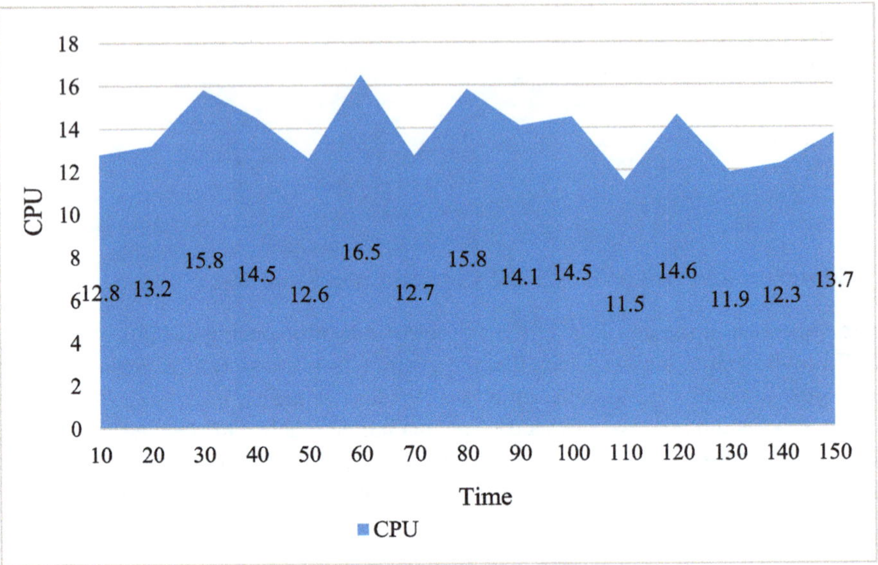

Fig. 2. Processor achievement test

As shown in Fig. 2, the average value of the processor in this test was 14.5%. This figure shows the approximate changes in the processor. By calculating the average value of the processor every 10 s, we can understand whether the processor's state is stable during operation. From the figure, it can be seen that the processor changes relatively smoothly and the server runs stably, thus verifying the effectiveness of the system in practical environments.

5 Conclusions

Starting from the current development status of underground environmental monitoring systems in coal mines, combined with the current applications of wired and wireless transmission, this article proposes a research and design of an intelligent detection and management system for coal mines based on Internet of Things technology. This paper designs the hardware of the system, and finally tests it in experiments to verify the feasibility of the system, which has high reliability in both data and system; it can effectively solve the monitoring problem under certain conditions. However, the data volume during the system testing process designed in the article is not large enough, and further optimization is needed in the future.

References

1. Ji, C.N., Huang, J., Huang, G.W., Rao, X.Z., Zhu, T.T.: Research on the transformation of coal exhausted cities based on the integration of landscape and ecological functions. J. Henan Polytech. Univ. (Nat. Sci.) **42**(3), 78–87 (2023)
2. Yang, C., Li, C.C., Wan, Y.Q.: Application of remote sensing technology in coalmine environmental monitoring and assessment. Coal Geol. China **34**(7), 61–66 (2022)
3. Sharma, R., Prakash, S.: Enhancement of relay nodes communication approach in WSN-IoT for underground coal mine. J. Inf. Optim. Sci.Optim. Sci. **41**(2), 521–531 (2020)
4. He, X.Q., Zhou, C., Song, D.Z., et al.: Mechanism and monitoring and early warning technology for rockburst in coal mines. Int. J. Miner. Metall. Mater. **28**(7), 1097–1111 (2021)
5. Jo, B., Khan, R.M.A.: An internet of things system for underground mine air quality pollutant prediction based on azure machine learning. Sensors **18**(4), 930 (2018)
6. Singh, A., Kumar, D., Hotzel, J.: IoT based Information and communication system for enhancing underground mines safety and productivity: genesis, taxonomy and open issues. Ad Hoc Netw.Netw. **78**(9), 115–129 (2018)
7. Choi, Y., Nguyen, H., Bui, X.N., Nguyen-Thoi, T., Park, S.: Estimating ore production in open-pit mines using various machine learning algorithms based on a truck-haulage system and support of internet of things. Nat. Resour. Res.Resour. Res. **30**(2), 1141–1173 (2021)
8. Yuan, L.: Scientific problem and countermeasure for precision mining of coal and associated resources. J. China Coal Soc. **44**(1), 1–9 (2019)
9. Hao, Y., et al.: New insights on ground control in intelligent mining with internet of things. Comput. Commun.. Commun. **150**(1), 788–798 (2020)
10. Jha, A., Verburg, A., Tukkaraja, P.: Internet of things-based command center to improve emergency response in underground mines. Saf. Health Work. Health Work **13**(1), 40–50 (2022)
11. Mishra, P.K., Kumar, S., Pratik, K.M., Kumar, J.: IoT based multimode sensing platform for underground coal mines. Wireless Personal Commun. **108**(2), 1227–1242 (2019)
12. Thirumal, G., Kumar, C.: Multilevel sensor deployment approach in IIoT-based environmental monitoring system in underground coal mines. Comput. Commun.. Commun. **195**(11), 1–13 (2022)
13. Ghosh, A., Mishra, P.K.: Underground coal mine environmental multi-parameter monitoring module with embedded sensors: application of UHF radio transceiver device. Radioelectron. Commun. Syst.. Commun. Syst. **65**(4), 200–212 (2023)
14. Dong, J., Chen, H.H.: Design of monitoring and early warning system for coal spontaneous combustion in GOAF based on optical fiber sensor. Coal Technol. **41**(8), 154–156 (2022)
15. Yan, J.W.: Research on environmental monitoring and early warning technology of intelligent coal mining face. Shanxi Metall. **45**(7), 191–192+206 (2022)

System Using Computer Combined Data Encryption Technology in the Security Management of Electricity Bills

Yue Zheng(✉), Xinxin Wang, Xin Wen, and Mengyu Tian

State Grid Shandong Electric Power Company Marketing Service Center, Jinan, Shandong, China
89539653@qq.com

Abstract. The electricity charge management of the power company is mainly to provide auxiliary services for the managers of the power company, which can not really change the operation mode of the power company, thus improving the operation efficiency of the company. The system of electric charge security management firstly uses the computer to establish the basic data, then inputs the electric charge data in MIS to form the electric charge receivable data, and performs the registration, business processing, electric charge accounting, account review, account book maintenance and other functions on the electric charge invoice, so as to achieve the automatic management of electric charge. Based on the above business information. The average value of security evaluation on electricity theft prevention of the electricity bill security management system is 4.44. Using computer combined data encryption technology to analyze customer electricity information can improve the quality and efficiency of electricity service.

Keywords: Electric Charge Security Management System · Computer Network · Data Encryption Technology · Electric Charge Accounting · Electric Charge Automatic Management

1 Introduction

The successful implementation of electricity fee management is an important guarantee for the sustainable development and expansion of reproduction of power supply companies. Since electricity is a commodity, there is also a lag in the recovery of electricity fees. Therefore, the power supply companies have launched "pre-purchase, pre-payment" and other electricity fee return strategies, which play a certain role in the prevention of security risks. With the development of market economy, the power market has also developed rapidly. Electric power companies are the basic service units of major enterprises. Their own social responsibilities and characteristics are facing new challenges and needs. However, a new normal has emerged in economic development, with the change of economic structure and speed, many unstable factors have begun to appear, and many large-capacity power generation enterprises have sent a red warning signal for the recovery of electricity bills. Large users are the main service objects of the power

Y. Pei et al. (Eds.): IC 2024, LNEE 1214, pp. 366–373, 2024.
https://doi.org/10.1007/978-981-97-4193-9_39

supply company. The number of users is large, and the management is difficult. The electricity fee is its main operating cost. The competition of various companies and the ups and downs of the industry make it possible for large users to shut down at any time, which will undoubtedly cause a huge risk of electricity fee recovery for the power supply company.

The electric charge management system of electric power companies is an organic combination of information technology and modern management technology, so that the electric charge management of electric power companies can be automated. Dinh H T believes that with the development of new technologies in the field of renewable energy and batteries, more and more houses are equipped with renewable energy and energy storage systems to reduce household energy costs. These houses usually have a home energy management system to control and dispatch each electrical equipment [1]. Teo T T believes that in the electricity market environment, the operating costs of consumers can be reduced by shifting consumption to a lower price cycle. He proposed the design of an advanced control strategy embedded in a grid-connected microgrid with renewable energy and energy storage capacity. The goal of the control strategy is to control the charging and discharging rates of the energy storage system, so as to reduce the operating costs of the end users through the arbitrage operation of the energy storage system, and reduce the power exchange between the main grid and the microgrid [2]. Kim H G believes that the standard consumer theory predicts that under commodity pricing, when the marginal price rises, the consumption should be lower than the bulk threshold. He examined whether consumers responded to the marginal price by using different forms of bulk pricing, including the increase of marginal price and fixed cost at the threshold [3]. Their research did not take into account the safety management of electricity bills, and the article will make further exploration later.

At present, the vast majority of power supply companies still use the method of "electricity first and then payment" to manage their electricity bills. Its biggest disadvantage is that it is difficult to recover, resulting in a serious "bad debt" phenomenon, which brings a heavy burden to the power grid, and even insufficient funds for updating technology. Therefore, the recovery of electricity bills has been an urgent problem for the management personnel of power companies. In recent years, the rapid development of information technology has put forward new ideas to solve the problem of power recovery in the power system. Therefore, it is the future development direction of the power company to use computer combined data encryption technology to transform the power system. According to the design of the system, the electric charge management of the power company is realized, and the main functions are given. The main structure of this paper is as follows:

The first part, the background and significance of electricity safety management system.

The second part studies the methods of electricity security management, including computer combination data encryption technology and electricity security management.

The third part, the design experiment of electricity safety management system, including the evaluation index of some power letters of credit.

In the fourth part, the safety management system of electricity fee is given and the future prospect is given.

2 Research Methods of Electricity Bill Safety Management

2.1 Computer Combined Data Encryption Technology

The encryption algorithm is to use a series of operations and transformations to change the original readable plaintext into an unreadable password. The key and algorithm are the keys of encryption, and the key is a special algorithm. In today's network society, cryptography is particularly important. In order to ensure the security of the network, effective mechanisms must be established to ensure the daily operation of the huge network [4].

At present, encrypting the information in the computer network is the most effective way to solve the problem of network communication security. Statistically speaking, when transmitting data on the network, the biggest event is the theft and destruction of data. Therefore, in practice, the commonly adopted method is to encrypt the data (i.e. plaintext) into the password that needs to be decrypted, and then transmit it through the network, so as to prevent the plaintext from being monitored during transmission, which can well protect the security of the data [5].

Although the software of electric power fee management system has its own particularity, its design idea is consistent with the management concept of modern enterprises. Early management software mainly used computers to adapt to the current management mode, replace manual labor, standardize the management process and improve management efficiency. Not to mention the power system, it lacks a better management mode, which can speed up the transmission of information, share data and make correct decisions for leaders. The electric power charge management system is designed under the guidance of modern MIS, and it is consistent with the development of power supply work at the grass-roots level.

2.2 Safety Management of Electricity Charges

Through systematic and systematic work and combined with the actual management of the power supply company, the system has been integrated [6, 7]. From the optimal point of view, the management of electricity charge accounting information of urban power companies is constructed. Power enterprises can collect, process and store all kinds of power data, status information and telemetry information by using power information acquisition system; These data can be counted, queried, calculated and analyzed; Calculation and analysis by region, time domain, voltage level, etc. [8]; The system can meet various requirements such as power demand analysis, electricity bill settlement, power metering equipment monitoring, anti-stealing analysis, load forecasting analysis, line loss statistical analysis, exception handling, demand management, etc. [9, 10]. The safety management mode of electricity charge is shown in Fig. 1.

The power supply company service and the power charge management system of the power supply company, the power supply company service and the office control management system all adopt the client/server architecture. The client uses standard SQL statements and other methods to access data in the database [11]. When the client sends the SQL statement, the database will return the dataset requested by the client to

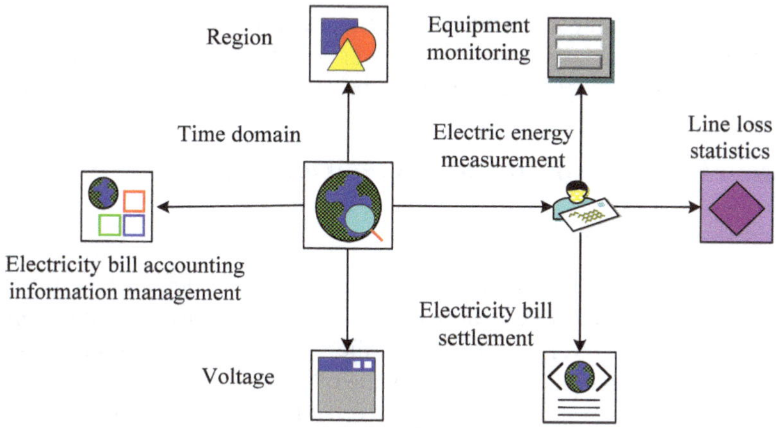

Fig. 1. Safety management mode of electricity charges

the client. Because this architecture centralizes all data on one server, which can ensure data security and maximize server performance [12, 13].

Because of its own advantages, the electric charge management system of electric power companies has gradually been adopted by various electric power enterprises and become the enterprise information management system. Now, with the increasing population of cities, there are more and more power users and power consumption [14]. Due to the particularity of the region where the power company is located, the collection of electricity charges is more complex and cumbersome. As operators and beneficiaries of the power industry, power enterprises have more and more opportunities for development of the power market due to the development of the power market. However, due to many fixed factors brought by the traditional human resource management mode, the development of power enterprises will also face greater challenges. At present, most power supply companies have introduced electricity fee management, but it is only an auxiliary means and cannot play a very good role. For example, meter reading, user query, analysis, processing and other power consumption need to be handled manually, which greatly reduces the efficiency and accuracy of the power supply department. In addition, the existing electricity fee management system only operates in the form of stand-alone software in the self-built local area network of the enterprise. Therefore, it is necessary to build a highly intelligent and efficient electricity fee management system that covers multiple power grids and power supply nodes and is convenient for users and enterprises.

In recent years, with the slow decline of China's economic growth rate, the tension between the supply and demand of electricity has been eased. The market-oriented reform of electricity must be gradually implemented, and the market-oriented competition mechanism must be introduced to regulate and develop the electricity market. In the external environment faced by power companies, the economic growth rate slows down, and the growth rate of power demand also decreases; The new technology will certainly have a certain impact on the power grid; The reform of power system and state-owned enterprises is also ongoing, and the power network is changing from "concept" to "practice", which has brought a severe test to the safety production of power enterprises.

3 System Design Experiment of Electricity Bill Safety Management

3.1 Hardware Platform

The system can work in any network environment. The system is based on the C/S architecture. It must have more than one database, and it must also be configured on the network. During the working process of the system, a large amount of data needs to be transmitted through the network. In view of the high utilization rate of the local area network of the power supply company, in order to solve the performance bottleneck of the server and client caused by the network problems, we must make full use of the existing network resources to make it run in the high-speed network. The network adopts IO/IOOM Ethernet and TCP/IP protocol. Database server: A server with a minimum configuration of hard disk no less than 50G and memory no less than 256M is required.

3.2 System Design

The system of electric charge security management firstly uses the computer to establish the basic data, then inputs the electric charge data in MIS to form the electric charge receivable data, and performs the registration, business processing, electric charge accounting, account review, account book maintenance and other functions on the electric charge invoice, so as to achieve the automatic management of electric charge. Based on the above business information, the system can query and analyze the accounts of customers, so as to improve the safety management level of electricity bills of enterprises. The database design of the system can meet tens of millions of levels of information. The parallel query method is adopted to improve the operation efficiency of the system and ensure the security and stability of the system.

4 Research Results of Electricity Bill Safety Management

Electric power is a special product, and its production, supply and sales are integrated. So far, there is still a traditional sales model of "use electricity first, pay later". Therefore, as a special product, electricity will be different from electricity in currency when it is traded between power companies and users, which creates conditions for some users with poor credit to default on electricity charges. Of course, there are many reasons why it is difficult to recover electricity bills. Although the major power supply companies are trying to collect electricity bills, there are still many problems to be solved. The electricity fee recovery of power companies is an important indicator to measure the operating performance of power companies. At present, major power grids and power supply bureaus attach great importance to the recovery of electricity bills, which is not only related to the fund operation of the power grid itself, but also related to the safety of social production and people's electricity use. The problem of electricity fee recovery of power users has increasingly become the main problem restricting the development and management innovation of power companies. Therefore, if power companies want to achieve greater development, they must find an effective solution.

The credit assessment of electricity charges is mainly based on incentives, and rewards users who pay better electricity charges to maintain a good situation of electricity charge recovery. The evaluation indicators of some electricity credit are shown in Table 1.

Table 1. Evaluation indicators of partial electricity credit

Index item	Standard score	Score range
Current rate recovery	80	70–90
Rate of repayment	10	5–15
Commitment rate	10	5–15

The electricity bill security management system is a relatively complex information system, especially in today's rapid development of information technology, the connection and cooperation between different services provide support for one or more enterprises. In this mode, an application is a collection of users, businesses and data services that meet specific needs. Therefore, it is very necessary to carry out scientific analysis and design of electricity charge management in the power market. As we all know, data is the core of the entire information system, and its definition is constantly changing. If we follow the conventional two-level architecture, a large number of interfaces and complex code will be generated, which greatly reduces the flexibility of the software, and even requires some small expansion of the database. Based on computer combined data encryption technology, this paper develops and implements a multi-level architecture, which greatly improves the scalability of the system. The performance test of the electricity bill security management system is shown in Fig. 2. The mean value of failure rate is 0.07, the mean value of correct rate of electricity charge display is 0.96, and the mean value of safety evaluation of electricity theft prevention is 4.44 (full score of 5 points).

The traditional operation and management mode of electric power industry is manual door-to-door meter reading. The workload is large, and the meter reading display is time-consuming and laborious. It is easy to cause employees to misread, miss and misread, and sometimes collude with users to steal from power supply companies, resulting in huge economic losses. With the increasing demand for electricity in society and the continuous development of modern intelligent technology, an efficient management system that can safely integrate hundreds of devices has become very necessary.

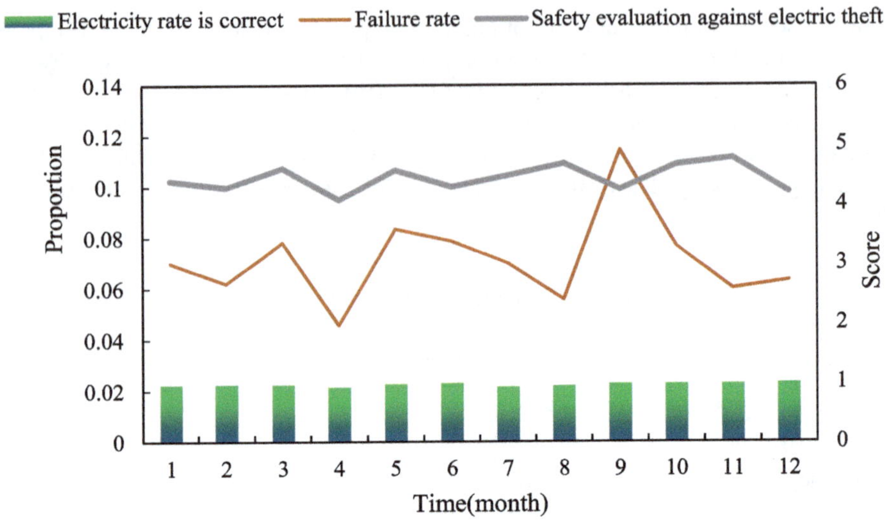

Fig. 2. Performance test of electricity bill security management system

5 Conclusion

In the process of developing the electricity bill security management system based on computer combined data encryption technology, we should fully understand the actual situation and needs of the enterprise, and reduce the workload of human resources by optimizing and integrating the system resources. The system has comprehensive functions, is easy to use, and is applicable to current employees and their conditions. At the same time, it meets the needs of large capacity and high speed, ensures the response speed of the system, and provides strong support for improving work efficiency and service quality. With the development of the electricity fee management work and technology of the power company, the requirements for its functionality, reliability, security and other aspects will become higher and higher, so the importance of the electricity fee management system of the power supply enterprise will continue to increase. On the basis of computer technology and network technology, the electricity fee management system of the power supply enterprise will also develop towards the direction of intelligence, generalization and ease of use. The article has done a lot of research on this, but there are still many problems and defects that need to be improved in the future.

(1) Combine the electric charge management system of the power company with other intelligent systems to form a complete information system.
(2) In the case of a large amount of data access, the power tariff management system of the power company should further consider the stability of the response time to the power user and ensure the security of the data.

Acknowledgment. State Grid Shandong Electric Power Company science and technology project. Research on iterative optimization of differentiated accurate payment strategy based on customer payment habits, SGSDYX00ZKJ2200096.

References

1. Dinh, H.T., Yun, J., Dong, M.K.: A home energy management system with renewable energy and energy storage utilizing main grid and electricity selling. IEEE Access **8**(99), 49436–49450 (2020)
2. Teo, T.T., Logenthiran, T., Woo, W.L.: Advanced control strategy for energy storage system in grid-connected microgrid with renewable energy generation. Iet Smart Grid **1**(3), 96–103 (2018)
3. Kim, H.G.: Estimating demand response in an extreme block pricing environment: Evidence from Korea's electricity pricing system, 2005–2014. Energy Policy **132**(SEP.), 1076–1086 (2019)
4. Kirubanad, V.B., Kumari, M.: Data encryption and decryption using graph plotting. Int. J. Civil Eng. Technol. **9**(2), 36–46 (2018)
5. Liu, J., Yang, D., Zhou, H.: A digital image encryption algorithm based on bit-planes and an improved logistic map. Multimed. Tools Appl. **77**(8), 10217–10233 (2018)
6. Koch, C., Hirth, L.: Short-term electricity trading for system balancing: An empirical analysis of the role of intraday trading in balancing Germany's electricity system. Renewable Sustainable Energy Rev. **113**, 109275.1–109275.11 (2019)
7. Sequeira, T.N., Santos, M.S.: Does country-risk influence electricity production worldwide? J. Policy Modeling **40**(4), 730–746 (2018)
8. Lucheroni, C., Mari, C.: Risk shaping of optimal electricity portfolios in the stochastic LCOE Theory. Comput. Oper. Res. **96,** 374–385 (2018)
9. Fathabadi, H.: Fuel cell hybrid electric vehicle (FCHEV): novel fuel cell/SC hybrid power generation system. Energy Conversion Manage. **156**, 192–201 (2018)
10. Rai, N.K.: A technical overview of wind power generation system. Int. J. Energy Environ. Econ. **11**(1), 1–5 (2021)
11. Moslemi, R., Mesbahi, A., Velni, J.M.: Design of robust profitable false data injection attacks in multi-settlement electricity markets. IET Gener. Transm. Distrib. **12**(6), 1263–1270 (2018)
12. Cronin, S.: Legal: lawsuits to continue despite swipe-fee settlement. Oil Express **41**(38), 3–4 (2018)
13. Lu, J., Wu, S., Cheng, H., Song, B., Xiang, Z.: Smart contract for electricity transactions and charge settlements using blockchain. Appl. Stochastic Models Bus. Ind. **37**(3), 442–453 (2021)
14. Lv, X., Li, M.: Application and research of the intelligent management system based on internet of things technology in the era of big data. Mob. Inf. Syst. **2021**(16), 1–6 (2021)

Author Index

Y. Pei et al. (Eds.): IC 2024, LNEE 1214, pp. 375–376, 2024.
https://doi.org/10.1007/978-981-97-4193-9

FSC
www.fsc.org
MIX
Papier | Fördert gute Waldnutzung
FSC® C083411

Kolibri 360 GmbH
Ferdinand-Jühlke-Straße 7
99095 Erfurt, Deutschland
produktsicherheit@kolibri360.de